Impact of Science on African Agriculture and Food Security

Impact of Science on African Agriculture and Food Security

Edited by

P. Anandajayasekeram
Manager, Capacity Strengthening Unit
International Livestock Research Institute, ILRI, Addis Ababa

M. Rukuni
Regional Director for African Programmes
W.K. Kellogg Foundation

S. Babu
Senior Research Fellow, Training Programme
International Food Policy Research Institute, Addis Ababa

F. Liebenberg
Project Leader, ARC

C.L. Keswani
Consultant

www.cabi.org

ILRI
INTERNATIONAL
LIVESTOCK RESEARCH
INSTITUTE

CABI is a trading name of CAB International

CABI Head Office
Nosworthy Way
Wallingford
Oxfordshire OX10 8DE
UK

Tel: +44 (0)1491 832111
Fax: +44 (0)1491 833508
E-mail: cabi@cabi.org
Website: www.cabi.org

CABI North American Office
875 Massachusetts Avenue
7th Floor
Cambridge, MA 02139
USA

Tel: +1 617 395 4056
Fax: +1 617 354 6875
E-mail: cabi-nao@cabi.org

A catalogue record for this book is available from the British Library, London, UK.

A catalogue record for this book is available from the Library of Congress, Washington, DC.

ISBN: 978 1 84593 267 1

Typeset by AMA DataSet Ltd, UK.
Printed and bound in the UK by Biddles Ltd, King's Lynn.

Contents

Contributors

P. Anandajayasekeram, *International Livestock Research Institute, ILRI, Manager, Capacity Strengthening Unit (CaSt), PO Box 5689 Addis Ababa, Ethiopia, Tel: +251 11 6463215 Ext. 2481, Fax +251 11 646 1252/646 4645, E-mail: p.anandajayasekeram@cgiar.org*

S.C. Babu, *Senior Research Fellow, International Food Policy Research Institute, 2033 K Street, Washington, DC 20006, USA, Tel: + 1 202 8625618, E-mail: s.babu@cgiar.org*

A. Bua, *Principal Research Officer, National Crop Research Institute, Namulonge, PO Box 7084, Kampala, Uganda, Tel: 256-772-461950, E-mail: abua@naro-ug.org*

Medson Chisi, *Golden Valley Agricultural Research Trust, PO Box 54, Fringilla, Zambia, E-mail: smip@zamnet.zm or medsonchisi@hotmail.com*

J.M.C. Esteurhuizen, *Agricultural Economist, Agricultural Research Council, PO Box 8783, Pretoria 0001, South Africa, Tel.: 27 12 4279700, Fax: 27 12 342 3948*

A.C. Isinika, *Institute of Continuing Education, Sokoine University of Agriculture, PO Box 3044, Morogor, Tanzania, E-mail: isinika@suanet.ac.tx, aidaisinika@yahoo.co.uk*

R.M. Aloyce Kaliba, *Research Associate/Policy Analyst, Aquaculture/Fisheries Center, University of Arkansas at Pine Bluff, 1200 North University Drive, Mail Slot 4912, Pine Bluff, AR 71601, USA, Tel: 870 575 8108; Fax: 870 575 4637; E-mail: akaliba@uaex.edu*

Daniel D. Karanja, *Senior Fellow, Partnership to Cut Hunger and Poverty in Africa, 499 S. Capitol Street SW, Suite 500B, Washington, DC 20003, Tel: 202-479-4501; Fax: 202-488-0590, E-mail: karanjad@msu.edu, www.africanhunger.org*

Carissa N. Marasas, *APHIS-PPQ-CIAO, 4700 River Road Unit 133, Riverdale, MD 20737, USA, E-mail: Carissa.Marasas@aphis.usda.gov*

E. Mazhangara, *5684 Shaw Street, Haslett, MI 48840, USA, E-mail: emazhangara@yahoo.com*

M.R. Mokoena, *Deputy Director, Commodity Marketing Directorate, National Department of Agriculture, Private Bag X15, Pretoria, 0007 South Africa, Tel: (012) 319 8080, Fax: (012) 319 8077, E-mail: MadimeM@nda.agric.za*

A.J. Moshi, *Zonal Director Research and Development, Eastern Zone, Ilonga Agricultural Research Institute, PO Box 33, Kilosa, Tanzania, Tel: + 255-023-2623201, Fax/Tel: + 255-023-2623284*

M. Mudhara, *Farmer Support Group, University of KwaZulu-Natal, Private Bag X01, Scottsville 3209, Pietermaritzburg, South Africa, E-mail: mudhara@ukzn.ac.za; mudhara@yahoo.co.uk*

M. Murata, *Project Monitoring and Evaluation Officer, Implementation and Coordination of Agricultural Research and Training (ICART) Project in the SADC Region, Food, Agriculture and Natural Resources*

Directorate, SADC Secretariat, P/Bag 0095, Gaborone, Botswana, Tel: (267) 3951863 ext. 5094, Fax: (267) 3924099, E-mail: mmurata@sadc.int

J.G. Niederwieser, *Plant Breeder, ARC – Vegetable and Ornamental Plant Institute, Private Bag X293, Pretoria 001, South Africa, Tel: 27 12 8419611, Fax: 27 12 808 1127.*

James F. Oehmke, *Professor and Liberty Hyde Bailey Scholar Department of Agricultural Economics, 317 AgH, East Lansing, MI 48824-1069, USA, Tel: 1-517-353-2981, E-mail: oehmke@msu.edu*

R. Randela, *National Treasury, Director: Justice and Secret Services, Pretoria, South Africa, E-mail: Rendani.randela@treasury.gov.za*

Rob Townsend, *Agricultural Economist, The World Bank, 1818 H Street, NW Washington, DC 20433, USA, E-mail: rtownsend@worldbank.org*

T. Zegeye, *Agricultural Socio-economics Research Department, PO Box 2003, Addis Ababa, Ethiopia, Tel: 251 11 6454434, Fax: 251 11 6461294, E-mail: socio_economics@eiar.gov.et*

Biography of Authors

Ponniah Anandajayasekeram
Senior Research Fellow, International Food Policy Research Institute

Ponniah Anandajayasekeram is a senior research fellow with the International Food Policy Research Institute. Before joining IFPRI, he was a senior research fellow with ISNAR in The Hague and managed its southern Africa office in Pretoria, South Africa. Previously, Ananda served as method specialist with Farm Level Applied Research Methods for Eastern and Southern Africa in Zimbabwe, as impact evaluation and policy analysis advisor for the Southern African Center for Cooperation in Agricultural Research and Training in Botswana, and as senior economist at the Australian Bureau of Agricultural and Resource Economics. Prior to that, he worked for CIMMYT in East Africa in various positions.

Ananda is professor extraordinaire in the department of Agricultural Economics, Extension and Rural Development at the University of Pretoria and has taught at Sokoine University in Tanzania and the University of Sri Lanka. He was president of the International Association of Farming Systems Research and the Southern African Association of Farming Systems Research and Extension. His research interests include agricultural marketing, farming systems, participatory research methods, R&D evaluation, impact assessment and agricultural innovation systems. He is the recipient of numerous awards and has been widely published, including 11 books, 37 papers and book chapters, 50 reports and 63 conference and workshop papers.

Mandivamba Rukuni
Regional Director for Africa Programmes, W. K. Kellogg Foundation

Mandivamba Rukuni is the current regional director for Africa programmes with the W.K. Kellogg Foundation based in Pretoria, South Africa. A graduate of the University of Zimbabwe (PhD) and the University of Reading (MSc), his career over the last 25 years has largely been as an academic and he has been associated with several universities (Zimbabwe, Michigan State, Pretoria). His work covers issues on food security in Africa, institutional development and transformation for agriculture and rural development, smallholder irrigation development in Africa, agricultural and R&D policy, land tenure, and community-based natural resources management. More recently, he has focused on business strategy models for rural

Africa as well as low-cost and effective education- and skill-enhancement alternatives for rural Africa.

Mandi's work in the area of land tenure and community-based natural resource management is well known. He chaired the Commission of Inquiry into Land Tenure Systems in Zimbabwe during 1993–1994. He was dean of agriculture at the University of Zimbabwe for 6 years, served as chair of the Agricultural Research Council and was consultant for a wide array of development assistance organizations including the World Bank, USAID and CIDA. He has served on several public sector boards in Zimbabwe including the Grain Marketing Board and the Agricultural and Rural Development Authority and on the boards of international development organizations such as CIAT and IFPRI. Mandi has published in all these fields, including 11 books, 15 book chapters, two monographs, 17 refereed journal articles, more than 70 conference and other peer-reviewed papers, and keynote addresses.

Suresh Chandra Babu

Senior Research Fellow, Head, Training Programme, International Food Policy Research Institute

Suresh Babu is a senior research fellow at the International Food Policy Research Institute (IFPRI). He is also the head of the Training and Capacity Strengthening Programme at IFPRI. His principal research areas include the effects of food policy on household food security and nutrition, and monitoring the impact of food and nutrition programmes on the nutritional status of the beneficiaries. Dr. Babu has conducted field research on food security and nutrition issues in South Asia and sub-Saharan Africa over the past 18 years. Before joining IFPRI, he was involved in implementing a food security and nutrition monitoring system for Malawi through a joint UNICEF/Cornell Food and Nutrition Policy Programme.

He has also been a senior food and nutrition policy advisor to the Government of Malawi and conducted evaluation of food and nutrition interventions as an evaluation economist for UNICEF-Malawi. At IFPRI, he has been a co-leader of a multi-country research project on Food Security and Nutrition Monitoring and conducted systematic client-consultations for food policy research in Ghana, Mali, Kenya and India.

The author of numerous articles, his recent papers have appeared in journals such as Food Policy, Food and Nutrition Bulletinand Social Sciences and Medicine. He has taught at Iowa State University, Cornell University, Bunda College of Agriculture and the University of Malawi, and conducted research for several international organizations including the FAO, UNICEF and the World Bank.

F. Liebenberg

Project Leader, ARC

Frikkie Liebenberg studied Agricultural Economics at Pretoria University and International Agricultural Marketing at the University of Newcastle-Upon-Tyne. He initially worked at the Directorate Marketing Department of Agriculture as a policy analyst and later the Western Cape Agricultural Development Institute as an agricultural economist. He joined the ARC DIA Unit in 1997 where his work focused mainly on research policy analysis, programme planning and M&E of research performance. The latter included assisting and guiding post graduate students on Impact Assessment studies, training researchers on M&E techniques and methodologies and coordinating the development of a corporate R&D Performance Evaluation System for the ARC. He also collaborated on an ongoing basis with ISNAR in the development of a training manual on M&E and IA of agricultural R&D Investments. He is currently serving as project leader in the ARC on a collation, analysis and synthesis of data for the R&D Policies and Strategies Project.

Prof. C.L. Keswani
Consultant

Professor Keswani is a plant pathologist and has served as a researcher at various universities in the USA. Between 1973–1985 he occupied progressive academic positions at the Sokoine University of Agriculture, Tanzania (formally University of Dar es Salaam) where he was involved in teaching research, and agriculture research administration (project planning, implementation, monitoring and evaluation) as well as institution building. Throughout 1985–1991 he served as technical advisor to the Plant Protection Research Institute at the Department of Research and Specialist Services (DR&SS) under the WB/IFAD project. During 1991–2002 he served as technical advisor to the DNAIDA Plant Quarantine Project and later as change process facilitator in DR&SS under the WB project. Presently, he is a freelance consultant involved in evaluation and impact assessment of agriculture, forestry and natural resource projects. He has numerous scientific publications and several consultancy reports to his credit.

Foreword

Resources for agricultural research are under stress in many developing countries, particularly in Africa. Yet without adequate investment in agricultural research, African and other developing countries cannot achieve productivity increases needed for food security, poverty reduction and sustainable management of natural resources. The challenge for policy makers is to identify the best investment options within agricultural research so that with limited resources, a high-level impact of productivity can be achieved. Such decisions require a complete set of information on the rate of returns on agricultural research investments. Impact assessment studies provide such information. Revitalized awareness of the importance of monitoring and evaluation (M&E) for project management, better use of scarce financial resources, increased focus of donors and other development partners on outcomes and impact, demand for accountability, and the recent changes in R&D paradigms are currently leading to a high demand for expertise in M&E, including impact assessment.

Impact assessment studies have been rare in Africa until recently for several reasons: paucity of data, lack of capacity and lack of demand for information-based decision making. This has resulted in very little documentation of impact of agricultural research and technology investments. This book, a joint effort of ILRI, IFPRI and the Kellogg Foundation, compiles available evidence of the impact of agricultural research in Eastern and Southern Africa. There is no doubt that the evidence assembled in this book will also facilitate the mobilization of much needed additional investments for agricultural research and development in the continent.

The authors of the chapters identify key research programmes and evaluate their production, income, socio-cultural, institutional and environmental impacts and spill-over effects. The methods and results used in these chapters will pave the way for future research on impact assessment and policy debate on 'Investment in agricultural research'. The methodological review in Part III of the book will be very useful for young researchers and academic purposes.

I congratulate the editors and the authors of the book for their dedicated effort in bringing out this volume. I have no doubt it will be of immense use to policy analysts, policy makers, and the research and development community at large.

Monty Jones (Dr)

Preface

Agricultural development has a crucial part to play in ending poverty, hunger and malnutrition in Africa. Throughout the 20th century investments in agricultural research and development have increased agriculture productivity and lifted millions from poverty and starvation. Poverty and hunger reduction are achieved through increased incomes of farmers who adopt new technology, expand income-generating opportunities and reduce food prices. Evidence also demonstrates that the rate of returns on investments in agricultural research and development (R&D) are comparable with many other public sector investments. However, investors in public R&D are no longer satisfied with activity-based progress reports. They look for outcomes and impacts on ultimate beneficiaries and overall growth and development rather than just outputs.

The decline in funding for agricultural research in Africa is exerting pressure on R&D managers to set priorities to maximize the social impacts of the relatively scarce research resources. Thus, the ultimate interest of those who invest in agricultural R&D is improvements in the lives of the poor, hungry and nutritionally insecure. Therefore, over the years, impact assessment (IA) has become a major tool in guiding investment strategies of donors, governments and financiers of agricultural and natural resources research.

Although IA of agricultural R&D programmes appeared in the literature in the mid-1950s, serious efforts to assess the impact of technology-based programmes in Sub-Saharan Africa started in the early 1990s. The University of Zimbabwe/Michigan State University Food Security Programme provided a foundation for impact assessment research in Eastern and Southern Africa. Since then much has been achieved through the efforts of the sub-regional agricultural research organizations (SROs). The Food and Agriculture and Natural Resources Sector (then the Southern African Center for Co-operation in Agricultural Research and Training, SACCAR) of the Southern African Development Community (SADC) and the Association for Strengthening Agricultural Research in East and Central Africa (ASARECA) in the Eastern Africa region institutionalized impact assessment and initiated training programmes to develop the necessary skills to conduct impact studies in the respective regions. These efforts were complemented by the International Agricultural Research Centers (IARCs) of the Consultative Group on International Agricultural Research (CGIAR) and the National Agricultural Research System (NARS). The International Service for National Agricultural Research (ISNAR), International Food Policy Research Institute (IFPRI), and more recently the International Livestock Research Institute (ILRI) played a significant part in this process. Emphasis was placed on developing national and regional capacity to conduct impact analysis, assisting in the institutionalization of the process as well as institutionalizing IA training. The activities of the SROs related to capacity development and institutionalization at the NARS level included:

- sensitization of NARS management on the need for IA as a management tool in decision making
- regional and in-country training on IA methodologies for social and biological scientists
- collaborative case studies by NARS scientists
- development of an IA procedural manual for NARS scientists
- assistance in the establishment of databases and monitoring and evaluation systems for evaluating R&D investments/technologies/programmes as an integral part of research management.

Many regional organizations implemented capacity development programmes in close collaboration with IARCs and universities in the region (University of Pretoria and University of Zimbabwe). About 400 scientists have been trained in IA methodologies through regional and national training workshops over the past 15 years. The ISNAR division of IFPRI, currently in collaboration with ILRI and selected national universities, continues to conduct an annual global workshop on Monitoring and Evaluation and Impact Assessment.

A core group of regional trainers participated in regional and national training workshops. The training strategy combined classroom teaching with field-level case studies. Participants as part of the workshops conducted national-level case studies with technical assistance from the trainers as part of the capacity development process. A number of these case studies resulted in doctoral and masters theses and are included in this book.

As part of this sub-regional initiative, a conceptual framework was developed to assess the impact of agricultural R&D investments and it enabled the analysts to examine the intermediate impacts, direct outputs, as well as economic, environmental, socio-cultural and spill-over effects using multi-criteria analysis. This framework was used in a number of studies reported in this book. Researchers have empirically verified the results of introduction of the various varieties and recommendations.

Information on returns to research guides future research priorities. Priorities for future research should depend on emerging issues, challenges and constraints that should be tackled for attaining agricultural growth targets. The chapters of this book have used primary and secondary data to measure the effects of various research programmes and present a comprehensive discussion of research results from various African countries in a single volume.

The book is organized into three major parts. Part I addresses challenges and contributions in African agricultural development, and presents available evidence to demonstrate the impact of R&D investment and discusses the role of science and technology in the transformation of African agriculture. Part II, Results of Impact Assessment in Eastern and Southern Africa, is a collection of impact study results from the region. These case studies deal with a wide range of enterprises including food crops, cash crops, livestock, floriculture and forestry products. Part III provides an overview of methodologies and best practices for conducting similar studies. The contents of the book emanate from individual and collaborative research by the authors through various research programmes over the past decades.

This book addresses African agricultural development problems, and it is aimed at researchers, policy makers, policy advisors and donors. It can be used as a primary or complementary textbook in postgraduate courses in agricultural development, agricultural economics, research management, crop science and agronomy. It will also be valuable for planners and donors interested in African agricultural development.

The editors of the volume are grateful to the organizations and numerous colleagues who directly or indirectly contributed to the publication of this book. Special thanks for the support from Dr Joachim von Braun and other colleagues at IFPRI, Drs Carlos Seré, John Mc Dermott and Bruce Scott of ILRI and the W.K. Kellogg Foundation. Sindu Workneh, Menberemariam Seyoum, Sengupta Debdatta and Elizabeth Carbone, all of the ISNAR Division of IFPRI, provided valuable assistance in preparing this volume for publication.

The cover is based on a design by Apollo Habtamu. The assistance provided by the Information Services at ILRI is much appreciated.

1 Challenges Facing African Agriculture

S.C. BABU, P. ANANDAJAYASEKERAM AND M. RUKUNI

Abstract: At independence in 1960, Africa was a modest food exporter while Asia was engulfed in a food crisis. The Green Revolution boosted food production in Asia and the global food problem shifted to Africa. However, science and technology have been promoted on an ad hoc basis in Africa's 45 years of independence. This chapter analyses why the Asian Green Revolution failed to take root in Africa, and why the average African grain yield has been flat since 1960. The chapter shows that the rate of return has been high on investments in research for a few commodities, such as hybrid maize, rice and flowers. However, most countries in Africa have a weak scientific and institutional foundation for transforming agriculture. Long-term investments are required to build the scientific and institutional foundation for a modern agriculture. This is a tall order, but this is what the USA accomplished from 1860 to 1912, what Japan did from 1868 to 1914 and what many countries in Asia and Latin America have accomplished over the past 40 years. The challenge now is to mobilize African political leaders and donors to make long-term accretionary (step by step) investments in science and technology that will boost cereal yields, increase agricultural productivity and drive down real (inflation adjusted) food prices. This is a proven way to reduce urban poverty and the poverty of rural families who are net food buyers.

This book brings together empirical evidence from 20 studies on the impact of investments in agricultural research and food security in Africa. Information on the returns to investments in research can help countries in setting priorities for agricultural research. The purpose of this chapter is to highlight the emerging issues, challenges and constraints on African agriculture that beg serious attention from the policy makers. It enumerates a broad set of food security challenges facing African agriculture. Finally, it identifies successes in African agriculture and recent efforts aimed at scaling-up the successes.

Introduction

Sub-Saharan Africa remains the most vulnerable region of the world in terms of achieving food security for all of its citizens. Recurring drought and crop failures have resulted in a high degree of food insecurity and malnutrition and kept a majority of the rural population in

 Impact of Science on African Agriculture and Food Security (eds P. Anandajayasekeram, M. Rukuni, S. Babu, F. Liebenberg and C.L. Keswani)

poverty (African Union, 2005). Yet agriculture remains a major contributor to rural incomes and the most important source of poverty reduction.

Poverty in sub-Saharan Africa has increased over the years. Over 300 million Africans, about 44% of the population, lived on less than a dollar a day in 2002 (World Bank, 2006). Although sub-Saharan Africa achieved an annual growth rate of 4.8% in 2004, the current levels of agricultural growth (1.6%) are inadequate to meet the poverty reduction objective of millennium development goals. Unless accelerated growth in agriculture is achieved in sub-Saharan African countries, the current projections indicate that by the year 2015, over 38% of the population will remain in poverty (World Bank and IFPRI, 2006). Efforts to increase agricultural growth require addressing several key questions that confront African policy makers.

- What can be done to speed up the growth of agriculture in sub-Saharan Africa?
- What are the emerging challenges in increasing the contribution of agriculture to rural poverty reduction?

Answers to these questions partly depend on understanding what has been achieved so far through research, technology development and policy reform.

This book brings together the results of 20 impact studies of investments in agricultural technology development and food security in Africa. Information on the returns to investments in research can help research managers in setting priorities for agricultural research. Future research priorities also depend on the emerging issues, challenges and constraints on agriculture growth and food security, and the aim of this chapter is to highlight those in African agriculture that beg serious attention from policy makers. It enumerates a broad set of food security challenges facing African agriculture. Finally, it identifies successes in African agriculture and recent efforts aimed at scaling-up the successes.

The problem

Africa's food crisis and drought captured world attention when a million people died during the 1985 famine in Ethiopia. However, Ethiopia's experience sparked a turning point in thinking about Africa's food problem. Instead of viewing periodic drought as the cause of food shortages and famine that could be ameliorated by national and regional grain reserves and by injections of food aid, the FAO, IFPRI and a number of scholars concluded in the early 1980s that Africa was facing a long-term structural food problem comparable with India's food crisis of the early 1960s. However, after four decades of independence, tens of thousands of donor-funded development projects and billions of dollars of foreign aid, most countries in Africa have been unable to generate a reliable food surplus and develop the capacity to manage their food economies in times of abundance and during times of scarcity. Without question Africa's food crisis has not been solved.

However, Africa's food crisis and the rapid spread of GM crops in Asia and Latin America in the past 5–7 years have sparked an intense wave of interest in the role of science and biotechnology in solving Africa's food problem. This renewed interest in science and technology is an acknowledgement that the gene revolution is bypassing Africa and the realization that South Africa is the only country in all of Africa that is currently producing GM crops commercially (Table 1.1). In short, not only did the Asian Green Revolution bypass Africa but the 'gene revolution' is currently bypassing Africa. This grim reality has prompted vigorous debate among African political leaders, policy makers, academics and members of the donor community. For example, the Minister of Agriculture in Tanzania had this to say about Africa being bypassed for the second time: 'Tanzania cannot afford to be left behind by technologies that increase yields, reduce farm costs and increase profits' (cited in Balili, 2005).

Table 1.1. Seven case studies: projections of the timeline for the release of GM crops to smallholder farmers in Africa.*

Crop	Target country/region	Problem addressed	Research started (year)	Projected time of delivering GM crops to smallholder farmers
Sweet potato	Kenya	Feathery mottle virus	1991	8 or more years
Potato	Egypt,† South Africa	Potato tuber moth	1993	4 or more years
Maize	Kenya	Maize stem borers	1999	4 or more years
Banana	Uganda	Banana weevil and diseases	2000	7 or more years
Cowpea	West Africa	Pod borer	2001	8 or more years
Cassava	Kenya, Nigeria, Malawi	Cassava mosaic virus	2001	8 or more years
Cotton	Major cotton growing countries	Cotton bollworms	2000	5 or more years

Source: Eicher *et al.* (2006)
*Excluding South Africa where GM crops are grown commercially.
†Michigan State University Bt potato research with Egyptian scientists was discontinued in 2001.

The problem driving this book is the stark reality that the long-term average food grain yield has been stagnant during Africa's 45 years of independence (Fig. 1.1). Research is needed to generate high yielding food crops that are profitable to smallholders on a recurring basis and at an acceptable level of risk. However, improved food grain varieties must be supported with massive investments in roads, irrigation and human capital, and with efficient input and marketing institutions. However, African policy makers and donors should not expect extension officers, NGO workers and farmers to develop high-yielding crop varieties because there is no known high-yielding food grain variety in Africa that has been developed by NGOs or extension workers. And because of the uncertainty, risk and the time involved in plant breeding (normally a decade or longer) to develop a new crop variety, private firms are reluctant to invest in plant breeding in poor countries. Therefore, long-term public sector investments in research will be needed to develop high-yielding crop varieties for smallholders. This is a proven way to reduce urban poverty and the poverty of rural families who are net food buyers.

This book brings together empirical evidence from 20 studies of the impact of technology investments in agricultural research and technology development in Africa. Information on the returns to these investments in research can help countries in setting priorities for agricultural research.

Africa's quest for a Green Revolution

Nigeria and many other African countries attempted to institutionalize the Asian Green Revolution model in the 1970s and 1980s. Nigeria made three bold attempts that provide insights for African nations and donors today. Nigeria embarked on independence in 1960 with a reliable food surplus, competitive exports (cocoa, palm oil and groundnuts) and a set of research institutions and trained agriculturalists that were the pride of West Africa. The reliable food

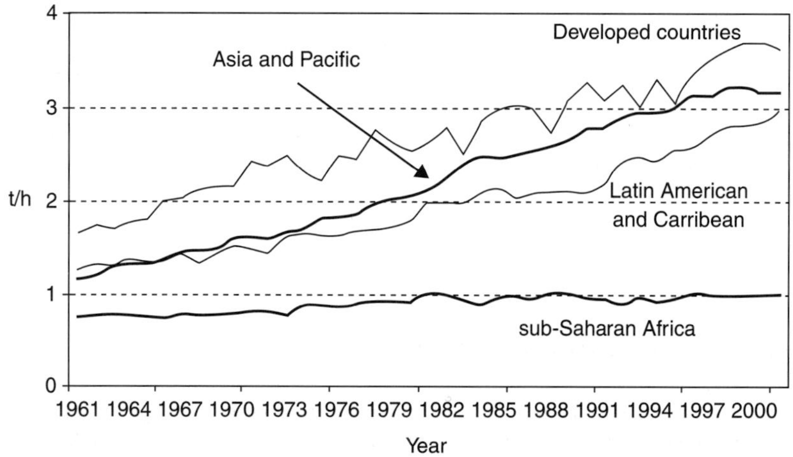

Fig. 1.1. Annual cereal yields by region, 1961–2000. Source: Eicher *et al.* (2006).

surplus explains why Nigeria's first Development Plan (1962–1968) concentrated on large-scale manufacturing and export crop production and devoted secondary attention to increasing food production. However, the combination of rapid population growth and a severe drought in the late 1960s led to a food crisis around 1970, forcing Nigeria to import food. Fortuitously, Nigeria's three decades of petroleum exploration led to commercial petroleum production starting in 1970. The petroleum exports generated $100 billion of foreign exchange earnings in the 1970s that provided more than ample resources to purchase food in international markets.

Flush with oil revenues and looking over its shoulder at Asia's Green Revolution, Nigeria launched three different crash programmes in the 1970s with one common aim: to create a Green Revolution in food production as quickly as possible. Because Nigeria has been ruled by military leaders for two-thirds of its 46 years of independence, it is instructive to assess how the Generals fared in their efforts to institutionalize the Asian Green Revolution model in Nigeria. Stunned by food deficits and rapidly escalating food prices, General Gowan launched the National Accelerated Food Production Programme in 1972. Gowan's crash programme was planned to achieve national self-sufficiency in six food crops by using improved technology that was assumed to be on the shelf; i.e. technology that was ready and awaiting diffusion to farmers. In practice, however, improved food production technologies were not readily available for local ecologies and consumer tastes.

Nigeria's second grandiose food production scheme was dubbed Operation Feed the Nation by the head of state, General Olusegun Obasanjo in 1976. This crash programme was personally spearheaded by Obasanjo, an agribusiness tycoon, and endorsed by other high ranking military officers who had become soldier-farmers. Operation Feed the Nation employed a military chain of command but it was scrapped in 3 years (1976–1979) because it did not contribute significantly to increased food production, to a drop in food prices or to the reduction of mounting food imports. Nigeria's third attempt to solve its food crisis was launched by a civilian ruler, Shedu Shagari, who became President in 1979. President Shagari launched a crash food production programme called the Green Revolution Campaign with the goal to make Nigeria 'self-sufficient in food by 1985' and a net exporter of food by 1987. Instead the overall rate of growth of staple food production fell 2.6% per year from 1980 to 1983.

Nigeria's three Green Revolution campaigns failed to achieve a tangible increase in national food production. Both military and civilian rulers used a top-down approach to bring about a

Green Revolution. They formed committees, task forces and issued directives but failed to develop a consistent policy package, economic incentives and rural service institutions to mobilize the energy of Nigeria's tiny family farms. Instead, each successive leader in the 1970s approached the chronic food-production problem with simplistic battle plans to win the war on food production in 3–4 years. Nigeria's annual agricultural growth rate of 1.7% from 1965 to 1980 was swamped by an annual population growth rate of 3.2%. The bottom line is that most African countries experimenting with Asia's Green Revolution found that they did not have high yielding food crop varieties that were profitable to farmers on a recurring basis and an acceptable level of risk.[1]

Challenges Facing African Agriculture

As two-thirds of the people in sub-Saharan Africa depend on agriculture for their livelihoods, rapid agricultural growth remains a major pathway to achieve the millennium development goal – eradicating extreme poverty and hunger in Africa. Agriculture will also continue to be the major source of rural income. Thus, removing the constraints on agricultural growth can increase the contribution of agriculture towards achieving the millennium development goals.

The productivity of the agricultural sector is in general low due to its dependence on erratic rainfall resulting in frequent droughts, a limited use of improved varieties of seeds and fertilizer, and poor marketing infrastructure. A better economic environment guided by appropriate policies and institutions is also essential for the smallholder sector to increase their productivity and income, thereby reducing poverty and food insecurity. Hazell and Diao (2005) confirmed that a major impact on poverty and food insecurity has to come from the improvement of staple crops that smallholders grow, because they employ the majority of the rural population.

Challenges to increase agricultural productivity

Science-based productivity enhancing technologies are the drivers of Africa's agricultural growth (Gabre-Madhin and Haggblade, 2003). Continued investment in agricultural innovation systems is fundamental for increasing agricultural productivity and lowering the cost of production in Africa. Average returns to past agricultural research investments have been variable because of institutional failures and poor performance of many national agricultural institutions in Africa, as shown by Anandajayasekeram *et al.* in Chapter 2.

Technology generation and dissemination

Investments in agricultural research must be increased to stimulate African agricultural productivity and translate the productivity gains from research into poverty reduction. In order to improve the benefits of research to the smallholder sector, researchers must first identify the farmers' technological needs in relation to the regional and global markets. Technological advances must go hand-in-hand with increased access to key inputs such as water, improved seeds, fertilizers and pest management practices. Successful agricultural research in Africa has produced several high-yielding crop varieties and technologies (Jones, 2005; FARA, 2006). Nevertheless, due to a lack of adaptive research, large-scale adoption has been limited.

Improving agricultural productivity of the smallholder sector in Africa will also require well-organized information management or knowledge sharing systems that provide price information, weather updates and natural disasters warnings. Another way to enhance the adoption of innovations is to improve the role of women who are key food producers. New innovations are needed to strengthen the capacity of the farmers and the extension workers in rural areas. In summary, increasing agricultural productivity of the smallholder sector will first require identification of readily available technologies and making them adaptable for various localities and agro-ecological systems and investments in research and extension.

Arresting natural resource degradation

Natural resource degradation has negatively impacted on African agricultural productivity over the past several decades. Continuous cultivation of cereal crops without adequate soil and fertility management has resulted in the depletion of soil nutrients and low productivity of farming systems. The soil infertility problem is exacerbated by growing rural population densities and by the high cost and inaccessibility of fertilizers (Borlaug, 2006). Research and innovation are needed to exploit mechanisms for smallholders who could take advantage of sustainable land and soil management technologies. While isolated successes have been achieved in improving the soil fertility through agroforestry techniques (Sanchez *et al.*, 1997), scaling up such soil management and soil fertility management technologies to a wide range of agro-ecological conditions remains a challenge (Pender *et al.*, 2006).

In order to boost innovation and adoption of soil fertility management technologies, community-based institutions should strengthen existing farmer associations and improve soil fertility. In recent years, NGOs have played a crucial part in improving soil fertility management by organizing farming communities for collective action. Yet additional institutional reforms are needed to create better incentives for rural communities to manage their land and soil resources in a sustainable manner (Levy, 2005). In summary, reducing the land degradation in Africa and increasing the adoption and use of soil fertility management technologies will improve the productivity of agro-ecological systems. Research is needed in order to identify cost-effective land and soil fertility management technologies that are easily adopted by smallholder farmers.

Prudent water management

A major constraint on the productivity of farming systems in Africa is the availability of water. Water is becoming a binding constraint for farming partly due to erratic rainfall and reoccurring droughts in various parts of the region. One way to increase the availability and the efficient use of water is through water saving technologies. Watershed development and small-scale irrigation systems are crucial for harvesting water for agriculture because large irrigation systems often result in water logging, salinization and sedimentation. Increased use of groundwater for crop production could be promoted in areas where groundwater sources are available, although water property rights and water prices remain contentious issues in African agriculture. A better understanding of the impact of water resource use in various agro-ecological systems and the conditions for successful water sharing will help to effectively manage water resources in agriculture (Shah *et al.*, 2002; SADC, 2006). Developing efficient water management technologies and disseminating them among smallholder farmers will increase productivity in African agriculture.

Strengthening marketing systems and trade liberalization

Strengthening marketing systems to reduce marketing costs and improve quality standards will be the cornerstone of improving rural incomes for a majority of African countries. Trade and market reforms carried out in the last 20 years as part of structural adjustment programmes have not fully benefited the smallholder farmers partly due to the inadequate market infrastructure such as roads, market information systems, and weak institutions governing standards and regulations (Babu *et al.*, 2002). In addition, poor development of market support services such as credit, transportation, refrigeration, storage and telecommunication networks have resulted in weak integration of smallholder farmers in rural areas to domestic and international markets. Although the trade and marketing reforms have improved the market performance in many African countries, their impact on smallholder productivity, market participation and poverty reduction has been limited (Bonger *et al.*, 2002).

Domestic market reforms

While the removal of parastatals as part of liberalizing domestic markets has helped to promote private-sector trading, a lack of capital in rural areas to initiate small-scale businesses, and thus entrepreneurship, has resulted in limited participation of the private sector in agricultural marketing. In recent years NGOs and community-based organizations have played important parts in facilitating effective marketing systems in remote areas. New innovations for improving the functioning of the markets and a new generation of policy reforms may be needed for making the smallholder sector competitive in regional and international markets (IDS, 2006).

Traditional export commodities such as cocoa, coffee and tobacco have also been affected by low productivity and quality, resulting in reduced competitiveness in international markets. Better organization of smallholders for improving the quality to meet international standards and strengthening the marketing institutions that can connect smallholder farmers to international markets will help in regaining such losses in competitiveness.

Effective domestic market reforms will be the precursor for successful participation of the smallholder sector in global trade. The productivity increases that could come from technological advances will increase the demand in the local markets, in addition to the demand that may emerge from rapid urbanization. Yet only 25–30% of agricultural production is currently marketed and the majority of market surpluses are not processed (InterAcademy Council, 2004). Marketing and agricultural processing costs continue to be high due to high transaction and transportation costs. This poses a major challenge for integrating smallholder farmers into the market economy and reducing the opportunity for increasing rural incomes. Improving access to domestic markets by infrastructure development could reduce transaction costs. Organizing market information systems to inform farmers and marketing agents will help to increase market competition. Better regulation of standards and quality of commodities will result in harmonizing market transactions and reduce exploitation of the middleman.

Participation in global markets

Trade liberalization and participation in global markets can generate income for the farmers who grow high-value crops for export. Yet such benefits could be realized only through the participation of African countries involved in WTO negotiations and would require a lot of technical and institutional support at the country and regional levels (Teunissen and Akkerman, 2005). Participation of the smallholder sector in international markets would also have an impact on their food security through world food prices. An understanding of the impact of trade liberalization on smallholder productivity and poverty levels remains minimal. It is important to know the differential impact of trade policies on commodities at the macro- and micro-economic levels. The role of regional markets in enhancing the benefits of regional trade among the countries in Eastern Africa, Southern Africa and Western Africa needs to be better understood. Finally, policy innovations are needed for better integrating smallholder farmers through vertical co-ordination of marketing systems (Scoones *et al.*, 2005).

Infrastructure, institutions and human capital

Institutional and infrastructure developments are key to agricultural growth in Africa (Rukuni *et al.*, 1998). Infrastructure development helps to reduce the prices farmers pay for inputs and to increase opportunities for selling their produce by connecting farmers in rural areas to market centres. Farmers located near areas with well developed infrastructure are likely to progress faster than those who live in remote areas with poor infrastructure, thus creating an income wedge between these two groups of farmers. Such inequity in access to infrastructure can also result in differences in access to health and educational services, which in turn can have a differential impact on agricultural productivity and poverty reduction (Babu *et al.*, 2002).

A major challenge for African policy makers is to decide on the priorities for various types of infrastructure as not all types of infrastructure investments yield the same benefits. Past research on the impact of infrastructure in China and India has shown that the largest productivity gains come from investment in agricultural research, followed by investments in education and rural roads (Fan and Hazell, 2000). Returns to investments in education and roads are particularly high in the marginal rainfed areas compared with irrigated high-potential areas. Such research results provide indications of what returns could be had from different types of infrastructure investments. There is a general agreement that even a modest increase in the infrastructure development in sub-Saharan Africa could bring in large returns, due to the low levels of the current infrastructure. Owing to low population density levels of production activities in many parts of sub-Saharan Africa, the per capita investment and maintenance cost of infrastructure development remains high. This has resulted in a widening gap in road density between African and other regions in the world over time.

Institutions that function effectively at the village, community, regional and national levels are fundamental for creating an enabling environment for agricultural development (Hall and Nahdy, 1999). At the village level, institutions that protect property rights and allocate agricultural resources such as land and water in an equitable manner will facilitate access of natural resources for the poor and landless households. Institutions that promote community management of natural resources are also essential for managing open access and common property resources. Collective action at the village and community levels is important for conserving natural resources and using them in a sustainable manner. At the community level, farmers' groups and women's self-help groups have been shown to be effective in organizing themselves for providing needed support for entrepreneurial activities. For example, improved micro-finance institutions and schemes have been successful particularly when they are built on local needs, knowledge and practice. However, formal financial institutions have not been effective in lending for farming activities, which has resulted in poor access of small-scale farmers to agricultural credit.

Developing human capital for agriculture is the key to poverty alleviation. Building the capacity of farmers, including the capacity of women through non-formal education can facilitate technology transfer and adoption. Formal training of extension workers and field level operators through investment in local training institutions is needed for energizing their role as promoters of new technologies in rural areas. Capacity strengthening of agricultural researchers both at the country and regional levels are important as well as efforts to retain them by effectively engaging them in problem-solving research.

African leadership

Scaling up and scaling out successes in African agriculture to benefit the smallholder farmers will require actions both at the country level and at the continental level. Recently, several African agricultural initiatives have been implemented to get agriculture moving in Africa. They include:

- The New Partnership for African Development's (NEPAD) Comprehensive African Agricultural Development Program (CAADP)
- The strategic plan of the African Union Commission on Rural Economy and Agriculture
- The United Nations' Inter-Academy Council study on science and technology strategies for improved agricultural production and food security in Africa
- The Forum for Agricultural Research in Africa

NEPAD has developed CAADP in order to generate momentum for continuing and expanding past successes in African agriculture. The broad objective of CAADP is to achieve food security by initiating action and associated investment on four broad priorities. They include:

1. Extending the area under sustainable land management and reliable water control systems, including increasing the access to irrigation.
2. Increasing market access to smallholder farmers through improved rural infrastructure and other trade-related interventions.
3. Increasing the productivity and supply of African agriculture across the region to reduce hunger, by effectively responding to food emergency crises.
4. Improving agricultural research funding for generating appropriate new technologies and identifying effective mechanisms for adoption of such technologies.

While these key areas have been long-standing issues for African agriculture, the CAADP initiative brings these challenges together in a focused manner with specific achievable targets. The initiative will scale up and document successes so that lessons learned could be shared among the countries. The CAADP initiative has broad political support among the key leaders in Africa and is African owned through the NEPAD initiative. The CAADP approach provides the opportunity for increased co-ordination among the partners and stakeholders in African agriculture development. It also brings together the donors and national governments, which will encourage a unified approach to agricultural development at the country and regional levels. Such harmonization will guide appropriate policies providing incentives for private farmers, teachers and other key actors in agricultural development. If implemented well, CAADP has the potential to increase welfare through enhanced productivity and integration of smallholder farmers into regional and global markets.

Conclusions

Africa is at an agrarian stage of history that is dominated by an institutional vacuum in the countryside and political neglect of the economic interests and the welfare of the rural majority. To be sure, there are encouraging agricultural reforms in a few countries such as Mali in West Africa and Rwanda in Central Africa. However, it will take decades of hard work in most African countries to craft a system of incentives, institutions and effective organizations to support a market economy and a productive agriculture. The time-consuming process of crafting incentives and institutions to develop a market economy was ignored in the 1960s and 1970s as African governments promoted state farms and parastatals followed by donor pressure in the 1980s and 1990s to privatize a wide range of government agencies. There is an urgent need for

African politicians and donors to acknowledge that Africa's success in getting agriculture moving is critically dependent on policy reforms, decentralization and massive public investments in roads, irrigation, agricultural service institutions and human capital improvement, and improving incentives and removing the heavy hand of the state. These are the same bread and butter issues that have preoccupied political leaders in India, Malaysia and Brazil over the past 40 years.

The severity of Africa's economic decline, its rural production crisis and the loss of African markets to Asian farmers is well documented. To reverse these trends requires political reforms and complementary investments in science, technology and human capacity-building initiatives. In short, strengthening the human capital and the institutional base of smallholder agriculture is essential for long-term agricultural growth and sustained rural poverty alleviation.

Several decades ago Professor Joseph Kizerbo of Burkina Faso reminded us that there is a growing realization that 'throwing billions of dollars' at Africa 'will not change anything.' The core problems today require African political leadership and African initiatives for generating a reliable food surplus, feeding people, regaining home markets and discovering how to be more competitive in regional and international markets.

The thesis of this book is that Africa's political leaders and policy makers need to craft a public sector-led agricultural strategy that is implemented by massive and long-term investments in strengthening S&T, and training in order to break the cycle of overseas training, technical assistance and the brain drain. Increasingly, comparative advantage is a function of investment in science, technology and the quality of people, rather than a function of the soil, sun and rainfall (InterAcademy Council, 2004). In the final analysis, there is a compelling, unavoidable need for S&T and human capital improvement to form the centrepiece of an Africa-wide strategy to boost cereal yields and generate a reliable food surplus over the coming decades. It took India 16 years to reach this goal and it may take most African countries one or two decades or longer.

Without a doubt, the development of a strategic plan to raise food grain yields on a sustained basis agricultural productivity should be grounded in a sense of Africa's history. The agricultural research history of Africa demonstrates that building national scientific research and training capacity is an incremental and multigenerational process that extends over decades and generations (Eicher and Rukuni, 2003). The historical record also documents the economic payoff to public investments in research on a wide range of commodities in Africa: hybrid maize, soybeans, tobacco and cotton in Zimbabwe, hybrid maize in Zambia, Tanzania and Kenya, hybrid sorghum in the Sudan, rust-resistant wheat in Kenya, improved tea clones in Malawi and Kenya, and cotton in Uganda.

The challenge for donors is to move beyond the resource transfer model that finances the construction of buildings and the purchase of equipment and vehicles for the National Agricultural Research System and universities. The challenge is to pursue a massive human capability-institutional building model that addresses the political and structural changes that Africa has to pursue in order to boost food grain yields that will reduce the real (inflation adjusted) price of food to urban consumers and net food buyers in rural Africa.

Substantial investments have been made over the last 50 years in getting agriculture moving in Africa. However, efforts to strengthen research institutions and to generate new technologies through agricultural research and development have produced mixed results (Waithaka and Minde, 2005). New approaches to public–private partnerships both at the country level and partnerships among the countries at the regional levels could substantially reduce the cost of research and increase the spill-in benefits of regional research to the individual countries. Capitalizing such new opportunities will also involve developing innovation systems that link universities of higher learning, national agricultural research systems, the private sector and civil society. Yet given the substantial past investments in agricultural research and development and their limited success in improving the welfare outcomes, justifying such investments in the future will require constant monitoring, evaluation and documentation of their impacts. This calls for institutionalizing impact assessment as part of the agricultural research planning process. Chapters 3–22 present the results of 20 case studies of the results of impact studies in Eastern and Southern Africa over the past 15 years.

Note

[1] Byerlee and Polanco (1986) found that farmers were often reluctant to adopt the entire package of farm inputs because of the risk involved. They found empirically that farmers adopted inputs in a stepwise fashion according to their relative profitability and risk.

References

African Union (2005) *Status of Food Security and Prospects for Agricultural Development in Africa.* Addis Ababa, Ethiopia.

Babu, S.C., Jagger, P., Hazell, P., Knox, A. and Rhoe, V. (eds) (2002) *Future opportunities for rural Africa.* Environment and Production Technology Division Workshop Summary Paper No. 11. International Food Policy Research Institute, Washington, DC.

Balili, D. (2005) *GM Crop Test Get Green Light in Tanzania.* SciDevNet, April 13.

Bonger, T., Gabre-Madhin, E.Z. and Babu, S.C. (eds) (2002) Agriculture technology diffusion and price policy. *Proceedings of a policy forum held by the International Food Policy Research Institute and the Ethiopian Development Research Institute,* March 25, 2002 in Addis Ababa, Ethiopia. 2020 Vision Network for East Africa Report 1. Ethiopian Development Research Institute and International Food Policy Research Institute, Washington, DC.

Borlaug, N. (2006) Importance of fertilizer in improving African agricultural productivity. Keynote address at the Africa Fertilizer Summit, June 6–9, Abuja, Nigeria.

Byerlee, D. and Polanco, E.H. (1986) Farmers' stepwise adoption of technological packages: evidence from the Mexican Altiplano. *American Journal of Agricultural Economics* 68(1), 519–527.

Eicher, C.K. and Rukuni, M. (2003) *The CGIAR in Africa: Past, Present and Future.* (OED Working Paper). World Bank Operations Evaluation Department, Washington, DC. Available at www.worldbank.org/oed/cigar/

Eicher, C.K., Maredia, K. and Sithole-Niang, I. (2006) Crop biotechnology and the African farmer. *Food Policy* 31, 6.

Fan, S. and Hazell, P. (2000) *Returns to public investment: evidence from India & China.* 2020 Focus 4 (Promoting sustainable development in less-favored areas). Brief 5 of 9. International Food Policy Research Institute, Washington, DC.

FARA (The Forum for Agricultural Research in Africa) (2006) *Framework For African Agricultural Productivity.* FARA, Accra, Ghana, 72 pp.

Gabre-Madhin, E.Z. and Haggblade, S. (2003) *Success in African agriculture: results of an expert survey.* International Food Policy Research Institute, Washington, DC.

Haggblade, S. (2005) From roller coasters to rocket ships: the role of technology in African agricultural successes. In: Djurfeldt, G., Holmen, H., Jirstrom, M. and Larsson, R. (eds) *The African Food Crisis: Lessons from the Asian Green Revolution.* CAB International, Wallingford, UK, pp. 139–159.

Hall, A. and Nahdy, S. (1999) *New methods and old institutions. The 'Systems context' of farmer participatory research in national agricultural research systems: the case of Uganda.* AgREN Paper No. 93. Overseas Development Institute, Agricultural Research and Extension Network, London.

Hazell, P. and Diao, X. (2005) The role of agriculture and small farms in economic development. In: *The Future of Small Farms.* Proceedings of a Workshop held by the International Food Policy Research Institute, Overseas Development Institute and Imperial College, June 26–29, 2005, in Wye, UK. International Food Policy Research Institute, Washington, DC.

IDS (Institute of Development Studies) (2006) *Achieving Food Security: What Next for Sub-Saharan Africa?* ID21Insights 61: 1–8. Available at: www.id21.org/insights/insights61/insights61.pdf (accessed 30 May 2006).

InterAcademy Council (2004) *Realizing the Promise and Potential of African Agriculture: Science and Technology Strategies for Improving Agricultural Productivity and Food Security in Africa.* The Royal Netherlands Academy of Arts and Sciences, Amsterdam, http://www.interacademycouncil.net/?id=9989 Accessed May 30, 2006.

Jones, M. (2005) Key challenges for technology development and agriculture research in Africa. *IDS Bulletin* 36(2), 46–51.

Levy, S. (ed.) (2005) *Starter Packs: A Strategy to Fight Hunger in Developing Countries? Lessons from the Malawi Experience 1998–2003.* CAB International, Wallingford, UK.

Pender, J., Place, F. and Ehui, S. (2006) *Sustainable Land Management: Lessons from the East African Highlands. Issue Brief* No. 43, 1–4. International Food Policy Research Institute, Washington, DC, pp. 1–4.

Rukuni, M., Blackie, M.J. and Eicher, C.K. (1998) Crafting smallholder-driven agricultural research systems in Southern Africa. *World Development* 26(6), 1073–1087.

SADC (Southern African Development Community) (2006) *Synthesis of Outcomes of the SADC Land and Water Management Scientific Symposium,* February 14–16, 2006 in Lilongwe, Malawi. Institute of Management, Lilongwe, Malawi.

Sanchez, P., Izac, A.M., Buresh, R., Shepherd, K., Soule, M., Mokwunye, U., Palm, C., Woomer, P. and Nderitu, C. (1997) Soil fertility replenishment in Africa: an investment in natural resource capital. In: Buresh, R., Sanchez, P.A. and Calhoun, F. (eds) *Replenishing Soil Fertility in Africa.* Soil Science Society of America (SSSA) Special Publication No. 51, Madison, Wisconsin.

Scoones, I., Devereux, S. and Haddad, L. (2005) Introduction: new directions for African agriculture. *IDS Bulletin* 36(2), 1–12.

Shah, T., van Koppen, B., Merrey, D., de Lange, M. and Samad, M. (2002) *Institutional alternatives in African smallholder irrigation: lessons from international experience.* Research Report 60. International Water Management Institute, Colombo, Sri Lanka.

Teunissen, J.T. and Akkerman, A. (eds) (2005) *Africa in the World Economy: the National, Regional and International Challenges.* Fondad, The Hague, The Netherlands.

Waithaka, M.M. and Minde, I.J. (2005) Striving for higher impacts in agricultural research and development. In: Omare, M.M., Makokha, M.O. and Oluoch-Kosura, W. (eds) *Shaping the Future of African Agriculture for Development: the Role of Social Scientists.* Proceedings of the Inaugural Symposium of the African Association of Agricultural Economists, December 6–8, 2004, in Kenya, Nairobi. African Association of Agricultural Economists, Nairobi.

World Bank (2006) *World Development Indicators.* World Bank, Washington, DC.

World Bank and IFPRI (International Food Policy Research Institute) (2006) *Agriculture and Achieving the Millennium Development Goals.* World Bank, Agriculture and Rural Development Department and International Food Policy Research Institute, Washington, DC.

2 The Role of Impact Assessment in Evaluating Agricultural Investments in Eastern and Southern Africa

P. Anandajayasekeram, S. Babu, M. Rukuni and S. Workneh

Abstract: Eastern and Southern Africa, in particular, have been burdened with chronic food insecurity, pervasive rural poverty and natural resource degradation. In many countries, agricultural research and development (R&D) has had limited success in improving agricultural productivity and the livelihoods of resource-poor households. Impact assessment (IA) can measure and quantify the benefits and consequences of R&D investments. Although IA of agricultural R&D gained momentum in the 1990s in Eastern and Southern Africa, relatively few of the studies in Africa have been published.

This chapter synthesizes the results of the IA studies included in this book and compares them with similar studies in other regions of the world. The available evidence indicates that rate of return for many R&D investments in Eastern and Southern African countries has been high for a wide range of programmes and commodities. Several challenges remain. There has been little methodological and practical work in assessing the economic impact of non-research outputs such as training, networking, advisory services, and policy and institutional reforms. Many assessments of the environmental impacts of R&D programmes are using qualitative methods because of the lack of data.

At present, IA results in Eastern and Southern Africa are mainly used to satisfy external accountability and reporting. Few agricultural research managers use evaluation results to support their resource allocation decisions. Fewer institutions have developed formal institutional evaluation systems to support decision-making and institutional learning. IA remains relatively new and externally driven. Institutional and human capacity is still lacking to assess impact regularly, incorporate results in a continuous process of planning research, update research outcomes and revise research plans. National R&D systems need to pay greater attention to internal learning, accountability and institutionalizing feedback mechanisms in developing IA procedures.

Introduction

In sub-Saharan Africa, 60% of the population live and work in rural areas. Recognizing the importance of agriculture in rural and economic development, the New Partnership for Africa's Development concluded that 'agriculture led development is fundamental to cutting hunger, reducing poverty . . . Agriculture must be the engine for overall economic growth in Africa'

 Impact of Science on African Agriculture and Food Security (eds P. Anandajayasekeram, M. Rukuni, S. Babu, F. Liebenberg and C.L. Keswani)

(NEPAD, 2002). This is evident also from studies showing that for every one dollar generated through agricultural production, economic linkages add another three dollars to the rural economy (Taylor and Howard, 2005). Thus productivity improvements in agriculture can contribute to increased human welfare, as demonstrated in the past century in several countries (UN Millennium Project, 2005). Investments in national, regional and global agricultural research have contributed to improved agricultural productivity. Yet decisions on how much to invest in agricultural research and in which enterprises to invest is crucially dependent on basic information on the returns to these investments. Impact assessment studies provide a basis for assessing the costs and benefits (both social and private) of investments in agricultural research and development.

Although IA of agricultural R&D programmes appeared in the literature in the 1950s, serious efforts to assess the impacts of technology based programmes in Africa started only in the early 1990s. In this chapter we shall synthesize the results from a set of IA studies reported in the chapters of this book. We present a basic analytical framework next. Then we review a broad set of studies and compare the evidence from the Eastern and Southern Africa (ESA) region with evidence from other regions in the world. Finally, we conclude the chapter with some suggestions to improve the use of IA results.

Conceptual Issues and Analytical Framework

A review of basic concepts and the analytical framework involved in IA of research provides the context for understanding the results reviewed in this chapter. IA is a special form of evaluation. In the literature, studies use the term 'impact' in many ways. It is sometimes taken to mean 'any effects' (both intended and unintended) that can be attributed to a specific action. In other cases, the concept of impact is used in a more restrictive manner referring only to the long-term outcomes. It could relate to results of a development programme on the people, economy, society or environment (Kumar, 1995) or to the ultimate effects on the country or organization (DANIDA, 1994). The concept of IA has been extended to look at the impact of research on the ultimate development goals – food security, protection of the environment and poverty alleviation (Cracknell, 1996). There is a logical evolution of IA of research from the relatively narrow focus on assessment of impacts of germplasm adoption and crop management research in the late 1970s and 1980s, to formal rate of return (ROR) and benefit distribution studies after that period. The next major methodological development was the work on spill-overs and inter-sectoral impacts. Finally, in the 1990s, IA studies expanded further into gender, environment and poverty. According to SPIA (2001) the term impact refers to the broad, long-term economic, social and environmental effects resulting from agricultural R&D.

Types of impact evaluations

Economic impact evaluations can be classified into two types: ex-post and ex-ante. Ex-post impact evaluation studies range in scope and depth – from simple story telling and anecdotal information, to partial (adoption studies) and comprehensive assessment of economic impacts (Maredia *et al.*, 2000). Ex-ante evaluations help to understand future and possible impacts of projects. In both approaches, the scope and coverage would vary depending on the objective of assessment. For example, one could assess the impact of a technological innovation on a research programme, include complementary services (such as extension, marketing, etc.) with the research programme, or assess the impact of innovation on the agricultural system as a whole. Impact can be measured at various levels of aggregation: the household, target population, national and regional levels. It could cover single, multiple sectors, or the overall economy. In contrast to single research project assessments, assessing full research programmes has advantages; it includes cost of all successful and unsuccessful projects and involves evaluation of one or more products of the research programmes.

No single IA tool is adequate to capture wide-ranging benefits of agricultural research. A 'multi-criteria analysis',[1] which may use a variety of methods, is often recommended. It is particularly useful in assessing programmes with multiple objectives. The label criteria covers a continuum, from relatively well-defined and easily measured quantities, such as yield gains, to less well-defined concepts, such as environmental quality.

Several considerations are important in assessing the impact of agricultural R&D on the welfare outcomes. A basic requirement is to ensure that the measured effects are, in fact, a result of the R&D programme under consideration and the need to verify that the farmers adopted the new innovation based on the recommendations from research. Capturing all costs and benefits of technology generation and dissemination is as essential as defining the 'with' and 'without' scenarios. Qualitative assessments using participatory methods can add value to quantitative data analysis. Finally, issues such as causality, attribution and incrementality also need careful attention.

Technologies released in the most recent 5–10-year period are ideal to study because adoption and diffusion of a technology is likely to continue for several years after its release. As the benefits from research require projections, impact studies normally include some elements of ex-ante estimation of technology adoption. Some of the studies reported in this book undertake such analysis.

A conceptual framework for impact assessment

In addition to technological improvements, R&D programmes invariably deal with non technological activities such as training, networking, development of techniques and methods, advisory services and organization and management, which may contribute to effective actions and institutional performance of research systems. These outputs affect the enabling environment of research organizations (both internal and external), ultimately impacting the overall research goals. Any comprehensive IA needs to address both technological and non-technological contributions of research and development programmes. Recognizing this need, a conceptual framework (see Fig. 2.1) was developed to guide the impact studies in the ESA region. This conceptual framework has been adopted and used in a number of empirical studies in the ESA region. Categorization of technologies into production technologies and R&D technologies is useful (Horton, 1990).[2]

A review and synthesis of impact assessment studies

In this section we shall first introduce previous reviews of agricultural IA studies and synthesis of results of the studies reported in this book. Over the last three decades, there were five major reviews of the economic impacts of R&D investments in Africa (Oehmke *et al.*, 1997; Alston *et al.*, 2000; Evenson, 2001; Thirtle *et al.*, 2002; Anandajayasekeram *et al.*, 2007). The results of these studies are summarized in Table 2.1. R&D investments assessed by these studies included a broad cross-section of the major types of research programmes. Most reported impact studies are ex-post in nature. The two key ex-ante studies reported are the ex-ante benefits from site-specific maize research in Kenya (Mills *et al.*, 1996) and the research priority setting under multiple objectives for Zimbabwe (Mutungadura, 1997).

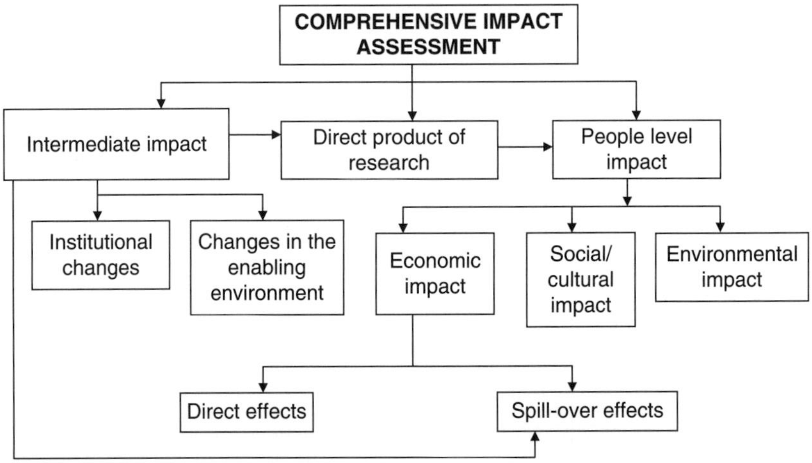

Fig. 2.1. Framework for comprehensive impact assessment. Source: Anandajayasekeram *et al.* (1996).

The unique feature of the study conducted by Thirtle *et al.* (2002) is that it looked at the ROR for the national investment using an econometric approach. According to this study the estimated overall ROR for the sample countries in Africa ranged between −12 and 58% (three of the 18 countries had negative ROR); for Asia the range was between −1 and 50% (only Sri Lanka had a negative ROR); and for Latin America, the range was between −22 and 44% (five of the 13 countries had positive RORs) as shown in Table 2.1. This study concluded that the investments in agricultural R&D raise agricultural value added for both Africa (22%) and Asia (31%) but not in Latin America (–6%).

The other four studies looked at the ROR for investments in specific commodities and/or projects and programmes. Evenson (2001) included a number of categories of investments namely extension, applied research, pre-invention science, private sector R&D and ex-ante research. Based on this review, the estimated median IRR for agricultural research for Latin America, Asia and Africa are 47%, 67% and 37%, respectively. Alston *et al.* (2000) assembled all reported (journals as well as grey literature) studies for the period 1953 to 1997. A total of 292 studies reporting a total of 1886 rates of returns was included in this analysis. The mean ROR for Asia and Pacific was 78.1 (222 observations); for Latin America and the Caribbean 53.2 (262 observations); for West Asia and North Africa 44.2 (11 observations); and for Africa 49.6 (188 observations).

The other two studies, Oehmke *et al.* (1997) and Anandajayasekeram *et al.* (2006), looked only at the studies completed in Africa. Oehmke *et al.* (1997) reviewed the economic impact studies across sub-Saharan Africa. Of the 27 RORs for the past investments in agricultural R&D, 21 showed RORs in excess of 12%. Examining the potential impacts of technologies released or still in development stage, 24 of the 30 forward-looking RORs show expected returns in excess of 12%. The review completed by Anandajayasekeram *et al.* (2006) included 81 reported studies. In estimating the ROR, 68% of the studies used the economic surplus approach and the rest employed econometric techniques. In 64% of these studies, the reported ROR exceeded 10%. In 39% of the studies, even under the worst scenario, the estimated ROR exceeded 40%. These are outstanding returns to investment by any criterion. Detailed investigation with the lower RORs suggest that researchers had not yet found the right mix of activities to produce cost-effective solutions in challenging agro-ecological environments (Oehmke *et al.*, 1997).

Table 2.1. Comprehensive ROR estimates by regions.

	Africa		Asia		Latin America		OECD	
Authors	No. of observations/ countries	Estimated ROR	No. of observations/ countries	Estimated ROR	No. of observations/ countries	Estimated ROR	No. of observations/ countries	Estimated ROR
Thirtle *et al.* (2002)*								
Overall agricultural R&D investment	17	22	11	31	13	–6	–	–
Evenson (2001)								
Extension	10	27	21	47	23	46	19	50
Applied research	44	37	120	67	80	47	146	40
Alston *et al.* (2000)	188	49.6	222	78.1†	262‡	53.2	990	98.2§
Oehmke *et al.* (1997)¶								
Ex-post	27	42.6	–	–	–	–	–	–
Ex-ante	17	34.9						
Anandajayasekeram *et al.* (2006)	81	38.0**	–	–	–	–	–	–

*ROR is estimated for overall investment in agricultural R&D at the national level.
†This includes Pacific countries.
‡This also includes Caribbean countries.
§This estimate is for all developed countries.
¶This covers the whole of Africa.
**This covers studies conducted in Eastern and Southern Africa.

Decomposition of measurements of African agricultural growth suggests that up to one-third of the growth in aggregate agricultural productivity is attributable to past investments in agricultural research (Oehmke *et al.*, 1997). This roughly corresponds to a contribution of agricultural research to economic growth of one-quarter of a percentage point. In other words, in the absence of agricultural research African economies would have grown a quarter of a percentage point slower than they actually did. In a recent study, Thirtle *et al.* (2002) concluded that a 1% yield increase reduces the number of people living on under $1 per day by 6 million and that the per capita cost of poverty reduction was $144 in Africa, $180 in Asia and $1140 in Latin America.

A synthesis of results of studies included in this book

A total of 21 impact studies in ESA are included in this book (see Table 2.2 for details). These studies were chosen to represent a wide range of commodities (food crops, cash crops and livestock) as well as approaches (econometric as well as non-econometric approaches). Over 80% of these studies used the comprehensive framework outlined earlier, followed some of the good practices (Maredia *et al.*, 2000) and assessed the impact of commodity programmes instead of individual projects or technologies. The studies include both successes and failures in estimating ROR. Thus the estimated ROR realistically represents the returns to R&D investment, minimizing the bias of the investigators. In addition, these studies also go beyond the estimation of economic returns to look at environmental, socio-cultural, spill-over as well as intermediate impacts of the R&D investments. Fifteen of the 21 studies reported in this book were completed as a follow-up of a Southern African Center for Cooperation in Agricultural and Natural Resources Research (and Training) initiated training programme conducted in collaboration with regional universities (University of Zimbabwe and University of Pretoria) as a part of capacity development partnership. These studies demonstrate the impact of the regional training initiatives. The results of the studies reported are presented in this section and the details can be found in the various chapters in this book.

Economic impact

In estimating the economic impacts of the R&D investments, 15% of the reported studies used an econometric approach and the rest used the economic surplus approach. The studies cover a wide range of commodities and agro-ecological conditions. Estimation included the costs of both success stories as well as failures. Owing to challenges in data availability, most studies estimated the RORs to research, extension and other complementary service costs such as marketing. In most cases, sensitivity analysis was performed to look at the effects of unreliable parameters.

The estimated ROR for the reported studies ranged between 0 and 167. Only three of the 19 studies reported had an estimated ROR of less than 10%. In over 50% of the studies the reported ROR exceeded 25%. About 20% of the reported studies had an estimated ROR of over 40%.

Although the research programmes evaluated in these studies may not be randomly selected, they include a broad cross-section of the major types of R&D programmes spanning from those that were heavily criticized to those considered highly successful. This compilation confirms that returns to research in ESA are similar to those found elsewhere, showing a high pay off for a wide range of programmes.

Table 2.2. Summary of impact studies reported in this book.

Country/level	Commodity	Time period	Rate of return (%)	Additional references*	Approach used
Ethiopia	Maize	1986–2000	29	Tesfaye (2001)	Surplus approach
Kenya	Wheat	1921–1990	0–12	Makanda and Oehmke (1996)	Surplus approach
	Maize	1955–1988	40–60	Karanja (1996)	Production function
Namibia	Millet	1988–1999	11	Anandajayasekeram *et al.* (1995)	Surplus approach
Tanzania	Maize	1965–1994	19–23	Moshi *et al.* (1997)	Surplus approach
	Livestock research	1966–1995	29–37	Kaliba (1995)	Supply response
	Agricultural research and extension	1971–1992	33.2	Isinika (1997)	Production function
Uganda	Cassava	1990–1996	167	Bua *et al.* (1998)	Surplus approach
Zambia	Sorghum	1983–2005	12–25	Chisi *et al.* (1997)	Surplus approach
Zimbabwe	Cotton	1970–1995	45–50	Mudhara *et al.* (1995)	Surplus approach
	Groundnut	1966–2000	59	Mazhangara *et al.* (1997)	Surplus approach
	Sunflower	1976–2000	20–23	Murata *et al.* (1997)	Surplus approach
	Sorghum	1980–1999	22	Anandajayasekeram *et al.* (1995)	Surplus approach
South Africa	Biological control of *Prosopis*	1986–2010	27–119†	Wessels (2000)	Surplus approach
	Lachenalia research and development	1965–2015	7–12	Niederwieser *et al.* (1997)	Surplus approach
	Proteaceae research and development	1974–2005	8–12	Wessels *et al.* (1997)	Surplus approach
	Livestock research	1970–1996	5–22	Mokoena (1998)	Surplus approach

(Continued)

Table 2.2. *Continued.*

Country/level	Commodity	Time period	Rate of return (%)	Additional references*	Approach used
	Russian wheat aphid integrated control programme	1980–2005	22–28	Marasas (1999)	Surplus approach
	Sorghum and millet improvement programme	1983–1993	5–14	Anandajayasekeram *et al.* (1995)	Surplus approach
	Disease control programme: tick and tick-borne disease	1997	1:08 to 1:1.2	Randela (2000)	Surplus approach
Enterprises	Wine grapes	1987–1996	40–60	Townsend and Van Zyl (1997)	Error correction model

Source: Adapted frcm Oehmke *et al.* (1997); Anandajayasekeram *et al.* (2007); Echeverria (1990); Marasas (1999); Thirtle *et al.* (1998); Thirtle and Townsend (1997). *Where the full information can be found.

†ROR was estimated for different rates of effectiveness of the control measures. For 100% effectiveness, the estimated ROR was 119; for 80, 50 and 10% effectiveness, the estimated RORs were 100, 71 and 27, respectively.

Environmental impacts

The adoption of modern agricultural technologies often results in both positive and negative externalities largely through its effects on the environment. Depending on the nature of the technology, the environmental impact can be on-site market impacts, on-site non-market impacts, off-site market impacts and off-site non-market impacts (Lubulwa and Davis, 1994). Quantifying and valuing the environmental impact of an agricultural R&D investment requires an understanding of the following: the source of the impact, the nature of the impact, the relationship between the impact and those variables that can affect stakeholders. They include current, potential and future producers, consumers and the society at large. Most empirical studies encountered two types of problems while attempting to assess environmental impacts – the data problem and the valuation problem. The major issue is how to generate information that reflects the physical, biological and economic diversity of the region/nation under study. A related challenge is to combine such information to yield reliable assessment about the region/nation. In the absence of data required for a complete analysis, it may be still possible to qualitatively identify the nature of the social benefits and costs together with the likely gainers and losers.

Though a number of studies reported in this book used the comprehensive framework, the environmental impacts were only identified and assessed qualitatively. Niederwieser *et al.* (1997) reported that in South Africa the *Lachenalia* R&D programme made a positive contribution to the preservation of biodiversity and increased the number of accessions from 17 in 1965 to over 1000 in 1997. The Proteaceae R&D programme also contributed significantly to the conservation of biodiversity in terms of gene bank accessions. The number of accessions increased from 139 in 1974 to 1774 in 1996. About 44 of these species have been classified as rare, endangered or vulnerable (Wessels *et al.*, 1997). Thus, there was strong evidence that the Proteaceae technology R&D programme was playing a crucial role in maintaining and conserving biodiversity.

The Russian wheat aphid control programme in South Africa (Marasas *et al.*, 1997) resulted in a decline in the area sprayed and the quantity of chemical used as well as the number of sprays per season. This was largely due to the use of resistant cultivars. In some studies the evidence was mixed. For example, in the Proteaceae study in South Africa, Wessels *et al.* (1997) reported that the fertilizer applied was relatively low, however the amount of chemical used to control pests and diseases was relatively high, compared with the alternative crops. A similar conclusion was reached by Mudhara *et al.* (1995) with respect to cotton in Zimbabwe. The recommendations on land preparation (reduced tillage), moisture management (tied ridges) and scouting had positive environmental effects, while the use of chemicals to control pests, diseases and weeds, and growing cotton as a mono-crop had negative effects on the environment.

Moshi *et al.* (1997) found a positive environmental effect of R&D investments in maize in Tanzania. The row planting of maize adopted by over 90% of farmers across all the seven agro-ecological zones was a good practice that controlled soil erosion. When maize is row planted across the slope, as in most parts of Tanzania, farmers mound the soil on to the seedling during weeding and small ridges are formed across the field, which help check the runoff of water. In addition, other maize management practices such as improved crop rotation, intercropping and ploughing crop residues will improve the soil structure and texture as well as the soil quality by adding more organic matter to the soil. Breeding for pest and disease resistance with flinty hard penicarp (varieties such as Kilima, Tmv-1 and Kito) reduced the need for chemical treatment both in the field and storage structures, implying a positive environmental impact.

In Zimbabwe (Mazhangara *et al.*, 1997), both large-scale commercial and smallholder farmers use groundnuts in their crop rotation to reduce soil nutrient mining in groundnut-based systems. Sunflower is used in crop rotation to control witch weed and reduce the incidence of pests and diseases. This has a positive impact on the environment (Murata *et al.*, 1997).

The biological control programme of *Prosopis* (Wessels *et al.*, 1997) had a positive environmental impact. The *Prosopis* invasion had a negative impact on the environment, in terms of land degradation, loss of biodiversity and water consumption (Harding, 1987; Verseveld *et al.*, 1998; Le Maitre, 1999). Any reduction in the rate of invasion of *Prosopis*, due to biological control, therefore had a positive environmental impact by reduction of the negative impact of *Prosopis* invasion on the environment. The biological control agents introduced in the programme were host specific and hence, have no negative impact on other indigenous plant and insect species.

In the case of Kenya (Makanda and Oehmke, 1996), the game park and the Masai grazing lands have been converted into wheat farms. The large-scale commercial farms have expanded into low-potential lands that tend to be more fragile, increasing the soil degradation or other resource problems. However, this problem was not serious because the large-scale farms had access to the funds to engage in resource conservation measures. With only 15% of the wheat area cultivated by smallholders, and some of these cultivating the less fragile, high potential land, production techniques that place greater emphasis on resource conservation can be employed on the vast majority of wheat farms without burdening the land-scarce farm family.

Intermediate impacts

There is ample evidence to show that the R&D investments have contributed significantly to institutional innovations in the form of methods and tools development, capacity strengthening as well as understanding of how these organizations interact with each other. Such impacts are the cornerstone of the emerging innovation systems perspective in agricultural R&D (Anderson *et al.*, 2005).

The Sorghum and Millet Improvement Programme contributed significantly to enhancing the enabling environment by assisting the development of national sorghum and millet research programmes, establishing research priorities, developing the necessary human capital, facilitating the development of physical infrastructure, establishing the necessary linkages (both within and outside the country) in a number of countries. Of the total R&D expenditure on this project, approximately 21% was utilized for long-term degree training, and trained a total of 89 graduates (approximately 30% were female) during 1984–1995. In addition, on the job and in-service training were also provided in research station management, breeding, plant protection, agronomy, economic and food technology (Anandajayasekeram *et al.*, 1995).

Niederwieser *et al.* (1997) reported that as a result of *Lachenalia* R&D activities, better links have been established between the research community, the propagator and the market (both local and international). The experience also showed the importance of researchers participating in the commercialization of the products and research priorities that are demand-driven based on the signals of the market. The R&D activities also enabled the development of indigenous human capacity to address various breeding and propagation aspects of bulb production. A number of other studies (Mudhara *et al.*, 1995; Marasas *et al.*, 1997; Mazhangara *et al.*, 1997; Moshi *et al.*, 1997; Murata *et al.*, 1997; Wessels *et al.*, 1997; Anandajayasekeram *et al.*, 2002) have provided evidence that R&D investments in agriculture have contributed significantly to human capital development including degree-oriented training and short-term skill building. Mazhangara *et al.* (1997) concluded that the success of the cotton industry in Zimbabwe was the result of carefully designed research, extension, marketing, credits and farmer training programmes. Chisi *et al.* (1997) reported that in Zambia the R&D investment in sorghum has contributed to a fairly sustainable programme in terms of staff, interdisciplinary approaches to sorghum research and strengthened planning, priority setting and resource allocation processes. A system of evaluating research projects and budgets has been initiated and the use of a logical framework has become an integral part of the system. Intermediate impacts emerging from R&D investment in agriculture could result in tangible benefits to the society. Ignoring such impacts could result in gross underestimation of the benefits leading to low RORs and misguided research priorities.

Socio-cultural impact

The socio-cultural impacts of agricultural R&D have been substantial. Mazhangara *et al.* (1997) reported that R&D efforts in groundnut have increased the income generated by smallholder farmers in Zimbabwe. Groundnuts are regarded as a women's crop in more than 80% of the smallholder households. The women, apart from selling groundnuts as nuts, also sell peanut butter and directly control the resulting income. The study also revealed that a significant proportion of the income generated through the sale of groundnuts in all three natural regions (NRII, NRIII and NRIV) is used to purchase daily household food needs. Peanut butter sauce added to vegetables and consumed with main meals is an important source of protein, which helped to prevent malnutrition among the rural poor, particularly among children.

Murata *et al.* (1997) concluded that sunflower is grown in the marginal environment in Zimbabwe and the income generated from it is used to purchase groceries to meet the daily needs of the households in the natural regions, having a significant positive impact on food security. Moreover, the domestication of *Lachenalia* (Niederwieser *et al.*, 1997) and expanded market opportunities in Japan and USA could create both direct employment in the commercial production units and indirect employment in the retail and wholesale markets. Similarly, the Proteaceae R&D programme also created significant employment opportunities, both full time and part time.

Randela (2000) concluded that mortalities resulting from tick-borne diseases in South Africa have a large impact on the livelihood of the rural households. The benefits derived from cattle farming by smallholders include food, income, manure and work (draught power). The income received from cattle or cattle products is largely used to meet basic needs. It is observed that the impact of cattle mortality due to tick-borne diseases on rural livelihood is both direct and indirect.

The maize R&D programme in Tanzania (Moshi *et al.*, 1997) has contributed to both household and national food security. The quantity of maize sold by the smallholder farmers increased over the years. As maize is the staple food, only surplus production is marketed. Hence, an increase in the quantity of maize sold shows an improvement in national food security. From the above cases, it is obvious that serious methodological efforts are needed to extend the standard IA procedures to account for broader socio-economic benefits resulting from agricultural R&D.

Spill-over effects

One of the key considerations in economic analysis of agriculture R&D investments is the technological spill-overs from one country to another or from one environment to another. Technological spill-overs in general increase the returns to research and can be 'spill-ins' or 'spill-outs'. The spill-overs of agricultural R&D results across geographical boundaries have implications for measures of impact on productivity, the implied rates of return to research, and for the national and international agricultural research policy.

Although the spill-over effects across commodities and geographical boundaries have been recognized in a number of studies reported in this book, very few attempts were made to quantify and incorporate them into the ROR estimates. This was largely due to the fact that the majority of studies estimated the ROR for the national investments.

In one of the studies dealing with sorghum and millet improvement in the SADC region (Anandajayasekeram *et al.*, 1995), an attempt was made to estimate the anticipated future returns to the network initiatives through the improved cultivars released in the region by incorporating the spill-over effects. A total of 26 sorghum varieties and 12 millet varieties has been released or pre-released during the period 1984–1995. Using the most conservative estimates given by scientists (of yield gains and adoption ceilings), and assuming a benefit flow period up to 2010, the estimated ROR for the total investment was 4%. Under the most optimistic scenario given by scientists, the estimated ROR was approximately 14%. This did not include the 17 sorghum varieties and 13 millet varieties that were being tested under the national on-farm verification trials.

The Russian wheat aphid-resistant cultivars produced in South Africa could also have resulted in a considerable amount of international spill-overs in Ethiopia, Egypt, Kenya, Tanzania and Lesotho where the prevalence of aphid was identified. Similarly, some of the sorghum varieties developed in Zambia are currently being used in Namibia, Botswana, Mozambique and Zimbabwe. Although this spill-over effect was recognized, this was not included in the analysis.

Previous to the *Lachenalia* programme, R&D to develop new floral crops was mainly restricted to countries in the northern hemisphere. The *Lachenalia* programme enabled South Africa to develop its indigenous capacity. The knowledge was disseminated to growers of other bulbs who were depending on the Dutch supply of propagation material. In addition, the programme is making approximately 250 crosses per year to enable the commercialization of the naturally occurring biological diversity of this flower bulb.

The foregoing results indicate that there is considerable evidence to indicate the existence of spill-over effects across commodities and national boundaries, which will increase the overall benefits of the R&D investments to the society. Owing to limited human capacity, the professional trained in one commodity and/or programme may also work in other related commodities, thus spreading the benefits.

The meta analysis performed by Alston *et al.* (2000) reported that only 12% of the 292 studies in their sample made any allowance for technology spill-overs; even fewer allowed for international spill-overs. An Australian Centre for International Agriculture Research study (Davis *et al.*, 1987) covering 12 different commodities, using a multi-country trade model, found that spill-over effects from regions where research is conducted to other regions with similar agro-ecologies and rural infrastructures ranged between 64 and 82% of total international benefits. Alston *et al.* (2000) concluded that:

- Intranational and international spill-overs of public agricultural R&D results are very important.
- Spill-overs can have profound implications for the distribution of benefits from research between consumers and producers and thus among countries, depending on their trade status and capacity to adopt the technology.
- It is not easy to measure these impacts and the results can be sensitive to the specifics of the approach taken, but studies that ignore spill-overs are likely to obtain seriously distorted estimates of ROR.
- More work is needed in this area to develop better methods to measure spill-overs and also to develop the necessary policy and institutional arrangements to harness the full potential of spill-over effects of R&D technologies.

Relative Performance of Public Sector Agriculture R&D Investments

Several recent International Food Policy Research Institute studies (Fan *et al.*, 2000, 2004a,b, 2005; Fan and Zhang, 2004) measured the effects of public spending on growth and poverty reduction in selected Asian and African countries using pooled time-series and cross-region data. To assess the impact of public investment on poverty, the number of poor people who will come out of poverty for a fixed investment (say one million shilling, 10,000 yuan, etc.) across the different sectors was estimated. Similarly, to estimate the economic benefit of the investment the benefit/cost ratios were estimated at the national level based on the increase in household income and/or productivity per unit of investment.

The ranking of the various public sector investments in relation to returns to investment and poverty reduction for the case study countries is presented in Table 2.3. In terms of returns to investment, except for Ethiopia, agricultural R&D ranked number 1, implying that it is the most efficient public sector investment possible. In terms of numbers of poor people out of poverty, agricultural R&D investments ranked among the top 3. Although limited, this evidence indicates that the investment in agricultural R&D performs as well or better than the other public sector investments and contributes significantly to poverty reduction. This demonstrates that there is opportunity to improve the growth and poverty impacts of total public investments through better regional targeting of specific types of investment (Fan *et al.*, 2005).

Summary of Studies of the Impact of Agricultural Research

The preceding review reveals that in the ESA region agricultural technology development and transfer (TDT) has had an impact across a variety of countries, commodities and agroclimatic conditions. Available ROR estimates indicate that African national governments have failed to invest adequate levels of resources in agricultural TDT, an issue currently being addressed through the New Partnership for African Development. Such resource allocation problems in TDT limit the potential impact of agricultural research to reduce food insecurity and pervasive rural and urban poverty in ESA. In a recent study, Roseboom (2002) concluded that under the assumption of full information and profit maximization, developing countries could have invested about 147% more in public agricultural R&D. In terms of agricultural R&D intensity, defined as the R&D expenditure as a percentage of agricultural GDP, they could have invested 1% rather than 0.4% during 1981–1985. Furthermore, the ESA IA results reveal that within the smallholder systems, the greatest benefits come from research leading to technologies that are simple, quickly adopted, widely applicable, result in large cost reduction and have a larger market. In addition, R&D efforts with longer gestation period and production cycles were associated with lower ROR.

A number of factors determine the type, intensity and focus of IAs (TAC, 2000). They include: purpose and objectives (learning lessons, accountability and future planning); impact dimensions of interest (environmental, economic and social); resources available to complete the assessment in a timely fashion; and previous experiences. Owing to data and valuation problems in a number of cases, the environmental impacts of R&D programmes were assessed in a qualitative manner.

There has been little methodological and practical work in the area of assessing the institutional impact of agricultural R&D. The results and impacts of institutional development activities may be difficult to quantify and measure and may take time to manifest. Paucity in coverage of institutional issues combined with preponderance of success stories or win–win cases means that the available IA efforts have denied lessons from failed cases. IA studies to date have not quantified the costs and benefits of regional programmes relative to costs and benefits that would be achieved if each country in the region carried out its own version of the

Table 2.3. Ranking of public investment effects in selected Asian and African countries.

	China	India	Thailand	Vietnam	Uganda	Tanzania	Ethiopia†
Ranking of returns in agricultural production							
Agricultural R&D	1	1	1	1	1	1 (52.46)	3
Irrigation	5	4	5	6			
Education	2	3	3	3	3	3 (9.00)	2
Roads	3	2	4	4	2	2 (9.13)	1
Telecommunications	4			2			
Electricity	6	8	2	5			
Health		7			4		
Soil and water conservation		6					
Anti-poverty programmes		5					
Ranking of returns in poverty reduction							
Agricultural R&D	2	2	2	1	1	3*	
Irrigation	7	7	5	6			
Education	1	3	4	3	3	2*	
Roads	3	1	3	4	2	4	
Telecommunications	4			2		1	
Electricity	5	8	1	5			
Health		6			4		
Soil and water conservation		5					
Anti-poverty programmes	6	4					

*The number of poor reduced per million shillings for education and agricultural research were 43.10 and 40.89, respectively.
†Data on returns in poverty reduction were not available for Ethiopia.

regional activity. Regional programmes increase the efficiency over separate national programmes in three ways: regional synergies, reducing duplication and improved ecosystem management (Oehmke *et al.*, 1997). The spill-over effects can substantially increase the total benefits of research investment (Evenson, 1987; Ewell, 1992; Sanders, 1994; Anandajayasekeram *et al.*, 1996; Alston *et al.*, 2000). Selective borrowing of technology can significantly improve the ROR for R&D investments. With the active involvement of International Agricultural Research Centres, development of subregional platforms to help co-ordinate research efforts and advances in communication technologies, it is likely that borrowing will be increasingly important in the future.

Factors contributing to limited use of impact assessment results

Despite the efforts made by the various regional and international institutions, progress in instilling IA mentality, culture and awareness within the various R&D institutes remains low. As most of the case studies are summative in nature, full potential of the impact studies has not been realized. A number of reasons can be advanced for this low impact and slow progress:

- Most studies ended up as post-graduate theses, reports, journal articles and/or conference papers and not much attempt has been made to translate them into a form and language that could be easily understood and acted upon by research managers and policy makers.
- Many IA results are used to satisfy external accountability and reporting requirements. There is limited evidence that the agriculture research mangers are beginning to use eval uation results in support of their decision making.
- Preponderance of success stories or win–win cases means the field of impact studies has been denied lessons from failed cases.
- Overemphasis on RORs and failure to highlight the impact of R&D technologies on national development goals make research mangers and policy makers sceptical of the data and methods used in IA. ROR studies cannot identify all causal factors and there is a tendency to attribute most returns to biophysical innovations thus underplaying the major contribution of institutional innovations.
- In contrast to the ideal Planning Monitoring and Evaluation (PM&E) system, few IA programme and project plans had clear objectives and indicators in practice. Plan implementation was seldom systematically monitored. Evaluations rarely depict information generated during implementation to compare achievements with objectives. PM&E is still being considered as a separate activity from plan implementation. None of the institutions has developed formal institutional evaluation systems that support decision making and institutional learning.
- Although ROR estimates are useful for making decisions on allocation of resources as well as justifying past expenditures of resources, they have limited predictive or prescriptive value. Thus emphasis should be based on both ex-ante and ex-post studies.
- Only a few impact studies provide clear guidelines for future action, making them less valuable as a research tool. It is not easy to discern meaningful patterns in the ROR, and to identify those factors that account for the systematic variation in the evidence. Many studies fail to analyse the factors contributing to impacts of R&D investment.
- Finally, the neglect of examining the apparent inconsistency between high RORs and poor aggregate performance has hampered the mobilization of funds. Studies have also failed to demonstrate the links between innovations at the farm level and improved macroeconomic performance.

These issues need to be addressed systematically in order to enhance the full benefit of IA studies.

The Way Forward

A number of areas should be addressed simultaneously in order to realize the full potential of IA as a management tool and to enhance the contributions of R&D efforts. These aspects are outlined in this section.

Methodological issues

Though a single ROR estimate is an aggregate measure of benefit from research and includes a number of socio-economic impacts, it is often unclear to policy makers that these effects are actually included in the ROR studies.

In order to enhance the impact of IA, a number of improvements are needed in the methods, analysis and presentation. These include:

- Incorporating environmental externalities and poverty related impacts.
- Institutional analysis to better understand the institutional innovations that contribute to successful R&D.
- Presenting information and policy briefs in an appropriate format and through the appropriate channel to the various stakeholders is critical to enhance communication and usefulness of results.
- Robust methodologies for assessing the institutional impacts and policy reforms are urgently needed. In recent years, the International Food Policy Research Institute has initiated assessing impacts of policy and management research (Anderson *et al.*, 2005). Much work is still needed in this area. There is a number of challenges for evaluating social and institutional impacts in a context in which partnerships and collaborative arrangements are critical to achieving programme objectives. The pooling of resources, sharing of responsibilities, and joint production and delivery of goods and services call for assessment methods to verify that collaborative mechanisms work as intended and do not produce perverse incentives and reduce rather than increase transaction costs.
- Assessment of complementary factors contributing to impact such as policy, input and output markets, price stability, infrastructure developments, institutional arrangements, extension and credit is equally important in realizing the full potential of the technology. This will enable the analysts to address the attribution problems, which is critical. Contribution analysis has been suggested as one of the approaches to address attribution problems (Mayne, 1999).
- Impact studies should look at the contributions of R&D investment to sectoral objectives such as sustainable agricultural transformation and aggregate agricultural growth and to social goals such as poverty alleviation, food security, improved nutrition, health improvements, gender consequences, employment creation and income distribution.

In pursuing research agenda for IA studies, it should be recognized that the National Agriculture Research Institutes (NARIs) in most developing countries do not have either the capacity or the resources to address these issues in a comprehensive manner. All stakeholders – NARIs, international agricultural research centres, and donors and academic institutions (both developed and developing countries) – should jointly pursue this agenda, sharing responsibilities based on respective mandates, comparative advantage and resource availability.

Institutional strengthening

The recent past has seen a dramatic change in the environment within which R&D institutions are operating in ESA. Most countries are undergoing drastic institutional reforms.

Pluralisms in service provision, decentralization of research and extension, greater involvement of private sector, non-governmental organizations and civil societies, and alternative funding mechanisms, are all important changes. At the same time the institutions are confronting a number of problems. These problems include falling research funding, low salaries, rapid turnover of staff, limited or non-existence of operating expenses, limited scientific interactions within and between organizations, lack of financial accountability and lack of accountability for impact. Priority setting exercises are relatively new and the allocation of resources to priority areas is hampered by political and institutional processes, lack of operating funds and tangible support from leadership. Institutional strengthening is needed to improve the efficiency of national and regional research organizations and systems. Improved TDT efficiency requires institutional innovations in research, extension, training and other organizations. Innovations in how these organizations interact with one another and function as one knowledge system in addressing the needs of the various stakeholders will determine the relevance of R&D institutes.

Facilitate institutionalization

The key issues that need to be addressed in moving ahead with the institutionalization of IA processes are capacity building, the interface between biophysical and social sciences, involvement of a broader range of stakeholders, the interface between NARIs and internal and external factors influencing impact, and an PM&E system for NARIs. Survey results of Anandajayasekeram and Martella (1999) clearly identify that the lack of capacity in African NARIs is one of the critical constraints for institutionalization. Thus, training scientists, in both biophysical and social sciences, is the most critical area requiring attention in moving forward. The agricultural higher learning institution should play a greater part in national capacity building.

Currently, the social science discipline takes the lead in most impact studies, yet senior and middle managers who make investment decisions tend to be biological scientists. Impact analysis should become an integral part of the programme planning to be effective and useful, and there is a need for active participation of biological scientists. Results of impact studies should be effectively communicated to policy makers, donor communities and other stakeholders to generate additional resources.

Perhaps the most critical issue in institutionalizing IA is the interface between NARIs and the internal and external factors that most influence the impacts of research (Oehmke *et al.*, 1997). In many instances the impact of research results are conditional on factors such as markets, inputs supply, policy and other institutional arrangements. Therefore, realization of full impacts may call for some strategic decisions regarding infrastructure and other institutional developments. Generating and incorporating IA results into strategic decisions in a selective way is the hallmark of a fully institutionalized IA programme.

Monitoring and evaluation systems for NARIs

At present most NARIs do not have a formal PM&E system to assess performance and impact. There is a need to establish need-based PM&E systems for agriculture and natural resource research. PM&E should become an integral part of project planning and implementation. Every research programme should include a logical framework, which should specify the indicators of success, data requirements as well as responsibilities for data collection. Nderitu (2000) argues that it is necessary to develop a harmonized planning, monitoring and evaluation system that vertically integrates organizational goals to programme, project and activity objectives and horizontally links research to key clients and stakeholders. Maredia *et al.* (2000) suggest that research organizations should have a panel of households that are regularly monitored for changes in key farm practices and productivity indicators related to the most important types of research outputs. Experience indicates that a fully integrated PM&E system should be viewed as a long-term goal for organizations to aspire to, not as something that can be implemented within a short period. Also this should be considered as a necessity, and not a 'nice to have' thing.

Conclusions

A number of studies have found that the net benefits of agricultural research are large and the ROR to R&D investments in ESA are similar to those found in other parts of the world. Evidence elsewhere in the world and the limited evidence from ESA countries suggest that investment in agricultural research has played a major part in helping to provide food for a large and expanding population (Pardey *et al.*, 2006). The productivity improvements in food crops have been closely linked to investments in agricultural R&D. The ROR for investments in agricultural R&D and the impacts of these investments are comparable with other public sector investments.

As funding agencies and policy makers demand better accountability for their investments, the importance of IA will continue to grow. They increasingly emphasize output and outcome focused evaluations. Given the limited and often declining resources for research, ex-ante IA will also play a crucial part in the allocation of scarce research resources. While there is a wide recognition that assessment of impacts can be a valuable guide for reorientation of plans and programmes, its use in the national R&D system, as part of the planning, monitoring and evaluation activities is less than effective.

Reaping the full potential of IA requires integrating and institutionalizing it in the overall research management and evaluation system of an organization. A self-evaluation culture needs fostering within the system, with R&D systems owning their assessment process and outcomes. The full involvement of decision makers and researchers in IA is key to the successful use of such impact studies. Yet IA is only a tool that supports decision making, it is neither a substitute for decision making, nor does it replace the judgement on the part of the decision makers.

Assessing the full range of the impact of R&D investments on agriculture is complex and expensive. It requires elaborate fieldwork and expertise. It is resource, data and skill intensive. Such resources and capacity needs must be recognized in planning IA activities. Assessing the economic impacts of non-research products such as training, networking, advisory services, policy and institutional reforms needs greater emphasis. Establishing relevant farm–household panel data sets for IA and maintaining institutional capacity to integrate IA results with resource allocation decisions will remain as key challenges for sometime to come.

Notes

[1] Multi-Criteria Analysis refers to a set of procedures designed to help the decision makers choose between alternative projects or options.

[2] For a detailed discussion, see Anandajayasekeram *et al.* (1996) and Marasas (1999).

References

Alston, J.M., Chan-Kang, C., Marra, M.C., Pardey, P.G. and Wyatt, T.J. (2000) *A meta analysis of rates of return to agricultural R&D: expede herculem?* Research Report No. 113. International Food Policy Research Institute, Washington, DC.

Anandajayasekeram, P. and Martella, D.R. (1999) Evaluation of agricultural research in Eastern, Central and Southern Africa. *Knowledge, Technology and Policy* 11(4), 13–41.

Anandajayasekeram, P., Martella, D., Sanders, J. and Kupfuma, B. (1995) *Impact Assessment of the SADC/ICRISAT Sorghum and Millet Improvement Program.* Vol. 1. SADC and SACCAR, Gabarone, Botswana.

Anandajayasekeram, P., Martella, D.R. and Rukuni, M. (1996) *A Training Manual on R&D Evaluation and Impact Assessment of Investments in Agricultural and Natural Resources Research.* SACCAR, Gabarone, Botswana.

Anandajayasekeram, P., Rukuni, M. and Mekuria, M. (2002) Why has impact assessment research not made more of a difference? Lessons from Eastern and Southern Africa. Presented at the International Conference on Impacts of Agricultural Research and Development, San José, Costa Rica.

Anandajayasekeram, P., Babu, S. and Workneh, S. (2007) *Impact of Research and Development in African Agriculture: Evidence from Eastern and Southern Africa.* IFPRI Discussion Paper. International Service for National Agricultural Research (ISNAR), Addis Ababa, Ethiopia.

Anderson, J., Bos, M.S. and Cohen, M.J. (2005) *Impact Assessment of Food Policy Research: a Stocktaking Workshop.* Synthesis Report, Impact Assessment Discussion Paper 25. International Food Policy Research Institute (IFPRI), Washington, DC.

Bua, A., Acola, G., Adupa, R.L., Otim-Nape, G.W., Baguma, Y.K., Sserunkuma, D., Manyong, V. and Coulibaly, O. (1998) *The Economic Impact of Investments in Cassava Research in Uganda.* Department of Research and Specialist Services, Ministry of Agriculture, Food and Fisheries, SACCAR, Republic of Zambia.

Chisi, M., Anandajayasekeram, P., Martella, D., Ahmed, M. and Mwape, M. (1997) *Impact Assessment of Sorghum Research in Zambia.* Department of Research and Specialist Services, Ministry of Agriculture, Food and Fisheries, SACCAR, Republic of Zambia.

Cracknell, B. (1996) Evaluating development aid: strengths and weaknesses. *Evaluation* 2, 23–34.

DANIDA (Danish International Development Agency) (1994) *Evaluation Report: Agricultural Sector Evaluation.* (Impact Studies, Vol. 2: Methods and Findings). Danish International Development Agency, Copenhagen.

Davis, J.S., Oram, P.A. and Ryan, J.G. (1987) *Assessment of Agricultural Research Priorities: an International Perspective.* ACIAR Monograph No. 4. Australian Centre for International Agricultural Research, Canberra.

Echeverria, R.G. (1990) Assessing the impact of agricultural research. In: *Methods for Diagnosing Research System Constraints and Assessing the Impact of Agricultural Research,* Vol. II: *Assessing the Impact of Agricultural Research.* ISNAR, The Hague, The Netherlands.

Evenson, R.E. (1987) *The International Agricultural Research Centers: their Impact on Spending for National Agricultural Research and Extension.* CGIAR Study Paper No. 22. The World Bank, Washington, DC.

Evenson, R.E. (2001) Economic impacts of agricultural research and extension. In: Gardner, B. and Rausser, G. (eds) *Handbook of Agricultural Economics,* Vol. 1. Economic Growth Center, New Haven, Connecticut, 84 pp.

Ewell, P. (1992) *The PRAPACE networks: CIP.NARS collaboration for sustainable agricultural production in Africa.* Papers presented at the Symposium on the Impact of Technology on Agricultural Transformation in Sub-Saharan Africa, Washington, DC.

Fan, S. and Zhang, X. (2004) Investment, reforms and poverty in rural China. *Economic Development and Cultural Change* 52(2), 395–422.

Fan, S., Hazell, P. and Thorat, S. (2000) Government spending, agricultural growth and poverty in rural India. *American Journal of Agricultural Economics* 82(4), 1038–1051.

Fan, S., Huong, P. and Long, T. (2004a) *Government Spending and Poverty Reduction in Vietnam.* International Food Policy Research Institute (IFPRI) Mimeo, Washington, DC.

Fan, S., Zhang, X. and Rao, N. (2004b) *Public Expenditure, Growth and Poverty Reduction in Rural Uganda.* DSG Discussion Paper No. 4. International Food Policy Research Institute, Washington, DC.

Fan, S., Nyange, D. and Rao, N. (2005) *Public Investment and Poverty Reduction in Tanzania: Evidence from Household Survey Data.* DSGD Discussion Paper No. 18. International Food Policy Research Institute, Washington, DC.

Harding, G.B. (1987) The status of *Prosopis* as a weed. *Applied Plant Science* 1, 43–48.

Horton, D.E. (1990) Assessing the impact of international research: concepts and challenges. In: Echeverria, R.G. (ed.) *Methods for Diagnosing Research System Constraints and Assessing the Impact of Agricultural Research.* ISNAR, The Hague, The Netherlands.

Isinika, A.C. (1997) *Evaluating Agricultural Research and Extension in Tanzania: the Production Function Approach.* Department of Research and Specialist Services, Ministry of Agriculture, SACCAR, Gabarone, Botswana.

Kaliba, Aloyce, R.M. (1995) The Socio-Economicas of Improved Cattle Raising under the Stall Feeding Management System in Semi-Arid Central Tanzania. Unpublished MSc Thesis, Department of Agricultural Economics, Kansas State University, Kansas.

Karanja, D.D. (1996) An Economic and Insitutional Analysis of Maize Research in Kenya. MSu International Development working paper No. 57. Department of Agricultural Economics, Michigan State University, East Lansing, Michigan.

Kumar, K. (1995) Measuring performance of agricultural and rural development programs. In: Picciotto, R. and Rist, R.C. (eds) *Evaluating Country Development Policies and Programs: New Approaches for a New Agenda.* Jossey-Bass, San Francisco, California.

Le Maitre, D.C. (1999) *Working For Water Programme: Prosopis and Groundwater: a Literature Review and Bibliography.* Environment ek Csir, Stellenbosch, South Africa.

Lubulwa, G. and Davis, J. (1994) *Inclusion of Environmental and Human Health Impacts in Agricultural Research Evaluation: Review and Some Recent Evaluations.* ACIAR Working papers Series No. 13. Economic Evaluation Unit, Canberra.

Makanda, D.W. and Oehmke, J.F. (1996) *The History of and Returns to Kenya Wheat Research.* Department of Agriculture Economics, Michigan State University, East Lansing, Michigan.

Marasas, C.N. (1999) Socio-economic impact of the Russian Wheat Aphid Integrated Control Program. PhD thesis. Department of Agricultural Economics, Extension and Rural Development, Faculty of Biological and Agricultural Science, University of Pretoria, South Africa.

Marasas, C., Anandajayasekeram, P., Tolmay, V., Martella, D., Purchase, J. and Prinsloo, G. (1997) *Socio-Economic Impact of the Russian Wheat Aphid Control Research Program, ARC.* LNR and SACCAR, Gabarone, Botswana.

Maredia, M., Byerlee, D. and Anderson, J. (2000) Ex-post evaluation of economic impacts of agricultural research programme: a tour of good practice. Presented to the workshop on 'The Future of Impact Assessment in CGIAR: Needs, constraints, and Options.' Standing Panel on Impact Assessment (SPIA) of the Technical Advisory Committee, Rome, 3–5 May, Rome.

Mayne, J. (1999) *Addressing Attribution Through Contribution Analysis: Using Performance Measures Sensibly.* Discussion Paper, Office of the Auditors General of Canada, Ottawa, Canada.

Mazhangara, E., Anandajayasekeram, P., Mudhara, M., Martella, D. and Murata, M. (1997) *Impact Assessment of Groundnut Research and the Enabling Environment in Zimbabwe: 1960–2000.* Department of Research and Specialist Services, Ministry of Agriculture, SACCAR, Gabarone, Botswana.

Mills, B., Hassan, R. and Pardey, P. (1996) Ex-ante benefits from site-specific agricultural research: maize in Kenya. In: *Contributed Papers, Conference Proceedings, Global Agricultural Science Policy for the Twenty First Century*, Melbourne, Australia, 26–28 August 1996, pp. 133–162.

Mokoena (1988) Assessment of Livestock Research in South Africa. Unpublished MSc Thesis, Faculty of Agriculture and Applied Biological Science, University of Pretoria, South Africa.

Moshi, A., Anandajayasekeram, P., Kaliba, A., Martella, D., Mwangi, W. and Shao, F. (1997) *Economic Impact of Maize Research in Tanzania.* SACCAR, Gaborone, Botswana.

Mudhara, M., Anandajayasekeram, P., Kupfuma, B. and Mazhangara, E. (1995) *Impact Assessment of Cotton Research and the Enabling Environment in Zimbabwe, 1970–1995.* SADC and SACCAR, Gaborone, Botswana.

Murata, M., Anandajayasekeram, P., Mudhara, M., Martella, D. and Mazhangara, E. (1997) *Impact Assessment of Sunflower Research and the Enabling Environment in Zimbabwe: 1976–2000.*

Mutungadura, G. (1997) Agricultural research priority setting under multiple objectives: the case of Zimbabwe, PhD thesis, Virginia Polytechnic Institute and State University, Virginia.

Nderitu, G.G. (2000) *A NARS Perspective on Impact Assessment in the CGIAR.* A paper presented to the workshop on The Future of Impact Assessment in CGIAR: Needs, constraints, and Options. Standing Panel on Impact Assessment (SPIA) of the Technical Advisory Committee, Rome, 3–5 May, Rome.

NEPAD (New Partnership for African Development) (2002) Comprehensive Africa Agriculture Development Programme. November. http://www.sarpn.org.za/documents/d0000488/ CAADP.pdf

Niederwieser, J., Anandajayasekeram, P., Coetzee, M., Martella, D., Pieterse, B. and Marasas, C. (1997) *Socio-Economic Impact of the Lachenalia Research Program.* ARC and SACCAR. Gabarone, Botswana.

Oehmke, J.F., Anandajayasekeram, P. and Masters, W.A. (1997) *Agricultural technology development and transfer in Africa: impacts achieved and lessons learned.* United States Agency for International Development Technical Paper 77, Washington, DC.

Pardey, P.G., Alston, J.M. and Piggott, R.R. (2006) *Shifting Ground: Agricultural R&D World Wide.* International Food Policy Research Institute (IFPRI), Washington, DC.

Randela, R. (2000) Socio-economic impact analysis of livestock disease control programme with special references to ticks and tick-borne diseases. MSc thesis, University of Pretoria, South Africa.

Roseboom, J. (2002) Underinvestment in agricultural R&D revisited. *Quarterly Journal of International Agriculture* 41(4), 297–316.

Sanders, J.H. (1994) Economic impact of the commodity research networks of SAFGRA. In: Sanders, J.H., Bezuneh, T. and Schroeder, A.C. (eds) *Impact Assessment of the SAFGRAD Commodity Networks.* US-AID/Africa Bureau, Washington, DC.

SPIA (2001) The future of impact assessment in the CGIAR, needs, constraints and options. *Proceedings of a workshop organized by the Standing Panel on Impact Assessment of the Technical Advisory Committee (TAC), 3–5 May 2000.* FAO, Rome.

TAC (2000) *Impact Assessment of Agricultural Research: Context and State of the Art, Impact Assessment and Evaluation Group (IAEG) TAC Secretariat.* FAO, Rome.

Taylor, M.R. and Howard, J.A. (2005) *Investing in Africa's future: U.S. Agricultural Development assistance for Sub-Saharan Africa.* Resources for the Future and Partnership to Cut Hunger and Poverty in Africa, USAID, Washington, DC.

Thirtle, C. and Townsend, R. (1997) *A review of returns to agricultural research expenditures.* Working document. Agricultural Research Council, Pretoria, South Africa.

Thirtle, C., Townsend, R.R., Amadi, J., Lusigi, A. and Van Zyl, J. (1998) The rate of return on expenditures of South African Agricultural Research Council (ARC). *Agrekon* 37(4), 621–631.

Thirtle, C., Lin, L. and Piesse, J. (2002) *The Impact of Research led Agricultural Productivity Growth on Poverty Reduction in Africa, Asia and Latin America.* Department of Environmental Science and Technology, Imperial College of Science, Technology and Medicine, London.

Townsend, R. and Van Zyl, J. (1998) Estimation of the rate of return to wine grape research and technology development expenditures in South Africa. *Agrekon* 37(2), 189–200.

UN Millennium Project (2005) *Investing in Development: a Practical Plan to Achieve the Millennium Development Goals.* UN Secretariat, New York.

Verseveld, D.B., Le Maitre, D.C. and Chapman, R.A. (1998) *Alien Invading Plants and Water Resources in South Africa: a Preliminary Assessment.* Council For Scientific and Industrial Research, Division of Water, Environment and Forestry Technology, Stellenbosch, South Africa.

Wessels, J. (2000) Impact Analysis of the Biological Control Program of Prosopis Species in the Britstown – De Aar District, South Africa. Unpublished MSc Thesis, University of Gent, Belgium.

Wessels, J., Anandajayasekeram, P., Littlejohn, G., Martella, D., Marasas, C. and Coetzee, C. (1997) *Socio-economic Impact of the Proteaceae Development and Transfer Program, ARC.* LNR, SACCAR, Gabarone, Botswana.

Zegeye, T. (2001) *The Impact of Technology Development and Transfer: the Case of Maize in Ethiopia.* Report No. 42, Ethiopian Agricultural Research Organization (EARO), Ethiopia.

3 Economics of Wheat Research in Kenya

DAVID W. MAKANDA AND JAMES F. OEHMKE[1]

Abstract: A net exporter of wheat in the 1960s, Kenya now imports the majority of its wheat requirements. Kenyan demand for wheat is rising with rapidly increasing population and higher incomes, leading to massive importation. Yet despite Kenya's investment in wheat research and extension over the past 90 years, the Green Revolution has had only a minor effect on Kenyan wheat production and yields. Consequently, Kenyans and donors are grappling with questions about appropriate investments in wheat research and technological development. This chapter combines an historical interpretation and quantitative evaluation of the Kenya wheat sector over the past century. Important findings include a more rapid increase in yields following independence in 1963, possibly attributable at least in part to investment in research; and a real rate of return to wheat research that is conservatively estimated to be in the range of 0–12%.

Introduction

The Kenyan wheat sector is unique in African agriculture (Heyer *et al.*, 1976). The Central and Rift Valley Provinces are one of the few areas that have the agroclimatic conditions necessary for successful use of modern, high-yielding, wheat varieties. Wheat is grown in the cooler and medium rainfall regions of Kenya at elevations over 5800 feet above sea level covering the Nakuru, Uasin Gishu, Trans Nzoia and Narok districts. Over 80% of the wheat is produced on large-scale, mechanized farms. Kenya has implemented mechanical, chemical and biological technology. Yet the Green Revolution that transformed Asian agriculture has had a minor effect on Kenyan wheat production. For example, Kenya was a net exporter of wheat throughout the late 1960s and early 1970s. By the end of the 1970s, Kenya was importing wheat to meet domestic demand.

This chapter has three objectives: (i) to present a chronology of Kenya's wheat sector for the 20th century, focusing primarily on yield, area and their determinants; (ii) to use this chronology to provide insights into what factors influenced wheat yields and output; and (iii) to calculate the rate of return (ROR) to investments in Kenyan wheat research.

Chronology of Kenya Wheat

Colonial establishment: 1895–1920

Experimental production of wheat by the Church Missionary Society started in 1895 in Machakos. In 1904, Lord Delamere, a pioneer farmer, began commercial production of wheat and by 1920 over 12,000 acres were planted to wheat. Most of the wheat output went to feed the urban populations in Mombasa and Nairobi.[2] Delamere promoted wheat among settlers and helped establish the British East Africa Maize Growers Association, which evolved into the Kenya Farmers Association 4 years later. In response to the 1908 devastation of his wheat crop by yellow rust (*Puccinia glumarum*), Delamere hired the wheat breeder, G.W. Evans.[3] Delamere established a wheat experimentation centre in Njoro, 120 miles west of Nairobi. In 1911, the government took over wheat research and established the National Plant Breeding Station at Njoro.

The post-World War I period: 1920–1929

From 1920 to 1928 the area planted to wheat increased over 500%, from 5547 ha to 33,595 ha. Yields were fairly constant over the period. Output changes mirrored area planted, increasing from 3357 metric tons (t) to 20,712 t, also an increase of over 500%. Increases in the area planted to wheat are attributable to the continued colonization and farming of alienated lands, primarily the highland areas, and increasing government support for wheat production, primarily through marketing policy.

In 1921, the colonial government responded to the low, recession prices for Kenyan exports by appointing the Browning Committee to protect and promote European production of a wide range of crops, including wheat. The Committee introduced producer-price supports, imposed an ad valorem tariff of 18–32% on wheat imports, and supported legislation to restrict the marketing of produce grown by Africans. Through a series of legislative and corporate manoeuvres, the Kenya Farmers Association effectively acquired monopoly power over wheat marketing.

The primary objective of wheat research was the development of rust-resistant varieties suitable for the agroclimatic conditions faced by settler farmers, with the first release of a Kenyan-developed variety in 1920. Real wheat research expenditures doubled from K7758 in 1921 to K15,596 in 1928. An improved, rust-resistant variety, Kenya Droop, was approved for release in 1929.

The depression era: 1930–1939

In addition to rapidly falling world wheat prices, the latter part of 1929 and all of 1930 were accompanied by a locust devastation and drought. Yields hit their nadir in 1931 and 1932, the area planted declined by more than one-half between 1928 and 1932, and the 1932 production of 5742 t was less than one-quarter of the 1928 production. While yield, area and output recovered somewhat in the following 7 years, previous highs were not attained in the depression era. Despite government and farmer efforts, stagnation in world markets limited the extent of the recovery in wheat production during the depression era.

World War II and recolonization: 1940–1954

The area under wheat expanded rapidly from pre-war levels of about 22,000 ha to 140,196 ha in 1955 and output rose more than fivefold from about 20,000 t in 1939 to 111,987 t in 1955.

The causes of this area expansion were increases in wheat prices and government subsidies during the war, and recolonization afterwards. Yields fluctuated during the period, rising to a peak of 1.24 t/ha in 1950, but declined to less than 0.8 t/ha in 1955. On average, yields were 0.10–0.20 t/ha higher than during the depression era.

World wheat prices tripled between 1942 and 1947 and Kenya became a net exporter of wheat from 1944 to 1952. In 1953, the world price of wheat began a long decline. During World War II, the British government promoted Kenyan agriculture as part of the war effort. Guaranteed prices and marketing control helped settlers increase food production for troops. Income subsidization of settler agriculture became firmly entrenched during the war years. In 1946, there was a new wave of colonial settlers, which further entrenched government subsidization of agriculture.

Research institutions were set up by the British colonial government after World War II to strengthen agriculture in East Africa. This was done under the auspices of the East African Services Organization, instituted in 1948.

Pre-independence: 1955–1962

In 1960, wheat covered a little over half of the total cereal area in the highlands (Odingo, 1971). Kenyan wheat research expenditures increased by an order of magnitude over the previous period: real research expenditures reached their all-time high in 1959. The increases went mostly to investments in land, laboratories and facilities, and fluctuated from year to year depending on capital purchases. This period also saw continued increases in varietal releases, with an average of six new wheat variety releases per year. Expenditures fell by nearly 50% between 1960 and 1962, due primarily to changing ownership of the research institutions.

Independence and re-institutionalization: 1963–1968

At independence, Kenya inherited well-established agricultural research institutions. These institutions were sustained throughout this period, but there was a sharp decline in funding from 1963 to 1965, as the colonial government pulled out of the country. However, real research expenditures recovered and by 1968 they had nearly reached the peak levels of 1959 and 1960. In 1967, agronomic research was institutionalized as a part of the wheat research programme. The research objective was to develop packages of agronomic practices, including tillage and soil moisture conservation practices, soil fertility control, weed control through efficient cultural practices and herbicides, and farm management and record-keeping practices (Makau, 1984).

Policy indecision: 1968–1977

Throughout this period, the new government struggled to determine what policies it should pursue, and changed policies in response to fluctuations in wheat output. Owing to mismatched producer and consumer prices, by 1971 demand exceeded supply and the Wheat Board had to import 59,500 t of wheat during 1972 to supplement local supplies. This was the start of Kenya's continuing dependence on imported wheat to meet domestic demand. The government attempted to increase wheat area in the mid-1970s, but by 1977 wheat imports equalled 50% of domestic production.

The Moi government

Since 1977 there have been drastic and fundamental changes in the organization of research and technology development in Kenya (Government of Kenya, 1981, 1986, 1988; McCarthy and Mwangi, 1982). The sudden collapse of the East African Community in 1977 led to the disintegration of a wide range of regional institutions that hitherto were jointly administered by Kenya, Uganda and Tanzania. The Science and Technology (Amendment) Act, 1979, provided the legal framework for the formation of the Kenya Agricultural Research Institute (KARI), among others, to replace the East African Agriculture and Forestry Research Organization and East African Veterinary Research Organization (Mazzucato, 1992).

Real wheat-research expenditures more than doubled in 1978. This is possibly an artefact of deterioration of regional institutions, so that the national system had to pick up expenses for some of the continuing programmes. Expenditures continued to climb until 1983, when they plummeted by 75%. They recovered slowly over the next 4 years to one-third their previous level, and then jumped by over 150% in 1987. Since that time they have fluctuated slightly below the 1987 level.[4] Until the 1980s most of the research at Njoro focused on technologies appropriate for large-scale production in the high-potential areas. However, the increase of smallholder wheat production and the increasing area under wheat in marginal areas has shifted the research priorities. Today, wheat research includes minimum and zero-tillage for the environmentally fragile marginal areas. The extension and demonstration components have been expanded to publicize small-scale wheat growing methods to the relatively new small-scale wheat farmers (Hassan *et al.*, 1992).

Trends in Wheat Yields and Research Expenditures

Visual observation of the yield data indicates a consistent, upward-trending path with random variation around this path of the degree expected in rainfed agriculture. Statistical decomposition of the series reveals an important subtlety in the data, specifically an increase in the rate of growth of yields following independence (Fig. 3.1). From 1921 to 1963, the predicted yields

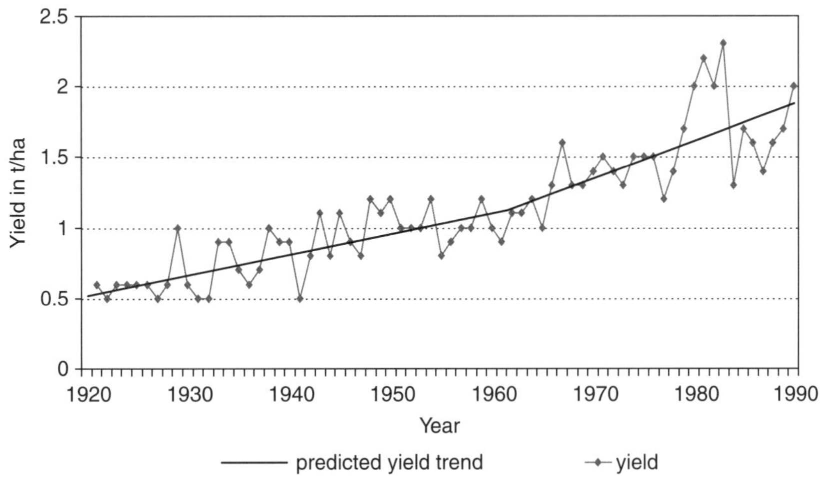

Fig. 3.1. Kenyan wheat yield and predicted yield trend: 1921–1990.

Table 3.1. Regression analysis of yield and research trends, and mean and trend shifts.

	Dependent variable	
Coefficient (*t*-statistic)	Yield	Research
Constant	0.537 (8.826)***	14,434 (1.208)
Permanent mean shift, 1953+		156,739 (5.313)***
One-time mean effect, 1953		–103,720 (–2.980)***
Trend	0.015 (6.836)***	
Trend shift, 1953+		–14,474 (–1.818)*
Second trend, 1963+	0.012 (2.385)**	
n	70	70
Adjusted R^2	0.78	0.59
Regression *F*	124.986***	26.084***

*, **, and *** denote significance at the 10, 5 and 1% levels, respectively.

increased at an annual rate of almost 1.5 kg/ha; from 1964 to 1990, the rate of increase is 2.7 kg/ha per annum – a change in the growth rate of 85%. Regression analysis shows that this change is statistically significant (Table 3.1).

This is a striking finding in the face of the pessimistic view that many African countries have not fulfilled the promises and expectations at independence. Of equal interest is the apparent robustness of the trend in yields, with the exception of the 1963 shift, to changes in the political–economic climate in Kenya. In a regression analysis, both the trend persisting throughout the sample period and the second trend starting in 1963 are statistically different from zero, and the predictive power of the equation is relatively good, as indicated by an adjusted R^2 of 0.78.

The Rate of Return on Investments in Wheat Research

Makau (1984) found that research had increased wheat yields sufficient to generate a 33% ROR on investment. Makau's finding is based on the regression of yields on area, lagged research, and rainfall over the period 1921–1984. In this section, Makau's analysis is extended in two ways: the availability of more recent data allows the sample period to be extended through 1990, and the regression model incorporates the trends and regime shifts found in the yield data series. The regression results are used to quantify the impact on production of a marginal increase in research expenditures. Valuing the increase in production at producer prices estimates the value of the research benefits; these benefits are compared with the costs (the increase in research expenditures) in the ROR calculation.

Makau's regression modelled wheat yields as a function of area, research lagged 12 years, and rainfall, using the natural logarithms of all variables. As a point of departure, this regression is reproduced using the current data set. The primary differences between the two data sets are: (i) the current data set extends an additional 8 years, to 1990; and (ii) in the current data set, the research has been deflated using Makanda's reconstruction and improvement of Makau's deflator (Makanda, 1985).[5]

Regression results show an adjusted R^2 of 0.68, substantially lower than Makau's reported R^2 of 0.965 (Table 3.2). This is probably caused by the greater volatility of research expenditures in recent years. The coefficient on area is positive and statistically significant at the 1% level, but an order of magnitude smaller than that found by Makau. The rainfall variable is negative, as is Makau's: this is to be expected, as average rainfall in the area is 930 mm/year and amounts greater than 950 mm/year are associated with increases in rust problems. The estimated coefficient on research expenditures of 0.191 is extremely close to Makau's estimate of 0.193; these coefficients indicate that an increase in expenditures of 1% will increase average yield by nearly 0.2%.

The second regression model estimated using the current data set adds to the model the two time trend variables from the yield regression presented in Table 3.1. Results include a small improvement in the adjusted R^2, to 0.75, and an F-test of the hypothesis that the coefficients on the trend variables are jointly zero rejects the hypothesis at the 1% level. The second trend (starting in 1963) is no longer positive, and not statistically significant. The area coefficient becomes negative and loses statistical significance. The research coefficient falls by nearly two-thirds, but remains significant at the 10% level. The reduction in magnitude is a direct consequence of the introduction of the trend variable. Interpretation of this finding is complicated by the fact that the regression analysis relies on the implicit assumption that the trend effect is exogenous, when in fact research may be causing some portion of the upward movement. Statistical analysis is further complicated by the correlation ($\rho = 0.55$) between the research variable and the time trend. Consequently, the estimated coefficient on research in the model with the trend may be biased downward.

The non-significance of the second trend, starting in 1963, is interesting because this is approximately the date at which the dramatic increase in research expenditures that started in the early 1950s would show up (with a 12-year lag). This indicates that the research variable does a better job, statistically, in explaining the increase in yield growth starting in 1963 than does an exogenous trend. However, there are a great many forces that could also explain this

Table 3.2. Regression models of yield as a function of research and other variables, 1933–1990.

	Dependent variable: ln(yield)	
Coefficient (t-statistic)	Makau's model	Current model
Constant	2.208	4.048
	(4.898)***	(5.501)***
Ln(Area)	0.167	−0.009
	(3.972)***	(−0.121)
Ln(Research lagged 12 years)	0.191	0.072
	(7.660)***	(1.976)*
Ln(Rainfall)	−0.161	−0.180
	(−1.480)	(−1.866)*
Trend		0.014
		(2.362)**
Second trend, 1963+		−0.001
		(−0.168)
n	58	58
Adjusted R^2	0.68	0.75
Regression F	40.5***	34.4***
F-statistic for both trend variables		8.4***

*, **, and *** denote significance at the 10, 5 and 1% levels, respectively.

yield increase: Kenya achieved independence in 1963, with a restructuring of the agricultural system and the emergence of the international research system. Consequently, circumspection is warranted when interpreting this statistical result.

An ROR on investments in wheat research can be calculated from the estimated coefficient on the lagged research variable in the regression equation. Makau calculates an ROR from the experiment of increasing research expenditures by 1% in each year in the sample. The annual benefits are then calculated by multiplying the coefficient (elasticity) on research in the yield regression by area by producer price, to obtain a value for the production increase attributed to research. The annual net benefits are the increase in production value less the increase in the cost of the research.[6] Makau's experiment is very similar to quantifying the value of increased production from a 1% increase in research when all variables are evaluated at the sample means.

Makau's calculations are reproduced using the regression results from Table 3.2. Based on the estimated coefficient of 0.19 from the reproduction of Makau's model, this results in an ROR of 12% (Table 3.3). As this represents the real returns to the investment (all prices and costs have been deflated), this number is acceptable by reasonable standards. The difference between this number and Makau's original 34% is due in large part to the different research deflator. When the research expenditure is increased by 1% in 1978 only (with a 12-year lag, 1978 is the most recent year in which the effect can be seen in this sample), the ROR remains essentially unchanged at 12%. Thus, recent research appears to be as productive as historical research.

The results using the estimated coefficient from the current model including the time trend are somewhat different. Increasing the research expenditures by 1% in every year results in a negative ROR.[7] When the 1% increase occurs only 1978, the ROR rises to 3%. The higher ROR is a result first of a greater area on which to apply the yield increase, and second of relatively high wheat producer prices in 1990. While an ROR of 3% is smaller than found in most studies for agricultural research, it does indicate a modest degree of success. However, the second regression model assumes that all of the recent upward trend in yields is exogenous and not due to research. Thus, the 3% ROR represents a lower bound on the real returns to research. As such, it indicates that Kenyan wheat research generates a low but acceptable real return.

Policy Implications

Wheat self-sufficiency

While self-sufficiency in wheat remains a stated goal of the government, this goal is impractical. The recent history of population growth, urbanization and income growth (the income of elasticity of bread estimated to be 0.97 and that of maize to be 0.70; McCarthy and

Table 3.3. Estimated RORs to Kenyan wheat research.

	An increase in real research expenditures of 1%	
	in each year, 1921–1990	in 1978 only
Regression coefficient from:		
Makau's model estimated from current data set	12%	12%
Current model estimated from current data set	< 0	3%

Table 3.4. Wheat consumption per capita per annum, by income group (kg).

	Income category		
Commodity	Low	Middle	High
Wheat	1.1	11.5	23.6
Maize	60.5	39.5	9.7

Source: McCarthy and Mwangi (1982).

Mwangi, 1982) suggests that demand will increase significantly in the absence of any relative price movements. None the less, maize is the principal food staple and is expected to remain so for the foreseeable future. Income growth will exacerbate the situation, with high-income groups consuming nearly 2.5 times as much wheat as maize (Table 3.4).

Self-sufficiency cannot be accomplished or maintained without large increases in wheat area. However, expanding wheat area will likely entail environmental degradation, loss of tourism or reduced livestock production. Moreover, the history of wheat area in Kenya indicates that expansion has occurred only when the government has provided significant subsidies. In the current budget situation, provision of subsidies does not seem to be a feasible policy.

Another way of achieving self-sufficiency is through intensification leading to higher yields. However, with no area expansion, yields would have to double or more in the next decade to meet projected consumption demand. The historical record on the growth of yields in Kenya shows that peak yields doubled between 1921 and 1950, and have come close to but not quite doubled since 1950. It seems unlikely that Kenya will be able to meet consumption demand by a doubling of yield increases within a decade.

Finally, the objective of self-sufficiency may not be desirable on economic grounds. The internal producer price of wheat in Kenya is approximately 125% higher than the world price (Makanda, 1995). While some of this price differential may be justified on the basis of transportation costs or positive externalities in supporting local farm income, it is clear that self-sufficiency in the near future will come only at a large cost. With in-kind food aid in the form of wheat available to Kenya, it seems unlikely that self-sufficiency is a cost-effective goal, at least for the near term.

Financial and economic profitability

Increases in area planted to wheat, and intensification of production, will depend on the economic and financial profitability of wheat farming. Wheat production in Kenya appears to have a comparative advantage over maize in the drier regions for mechanized wheat production in farms of 4 ha or more (Table 3.5). Smallholder wheat production using labour-intensive methods on farms of about 1 ha (or possibly in mixed farming systems with 1 ha of wheat) is more profitable than maize production, but financially maize is preferable: 'From the farmer perspective, however, using more labour-intensive technologies to produce wheat appears to be no more profitable than growing maize on small holdings' (Longmire and Lugogo, 1989). The effect of recent adjustment programmes is unclear: normally this would be expected to bring the financial incentives more in line with the economic incentives and thus give a comparative advantage to wheat. However, such programmes more effectively transmit world prices to farmers, and as long as developed countries are subsidizing wheat production and export, it is uncertain if the domestic resource costs justify Kenya wheat production.

Table 3.5. Financial and economic returns to wheat production in Nakuru District, Kenya, by farm size, 1987.

Crop/practice	0.4 ha	1 ha	4 ha	10 ha	40 ha
Maize/oxen					
Financial returns, KS/ha	3639	3652	3661	3664	3667
Economic returns KS/ha	301*,†	–114†	589	–730	–829
Wheat					
Oxen/labour intensive					
Financial returns, KS/ha	3632	3880	3997	4002	4006
Economic returns, KS/ha	–301	39	–544	–683	–781
S. Asian system					
Financial returns, KS/ha	2885	3332	3625	3703	3758
Economic returns, KS/ha	–933	–622	–854	–924	–974
Full mechanization					
Financial returns, KS/ha	2705	3672	4286	4405	4566
Economic returns, KS/ha	–774	–39	589	683	781

*For 0.5 ha plot.
†Hand cultivation.
Source: Longmire and Lugogo, 1989.

Research policy and research targets

More reasonable objectives can be stated in terms of yields or multifactor productivity, a measure of how much output is generated per aggregate amount of all measurable inputs. To improve the individual welfare of a population growing at about 3% per annum, increases in production of close to 4% per annum are necessary. For example, achieving increases in average yields of 4% per year would lead to a doubling of output in two decades with no increase in land area, substantially faster than the historical record. Achieving such growth may require a combination of further improvements in varieties available, agronomic and mechanical recommendations, and greater use of purchased inputs such as fertilizer, and possibly substantial investments in irrigation.

It is unclear how heavily KARI should emphasize wheat productivity increases generated from internal research. Maredia (1994) argues that varieties from the International Centre for Maize and Wheat Improvement (CIMMYT) outperform nationally developed varieties in most macro-ecologies. This suggests that CIMMYT should de-emphasize breeding and focus on KARI to facilitate testing, adaptation and transfer of CIMMYT wheat varieties. The success of such a strategy depends in part on effective varietal release regulations and policies, and on the ability of CIMMYT to maintain desirable varietal characteristics such as rust resistance.

Smallholders or commercial farms?

KARI has redirected some of its wheat research efforts to address the problems faced by smallholders. However, smallholders account for approximately 15% of wheat area, and the traditional comparative advantage of maize over wheat on small farms suggests that this proportion will not grow dramatically. In addition, there seems to be some preference by small farmers towards maize for food security reasons. Thus, in order to achieve average wheat yield increases of 4% per year, KARI will have to continue to focus on developing new technologies for use on commercial farms. For example, if yields on large farms did not increase, then smallholder yields would have to increase by more than 25% per year in order to achieve the 4% aggregate yield growth target. In contrast, if smallholder yields are stagnant, then large-farm yield increases of 4.7% would achieve the aggregate target. The research policy implication is to continue to focus on technical improvement for large farms, with adaptive research for transmitting these improvements to smallholders and some research on problems specific to smallholders.

Environmental concerns

Environmental concerns are nowhere more evident than in Narok district, where game park and Masai grazing lands have been converted to wheat production. Large-scale commercial farms have expanded on to low-potential lands that tend to be more fragile, increasing the risk of soil degradation or other resource problems. Fortunately, the trade-off between household food security and environmental sustainability is less pronounced in Kenyan wheat than in other parts of Africa. With only 15% of wheat area cultivated by smallholders, and some of these cultivating the less-fragile, high-potential land, production techniques that place greater emphasis on resource conservation can be employed on the vast majority of wheat farms without hurting the land-scarce farm family. Large-scale farms may have access to the funds necessary for significant investments in resource–conservation measures. Commercial farmers may choose to make such an investment if it will enhance the sustainable, productive capacity of the land. There is scope for investments that will lower unit costs at the same time as sustaining the agricultural resource base. Current research is addressing this issue.

Conclusions

The history of Kenyan wheat is dominated by government subsidies. For those few periods in which subsidization declined, such as the late 1960s, area and production declined significantly. The current budget situation will make it difficult to maintain subsidies at historical levels. Perhaps the most intriguing findings are: (i) that the rate of yield growth increased in the early 1960s; and (ii) that prior increases in research expenditures provide at least a statistical explanation of the more rapid yield growth.

For recent years, the estimated RORs indicate that investments in Kenya wheat research are generating at least a modest return. If the higher rate of yield since the 1960s is in fact attributable to research, then the RORs may be substantially higher than those estimated in this paper. The returns may be expected to increase further if area planted increases and/or if domestic wheat prices increase.

Appendix: Data Sources

For the period 1921–1982 data on wheat output, area, yield and research expenditure were obtained from Makau. The following quotations delineate his sources: yearly wheat production data and area planted with wheat was obtained from Pinto and Hurd, 1970, Annual Reports of the former Kenya Wheat Board and the current records of the National Cereals and Produce Board. Gaps for the war years 1939 and 1940 were filled with estimates from the Kenya Statistical Abstracts on production and records of correspondences in the Kenya National Archives on yields per acre (p. 196).

Research expenditures were determined from the following main sources and the best likely figures chosen:

- Kenya Department of Agriculture Annual Reports: 1921–1937. The annual reports stopped reporting on expenditures from 1938 onward.
- Colony and Protectorate of Kenya Estimates of Revenue and Expenditure: 1926–1954. From 1955 onwards expenditures for Plant Breeding Services and hence wheat research expenditures were not recorded separately in the estimates.
- Development and Reconstruction Authority Annual Reports: 1945–1951. This was an authority started in 1945 after World War II for reconstruction against the effects of the war. Apparently it disappeared around 1951 after it outlived its usefulness.
- The Colony Development and Welfare Fund Annual Reports and correspondences in files at the Kenya National Archives: 1954–1956.
- NPBS, Njoro records of expenditure and Authority to Incur Expenditures: 1957–1980. It is reported in archival records that there was a fire outbreak at the NPBS, Njoro in 1953/54 and therefore the records at the station before 1955 were destroyed.
- Wheat Board of Kenya Annual Reports: 1963–1976.
- End of Year Ledgers of the Ministry of Agriculture: 1974–1982. It is unfortunate that these ledgers are destroyed after a few years: they contain accurate information on the actual expenditures on every research station under the Ministry. It is not possible to get such accurate information from stations themselves because they control only part of their budget, the rest of the budget salaries and allowances being controlled at the Ministry Headquarters. Such information is easily lost after a few years when the Ledgers are destroyed.
- Danish International Development Agency, CIDA, NPBS Njoro records and the 1976 UNDP Compendium on Development Assistance to Kenya.
- Kenya Government Appropriation Accounts and Audit Reports (pp. 197–198).
- For the period 1982–1990, Makanda extended the data based on records kept at the National Plant Breeding Center, Njoro.

Notes

[1] Department of Agricultural Economics, Michigan State University, East Lansing, Michigan. The authors acknowledge support from USAID and the Rockefeller Foundation. Responsibility for errors remains with the authors.

[2] The population of Nairobi was 4950 in 1909.

[3] Makanda (1995) reports that Delamere hired a breeder in 1907.

[4] For this time period, the reader should not attach much importance to any particular expenditure figure because changes in exchange-rate policy, the availability and allocation of donor funds, and it was a period of fluctuation and change in wheat research.

[5] Makau (1984) constructed a Capital Deflator Index to translate nominal research expenditures into real research expenditures, using regression analysis of various price indexes to interpolate missing values in early years. After extending Makau's data set to 1990, Makanda (1995) reconstructed Makau's deflators using the longer data series.

[6] This method implicitly assumes that the farmer incurs no additional costs from implementing the research results. As the majority of wheat research in Kenya is varietal development and testing, there should be only minor costs associated with adopting a different variety. Therefore this assumption is deemed acceptable.

[7] For an elasticity of yield with respect to research of 0.7, there are at least two sub-periods in which the net benefits are negative. Consequently, there will be multiple ROR solutions, but numeric search indicates that the reported results are the only ROR solutions found in the relevant ranges.

References

Farm Economics Survey Unit (1964) *Farm Production Costs in the Njoro Area 1958–61.* Ministry of Finance and Economic Planning, Economics and Statistics Division, Report no. 20. Nairobi, Kenya.

Government of Kenya Colony Development and Welfare Fund. *Annual Reports, 1930, 1936.* Government Printer, Nairobi.

Government of Kenya (1981) *Food Policy Paper of 1981.* Government Printer, Nairobi.

Government of Kenya (1986) *Sessional Paper No. 1 of 1986, Economic Management for Renewed Growth.* Government Printer, Nairobi.

Government of Kenya (1988) *1989–93 Development Plan.* Government Printer, Nairobi.

Hassan, R., Mwangi, W. and Karanja, D. (1992) *Wheat Supply in Kenya: Production Technologies, Sources of Inefficiency and Potential for Productivity Growth.* KARI/CIMMYT Project Report, Njoro, Kenya.

Heyer, J., Maitha, J.K. and Senga, W.M. (1976) *Agricultural Development in Kenya.* Oxford University Press, Nairobi.

Longmire, J. and Lugogo, J. (1989) *The Economics of Small-Scale Wheat Production in Kenya.* CIMMYT Economics Working Paper 89/01. CIMMYT, Mexico D.F.

McCarthy, F.D. and Mwangi, W.M. (1982) *Kenyan Agriculture toward 2000. International Institute for Applied Systems Analysis,* Research Report 82-19. Laxenburg, Austria.

Makanda, D.W. (1995) Wheat policy in Kenya. PhD thesis, Department of Agricultural Economics, Michigan State University, Michigan.

Makau, B.P. (1984) Measurement of economic returns to research and development: the case of wheat research in Kenya. MSc thesis, Department of Economics, University of Nairobi.

Maredia, M. (1994) The economics of the international transfer of wheat varieties. PhD thesis, Department of Agricultural Economics, Michigan State University, Michigan.

Mazzucato, V. (1992) Non-research policy effects on the rate of return to maize research in Kenya: 1955–1988. MSc thesis, Department of Agricultural Economics, Michigan State University, Michigan.

Odingo, R.S. (1971) *The Kenya Highlands.* East Africa Publishing House, Nairobi.

Pinto, E.F. and Hurd, E.A. (1970) Seventy years with wheat in Kenya. *East African Agricultural and Forestry Journal* xxxvi, special issue.

4 Impact of Sorghum Research and Development in Zimbabwe: the Case of SV 2

P. Anandajayasekeram, D.R. Martella, J. Sanders and B. Kupfuma

Abstract: Sorghum and millet are important for human consumption in marginal rainfall regions, for low income consumers in other regions and for beer production in Zimbabwe. After independence the research and policy emphasis on sorghum shifted towards communal farmers, from red-seeded to the white-seeded sorghum varieties. The variety SV 2 was identified from an International Crops Research Institute for the Semi-Arid Tropics (ICRISAT) regional trial in 1980 released in 1987 and subsequently commercialized by Seed Co-op, a private seed company. The diffusion of SV 2 was accelerated during the dry years of the 1990s. In this study an attempt was made to assess the impact of R&D investment on SV 2. Farm surveys indicated that in a normal year there were substantial yield gains from the introduction of SV 2 alone without and with the use of fertilizer. The benefits of the introduction of SV 2 are calculated with two different assumptions: with and without fertilizer. The estimated net present value for 10, 15 and 20% discount rate was positive indicating that the development and transfer SV 2 generated benefits that are over and above the amount invested. The estimated internal rate of return was about 25%. The effect of use of fertilizer on IRR was modest. Sensitivity analysis indicated that omission of administration, overhead and depreciation costs, and the extended benefit flow period significantly affect the rate of return to technology development and transfer.

Introduction

The focus of agricultural policy and research in Zimbabwe prior to independence in 1980 was on the white, large farm sector. Then an estimated 6500 farmers held 50% of the land and produced 94% of the market surplus (IFAR, 1993). The emphasis of the national research programme on sorghum in the 1970s was on short season, early cultivars for brewing. After independence the research and policy emphasis shifted to the communal farms sector. Research attention changed from the red-seeded to the white-seeded sorghum cultivars preferred for human consumption.

The variety SV 2 was identified from an International Crops Research Institute for the Semi-Arid Tropics (ICRISAT) regional trial in 1980, released in 1987, and diffusion accelerated during the dry years of the 1990s. This chapter documents the introduction of SV 2 and synthesizes what one could learn from this experience. It is important to keep in mind that SV 2 is largely a product of the national research services of Zimbabwe. With the available data, it was difficult to estimate the rate of return (ROR) for research alone. Therefore, the ROR was estimated for research as well as the complementary services.

The Importance of Sorghum and Millet

Maize is the predominant staple food in Zimbabwe. Wheat production and consumption increased rapidly during the 1980s as well as the consumption of rice. The consumption of maize, sorghum and millet has been declining, especially in urban areas and among middle and upper income sectors.

Sorghum and millet are not major crops in Zimbabwe. Over the 1989–1991 period the combined sorghum-millet production was only 12% of maize production (Table 5.1). Nevertheless, sorghum and millet are important for human consumption in marginal rainfall regions, for low income consumers in other regions and for beer production. These coarse grains will become important in animal feed as incomes increase to create demand increases for animal protein.

There are two sorghum production sectors in Zimbabwe: the commercial sector on good land with high input use and production principally for the breweries, and the communal land with little input use besides labour and production principally for human consumption. Average yields in the commercial sector are 2 t/ha and above and reflect the use of inorganic fertilizers. Communal yields, however, are about 500 kg/ha. For sorghum area (Fig. 4.1) and production, the communal sector is clearly the dominant sector.

The production of sorghum and millet is concentrated in the more marginal and lower fertility soils. The use of inorganic fertilizer on these crops in the communal areas is minimal so yields are low and declining as low fertility soils are further depleted. A sustainable cropping system on these low fertility soils will understandably require higher nutrient levels.

The Development and Diffusion of SV 2

The introduction of any new cultivar is always a joint effort and separating the individual contributions is extremely difficult. The DR&SS Sorghum Program identified SV 2 from a Sepon

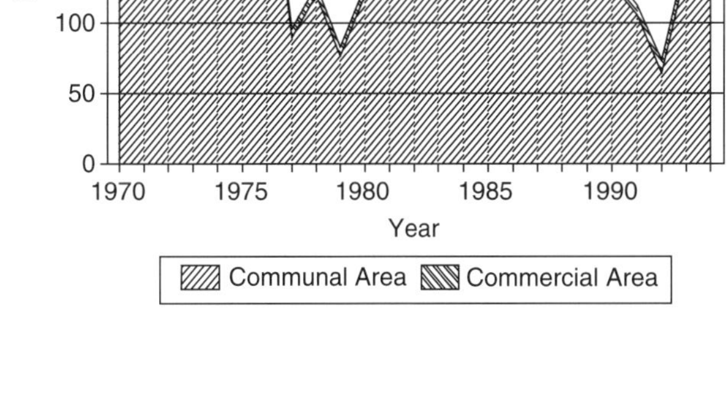

Fig. 4.1. Zimbabwe sorghum and millet area 1980–1994. Source: Zimbabwe Central Statistical Office.

nursery of ICRISAT in 1980. After experiment station testing, regional testing and farm trials in collaboration with the national extension service (AGRITEX), SV 2 was released by the Zimbabwean government in 1987. Breeder seed was produced in 1988 and provided to the Seed Cooperative for multiplication and distribution. Sorghum and Millet Improvement Programme's (SMIP) principal contribution was emergency seed production and distribution after the catastrophic drought of 1991–1992. Moreover, SMIP helped with collaborative on-farm trials of SV 2 in the 1990s along with the national programme and the extension service (AGRITEX). SMIP and the DR&SS provided seed to various organizations including NGOs for on-farm testing and distribution to farmers.

Farm Trials of SV 2

Most of the on-farm trials were conducted using 100 kg/ha of compound fertilizer D (7, 14 and 8 kg/ha of the three principal nutrients) and 50 kg/ha of ammonium nitrate (34.5%) side-dressing after 3 weeks (Table 4.1).

On-farm trial results at various communal areas and the farmer yields in selected areas are presented in Table 4.2. The same moderate levels of fertilization were apparently utilized in all the on-farm trials. There is substantial variability in these results with a 17% mean difference for a normal rainfall year with fertilization. In the absence of the new cultivars, yields would have been 14% lower. These estimates were obtained from 180 farmers participating in the emergency seed impact assessment study (Friis-Hansen and Rohrback, 1993). From the farmers' yield levels, it appears that in half of the sites farmers were also provided with inorganic fertilizer, as that was the prevalent practice in AGRITEX when the SV 2 and other new cultivars were distributed. In Hwange, Kezi and Mwenesi yields were much lower so perhaps there was no fertilization or much lower initial soil fertility.

The farmers' absolute yields were lower than those of the researcher managed trials. On average there was a 54% increase in yield with SV 2 in a normal rainfall year or, in the absence of the new cultivars, yields would have been 35% lower during a normal rainfall year.

The farm trials of the Zimbabwe national sorghum programme showed mean yields for 1992/93 of 2.15 t/ha for SV 2, and 1.56 t/ha for the local variety with researcher managed, farmer implemented trials. In farmer managed and farmer implemented trials for SV 2 and local landraces, yields were 2.05 t/ha and 1.76 t/ha respectively. These were proportional yield gains of 38% and 16%, respectively. Or in the absence of these new cultivars yields would have been reduced by 27% and 14%. The standard fertilizer recommendation indicated above was apparently used on these trials.

Table 4.1. Yields with fertilization in the Tsholotsho District of Zimbabwe.

	Yields (t/ha)	
Crop year	SV 2	Local
1989/90	2.16	1.5
1990/91	1.6	1.2
1991/92	0.51	0
1992/93	3.0	1.9
1993/94	1.6	1.1

Source: Farm records of AGRITEX.

Table 4.2. Sorghum yields in a normal rainfall year, 1992/93.

	SMIP on-farm yield trials (kg/ha)		Farmers' yields (kg/ha)	
Communal area	SV 2	Landrace	SV2	Landrace
Binga	2080	2130		
Hwange	3300	2220	234	59
Lupane	1860	1690		
Tsholotsho	1520	1060	1853	1065
Plumtree	1440	1260		
Gwanda	970	920		
Mutoko	3380	2450		
Marange	1430	1140		
Gutu	1330	1460		
Chivi	2850	3110	1514	788
Muzarabani 1	2790	1980		
Muzarabani 2	1780	1750		
Gokwe			1459	1309
Kezi			406	188
Mwenezi			534	482
MEAN	2061	1764	1000	649
Binga	2080	2130		

Source: SMIP Seed Impact Assessment Survey, 1993.

The yield information obtained in an adoption survey conducted by SMIP in 12 different communal areas is presented in Table 4.3. The amount of fertilizer used by farmers was negligible. In 2 of the 3 years, farmers faced adverse or low rainfall conditions. For 1991/92 and 1993/94 period without fertilizer and at low rainfall levels, the local landraces out-yielded SV 2. SV 2 only did better than the locals in the normal rainfall year. This reflects farmers' rationale for choosing low yielding but more stable local cultivars at zero fertilization levels. The SMIP survey reveals that when SV 2 is fertilized even in the low rainfall years, such as 1991–1992 and 1993–1994, it out-yielded local sorghum. To exploit the potential of the cultivar, the critical input combination is to introduce both SV 2 and inorganic fertilizers.

Diffusion and yield differences

One basic problem in estimating the diffusion of varieties is that once the seed becomes available to farmers, they can also produce seed for themselves, their neighbours and relatives and for sale. Therefore, estimates of the extent of diffusion cannot be made from seed sales except for the first year of release and under exceptional circumstances such as after the drought of 1991–1992.

If all the seed produced by the Seed Cooperative were sold domestically, then 3.2% of the area in 1988 would have been under SV 2. During the late 1980s and early 1990s, many NGOs were buying seed in Zimbabwe for distribution in Angola, Mozambique and Namibia. However, in the first year of the availability of a new cultivar we assumed that most of the seed would be distributed locally. Thus, an estimated 3% of the communal sorghum area was planted under SV 2 in 1988 (Table 4.4).

Table 4.3. Yields in the communal areas with farmer practices: 1991/92–1993/94.

	Yields (kg/ha)	
Year	SV 2	Landraces
1991/92	360	375
1992/93	733	435
1993/94	283	323
Average	459	378

Survey Source: SMIP Adoption.

Table 4.4. Diffusion estimates for SV 2 in the communal areas, 1987–1998.

		Percent diffusion based on		
Year	Seed sales (metric tons)	Seed sales*	Survey data†	Estimation
1987	58.5			
1988	61.9	3		3
1989	48.1			(5)‡
1990	69.3			(8)
1991	93.8			(12)
1992	435.2		17	17
1993	114.8	29	30	30
1994			37	37
1995			36	36
1996				(38)
1997				(40)

Source: *Zimbabwe DR &SS; †SMIP Adoption Study.
‡The data in parentheses represent interpolation before 1992 and conservative projections after 1995.

In 1991/92, Zimbabwe had a serious drought with many farmers experiencing complete crop loss. There was an emergency seed programme to provide farmers with seed for 1992/93 crop season. Using the seeding rate of 10 kg/ha, sufficient emergency seed was produced for 29% of the communal sorghum area. Assuming that there was some minimal seed retention and utilization by farmers, 30% of the area was estimated to be in SV 2 in 1992/93.

These seed sale estimates were complemented with 1994/95 survey results from a sample of 256 farmers from 12 different communal areas. Recall data were utilized for the estimates of the sorghum area in the new cultivars from 1991 to 1994.

Finally, for the missing years the impact assessment team interpolated the adoption based on the discussions with research and extension staff. These estimates were very conservative for the future years. The adoption rate declined in 1995 and farmers claim that the principal reason was the difficulty in obtaining seeds. Seed and fertilizer distribution was assumed to improve but the adoption rate was assumed to slow down and peak at about 40%, i.e. the adoption ceiling was assumed to be 40%.

Various sorghum breeders were also asked for their subjective views of the extent of diffusion of SV 2. DR&SS estimated that 10–15% of the area of communal sorghum was in SV 2 in 1991/92 and 33% in 1992/93. SMIP breeders thought that 33% of the communal sorghum area was in SV 2 in 1994/95 and in another 5 years diffusion would peak out at 43%. These estimates are consistent with the diffusion estimates used in this study. According to this survey the farm area in sorghum remains constant over the period studied, but there is a gradual shift to SV 2. The sharp increase in adoption during 1993/94 was largely due to the emergency seed distribution. Farmers explained that there was a shortage of seeds in 1994/95 constraining them to reduce the size of the area desired for SV 2. The 1993/94 crop season was a dry year and the extension service provided SV 2 and fertilizer in 1994/95.

The social benefits of SV 2

The international price for sorghum was $100 per metric ton (+) in 1993 and $104 in 1992. Given the fact that maize is usually imported to offset shortfalls in sorghum and millet production, the international reference price might alternatively be the maize price. As the international price for sorghum and maize are in the same range this will not affect the analysis. Adding a $40/t transport and handling charge (ocean and railroad) gives an import parity price of approximately $140/t to the urban areas of Zimbabwe, where the competition for the meal and flour market takes place. This constant import parity price was assumed over the period for the purpose of this analysis.

SMIP and other survey data show that without fertilization there is very little advantage to SV 2 over local landraces in dry years. However, in farm-level experiments during the dry years, 1991/92 and 1993/94 crop seasons, the fertilized SV 2 out-yielded the fertilized local landraces. Farm surveys indicated that in a normal rainfall year there was a substantial yield gain from the introduction of SV 2 alone without fertilization. In the normal year, 1992/93, the local variety without fertilizer yielded 435 kg/ha and SV 2 yielded 733 kg/ha. The trial data show that with fertilizer yields ranged from 1 to 3 t/ha. The local sorghum also yielded much more with fertilization in these farm trials, but the combined effects of SV 2 and inorganic fertilizers were substantial.

The benefits of the introduction of SV 2 are calculated with two different assumptions. In the first assumption, only the new cultivar is introduced without fertilization. In the second only the new cultivar is introduced up to the present (1994/95 crop year) without fertilization. Then over the next 5 years, from the 1995/96 crop season onward, the standard fertilizer levels utilized in the farm testing of SV 2 are adopted on a gradually increasing percentage of farmers' fields. The addition of 100 kg/ha of the compound D (7, 14 and 8% of N, P_2O_5 and K) and 50 kg/ha of ammonium nitrate increases SV 2 yields to 1500 kg/ha in normal years and 750 kg/ha in dry years when the unfertilized local cultivar yields decline to 335 kg/ha. In the experiment/trial data, only two levels of fertilization were indicated. The high level of fertilization was used in the costs estimates for this study, although further fertilizer experiments may suggest a different level that could yield the maximum benefit for investment in fertilizer. Unfortunately, more data were not available to make this adjustment.

The yield data used to estimate the actual and the potential benefits to the introduction of SV 2 with and without the future introduction (after 1995) of inorganic fertilizer are summarized. The impact assessment team assumed that if the planting time were correct, then even in the dry years there would be an advantage to SV 2. In dry years the yield decline in the absence of the new cultivar would have been 10%. With these yield data, the benefits of the introduction of SV 2 are estimated (Table 4.5). These are economic benefits because the study used the border price of sorghum and did not include the subsidies on the seeds and fertilizers.

Table 4.5. Calculated benefits accruing to improved varieties, 1998–1999.

	Benefits (1993 US$–1000)	
Year	Observed*	Potential†
1988	282	282
1989	186	186
1990	341	341
1991	379	376
1992	179	179
1993	1285	1285
1994	244	244
1995	1319	1319
1996	1405	1408
1997	1493	1501
1998	219	333
1999	1582	1607

Source: SACCAR impact study estimates.
*Cultivar alone, no fertilizer.
†Cultivar alone until 1994/95, then moderate fertilization.

The demand and supply elasticities are both assumed to be 0.4. The ROR calculations are made using the Akino–Hayami method for calculating social benefits.

Costs

In any benefit–cost analysis, it is important to estimate and account for the costs involved in the development and transfer of the technology, including the adoption costs to the farmers. The costs of developing the original ICRISAT material was assumed away as these are considered as global public goods. The various cost components that need to be included in the analysis are: research costs (including on-farm trial costs), extension costs, cost of seed multiplication and distribution, as well as the adoption cost to farmers. These cost components are discussed in the following sections.

SV 2 adoption costs

The main cost farmers incur by adopting SV 2 is the purchasing of seed, and to a limited extent, fertilizer. The price charged for sorghum seed by the Seed Cooperative was not available and therefore estimates had to be made. Using the farm-level seed price of 1992, a seeding rate of 10 kg/ha, and the official exchange rate in 1993, seed costs were estimated to be US$2.50/ha (1993 dollars).

For fertilizer the standard recommendation of 100 kg/ha of compound D and 50 kg/ha of ammonium nitrate was costed in 1993 prices and converted to 1993 dollars with the official exchange rate to give fertilizer costs of US$23.40 (1993)/ha. The SV 2 area fertilized was projected to be 5% in 1996, 10% in 1997, 15% in 1988 and 20% from 1999 onwards.

SV 2 research and extension costs

Although the introduction of SV 2 in Zimbabwe was a joint effort, DR&SS played a lead role with vital support coming from AGRITEX, SMIP and ICRISAT. While it is possible to calculate the benefits of such joint programmes, it is difficult to estimate the costs incurred by each institution. In the case of sorghum research and development, DR&SS and AGRITEX do not have commodity based accounting systems. SMIP also, being a regional programme, does not have a system of apportioning costs on the basis of commodity by country and by activity. This means that the research and extension costs for the introduction of SV 2 are at best estimates.

There are three components of research and development costs. First are the adoption costs, which have already been accounted for in the form of seed and fertilizer costs. The second component is the research costs, and the third component is extension or transfer costs. Research and extension costs should be specific to the technology in question. For this study, the costs of interest are only those that the various research and extension institutes incurred in developing and introducing SV 2. For both research and extension, the costs should be broken down into: research running expenses, staff salaries and benefits, administration and overhead costs, and capital assets depreciation (buildings and laboratories). If the research and extension cost data are not available in this form, then data on professional staff establishment and responsibilities, and total expenditures can be used to estimate the proportions of each cost item that can be attributed to the development and transfer of a technology. For the introduction of SV 2, the costs of interest are those incurred by DR&SS, AGRITEX and SMIP.

DR&SS researchers provided estimates of sorghum research running costs over the years. For the years when data were unavailable, such as in the future, extrapolations were made on the basis of past trends. There were no data on salaries and benefits for staff. The salaries and benefits for staff were estimated assuming a ratio of 2.5 times the operation costs, as on average, operation costs are assumed to constitute 40% of the research running costs and staff salaries budget. There was no basis on which the research administration and overhead and depreciation costs could be estimated given the available database. Thus administration and overhead expenditures and depreciation costs are not included in this analysis and represent an underestimation of the costs of sorghum research. This issue is addressed later through sensitivity analysis.

Adding the running costs and salaries and benefits estimated, gave the total sorghum research budget and not the budget used to develop and transfer SV 2. The breeder spends 30% of the sorghum budget on non-hybrid varieties. Although SV 2 is just one of the non-hybrid varieties, the assumption was made that from 1979/80 to the time when SV 2 was released in 1987, the non-hybrid sorghum breeding budget was equally split between SV 1 and SV 2. Thereafter the breeding programme had to do maintenance research for SV 2 at one-third of the non-hybrid breeding budget. The agronomy and plant protection researchers only started working on SV 2 after it was released in 1987. One-third of their sorghum budget was assumed to have been spent on SV 2.

The extension workers also started to work on transferring SV 2 to communal farmers after it was released. In the absence of a breakdown of extension costs by commodity and on the basis of a total research to extension budget ratio of approximately 1:3, the sorghum extension budget was assumed to be about three times that of sorghum research. As in the case of sorghum agronomy and plant protection, the assumption was made that a third of the expenditures are attributable to the promotion of SV 2. As the research budget does not include administration and overhead costs and depreciation, one can assume that the extension expenditures are underestimated and therefore the ROR will be biased upwards.

SMIP also played a part in the introduction of SV 2, especially on the transfer of the variety. They achieved this through on-farm trials, material support to the DR&SS sorghum research team, and through the 1992/93 emergency seed multiplication and distribution activity.

The information on the time that SMIP sorghum scientists spent working in Zimbabwe on SV 2 was available. For this study, 10% of the assistance to DR&SS was assumed to be related to the introduction of SV 2.

The costs of the emergency seed production and distribution can be broken down on the basis of variety. The amount spent on the production of SV 2 seed earmarked for Zimbabwe is simply the amount of seed produced (399 t) multiplied by the costs of production (US$2216/t). This seed was cleaned and packed at the Seed Cooperative before being distributed by the government through AGRITEX and NGOs. The costs of these operations were not available. Therefore, US$250/t was used as an approximation of the cleaning, packing and distribution costs.

As in the case of DR&SS and AGRITEX, there is no information on administration and overhead costs and depreciation of capital assets for SMIP. There is no easy way of getting around this except to obtain this information. In this study, the effect on the ROR of assuming that these costs for all three institutions make up over 25% of the total costs is examined through sensitivity analysis.

The Returns to SV 2

With cost estimates for SMIP, AGRITEX and DR&SS, development and transfer expenditures and the returns to SV 2 introduction were estimated. The ROR is biased upwards in the absence of administration and overhead expenditures and depreciation. As it was not feasible to separate the effects and the costs of different institutions involved in the development and introduction of SV 2, the ROR estimate is for the combined efforts of all the institutes involved.

The benefits after research and transfer costs (these do not include administration and overhead expenditures and depreciation) are deducted. The benefits are dependent on the quality of the season with low or negative net benefits during drought years. In 1992 the net benefits were negative reflecting the severity of the drought. The next year was generally a good year, but this does not show up in higher net benefits because of the high costs incurred in the provision of seed (almost US$1 million).

The assumption on what the social discount rate was in Zimbabwe during the period 1980–1999 determines whether the expenditures incurred in the introduction of SV 2 were worthwhile. There is a school of thought that believes that the discount rate should be the real interest rate on long-term government bonds. This reflects the social opportunity cost of public funds. Another school of thought believes that the social time preference (the rate at which society is collectively indifferent between present and future consumption) should be used. This only can be used practically if proxies are found and the best proxy is the real rate of interest. The current nominal bank lending rates in Zimbabwe are about 33% and the inflation rate is about 22% giving a real interest rate between 9 and 10% at the time of the study.

Table 4.6 shows the net present values at 10, 15 and 20% and the internal ROR to SV 2 introduction with and without the assumed adoption of fertilizer. The net present value at all discount rates is positive indicating that the development and transfer of SV 2 generated benefits are over and above the amount spent. The internal rates of return for with and without fertilizer cases are almost the same. Assuming that the social discount rate in Zimbabwe is 10% (the real interest rate), and administration and overhead costs and depreciation are negligible (an unrealistic assumption), then the introduction of SV 2 is a socially profitable programme.

The effect of fertilizer adoption is modest. This could be due to the fact that fertilizer adoption is assumed to occur during the last 4 years of the study period and/or the rates used in

Table 4.6. Zimbabwe: returns to SV2 introduction.

		Measure of return		
		Net present value*		
Treatment	Internal rate of return (%)	10%	15%	20%
With fertilizer	25.9	1,003,656	507,496	173,911
Without fertilizer	25.8	982,849	496,952	169,220

Source: SACCAR impact study estimates.
*Net present values are in US$.

the experiments may not be the most economic levels. If the period was made long enough for the fertilizer adoption ceiling to be reached, the difference in returns would become more apparent. The full impact and the returns to the introduction of SV 2 should be determined using a longer time period than that used in this study. The decision to stay with the 1980–1999 period is due to the desire to minimize errors that emanate from the use of predicted values of such variables as production, prices, adoption rates, etc. The effect of lengthening the study time period is examined as part of sensitivity analysis.

Sensitivity analysis

In this section two critical variables are changed in the analysis, and their effect on the ROR is assessed. The first is the research and transfer cost estimate, which is increased to allow for the effect of including administration and overhead expenditures and depreciation costs in the analysis. Such fixed costs are usually assumed to make up about 25% of the total costs. The research and transfer running costs and staff salaries and benefits are therefore increased by a third.

The second variable to be altered is the length of the period of the study. The analysis period is extended by 6 more years. The sensitivity analysis looks at a longer period with the current cost levels and with and without the overhead and administration costs and assesses the effects of this on the ROR estimated.

The effect of increasing research and transfer costs to allow for administration, overhead and depreciation costs is the reduction in the ROR by four percentage points, which is from about 26% to 22% (Table 4.7). This amounts to a 15% decline in the ROR estimated. With the social time preference rate assumed to be about 10 %, the introduction of SV 2 has been profitable. If 6 more years of benefit flow are added, the ROR increases by less than three percentage points to about 24%, which is above the assumed social time preference rate. This demonstrates that the omission of administration, overhead and depreciation costs affect the ROR to technology development and transfer significantly.

The primary objective of increasing the period of analysis by 6 years was to examine if the benefits with fertilizer would be different from the benefits calculated without assuming the fertilizer adoption. The ROR for the two cases is clearly not very different from each other. It is interesting to note that the ROR for the fertilizer case is lower than the without case for longer periods. This result could be due to the high cost of fertilizer. In addition, the fertilizer levels assumed in this analysis may be too high and therefore not optimal at the farm level. This is an area for further investigation in fine tuning sorghum production practices, especially fertilizer recommendations.

Table 4.7. Return to SV 2 introduction: sensitivity analysis.

	Internal rate of return (%)			
Treatment	Running costs and salaries only 1980–1999	Running costs and salaries* only 1980–2005	Running costs, salaries and fixed costs 1980–1999	Running costs, salaries and fixed costs* 1980–2005
With fertilizer	25.9	27.5	21.9	24.0
Without fertilizer	25.8	27.6	21.8	24.1

*Period of analysis extended by 6 years.

Conclusions

The improved white sorghums of the SV 2 type are considered superior for food and have the potential for drought escape due to their shorter growing period. They also respond well to fertilization when compared with local landraces. However, to determine whether the introduction of SV 2 has been economically beneficial, several issues need to be resolved. According to farmers interviewed, SV 2 is more susceptible to bird attack than local varieties, and more vulnerable to storage insects and striga problems.

The determination of the ROR to the introduction of SV 2 will ultimately depend on the availability of quality data to facilitate a more accurate estimation of both benefits and costs. In this study the unavailability of cost data related to development and diffusion of the variety has been the major limiting factor in estimating a relatively accurate ROR, particularly the data on administration and overhead expenditures and depreciation costs. The ROR estimated in this study is, therefore, biased upwards and the effect of including these fixed costs has been demonstrated in the section on sensitivity analysis.

References

IFAR (1993) *Agriculture in Zimbabwe, the Role of International Research Centers*, Country Report No. 3. IFAR, Arlington, Virginia.

Friis-Hansen, E. and Rohrbach, D.D. (1993) *Drought Relief Emergency Production of Sorghum and Pearl Millet Seed, Impact Assessment*, ICRISAT Southern and Eastern Africa Region Working Paper 93/01. ICRISAT, Bulawayo.

5 Ex-ante Analysis of the Sorghum and Millet Improvement Program

P. ANANDAJAYASEKERAM, D.R. MARTELLA, J. SANDERS AND B. KUPFUMA

Abstract: The Southern African Development Community (SADC)/International Crops Research Institute for the Semi-Arid Tropics Sorghum and Millet Improvement Program (SMIP) was started in 1983 as a regional initiative of the SADC in collaboration with one of the International Agricultural Research Centres. This chapter deals with the estimation of the rate of return for the combined investments of the National Agricultural Research System and SMIP. By necessity this study is both ex-ante and ex-post. Two scenarios were considered in the analysis: optimistic and conservative. The differences between the two scenarios were based on the levels of adoption ceiling and yield gains assumed. Seven countries released sorghum varieties and five countries released millet varieties under this collaborative programme. Both the Akino–Hayami method and simple benefit–cost analysis were used. The results revealed that if the appropriate discount rate is 10%, investments in sorghum and millet R&D only paid off if the optimistic estimates are the true representation of the adoption of varieties, the yield gains and the research and transfer cost. The rate of return for this scenario is 14%, and that for the conservative scenario is about 5%.

Background

The SADC/ICRISAT Sorghum and Millet Improvement Program (SMIP), conceived as a 20-year activity, was begun on 15 September 1983 as a response to concerns of the Lusaka Declaration (1 April 1980) for environmental degradation and food security of poorer people in the marginal rainfall areas of the SADC countries. SADC invited ICRISAT to establish a research station at Matopos, Zimbabwe, to assist with research on sorghum and millet in order to meet the food security needs of rural populations living in drought-prone areas. Canada, Germany and the USA helped fund the SMIP, including the establishment of adequate regional research facilities at Matopos.

Start up: SMIP: Phases I and II (1983–1993)

Phases I and II of SMIP have been successful in building capacity for conducting sorghum and millet research, constructing the regional research facilities, assisting several of the National

 Impact of Science on African Agriculture and Food Security (eds P. Anandajayasekeram, M. Rukuni, S. Babu, F. Liebenberg and C.L. Keswani)

Agricultural Research Systems (NARS) with construction and start-up planning, including station development, providing short- and long-term training for NARS staff, collecting and characterizing germplasm, and collaboration with the NARS to develop improved open-pollinated and hybrid varieties of sorghum and millet.

During the initial years, priority was given to the development of facilities at the SMIP centre at Matopos, which today possesses a physical facility that is conducive to high research performance. After 8 years of regional and collaborative national research efforts, the technical advantages of improved sorghum and millet technologies are no longer doubted. Improved cultivars of sorghum and millet have offered grain yield advantages of 50–200% under researcher managed conditions. On average the drought-tolerant materials developed by SMIP perform better under very harsh conditions.

Phases I and II were also marked by major investments in human resource development, including the provision of long-term degree training to over 90 national scientists in a range of disciplines. Many more scientists and technicians have received in-service training. Workshops, regional trials and monitoring tours have created a regional sorghum and millet research community linked to an international germplasm base. The network has shared research ideas and technologies across national boundaries.

Sorghum and Millet Improvement Program: Phase III (1993–1998)

The payoff to the building of institutional capabilities to conduct sorghum and millet research derives from the development and ultimate adoption of improved cropping technologies designed to increase household food security and farm incomes. In recent years, an increasing number of sorghum and millet technologies – both improved cultivars and improved management practices – have become available to small farmers. However, the adoption of improved seed technologies has been variable. In most SADC countries, adoption rates of improved cultivars have been low.

The strategy of Phase III of the project was based on the assumption that: (i) excellent improved sorghum and millet cultivars have been developed and hold sufficiently promising yield increases for rapid adoption by farmers; and (ii) that a combination of public and private input distribution and information dissemination channels exists to extend available technologies to a substantial number of sorghum and millet producers in the SADC region. SMIP was expected to play a catalyst role of moving sorghum and millet technologies to farmers, particularly after a severe drought, by helping to develop regional networks for SMIP and technology transfer. Through its participation in such an effort, SMIP worked with a number of regional and national organizations to strengthen networks that can then operate effectively and more efficiently than the present structure existing within the SADC region for technology generation and transfer.

The project goal was the stabilization of food supplies in the region, leading to improved nutrition and income for poorer people farming in drier areas. The purpose of this project is to increase the production of sorghum and millet with good consumer acceptance, local adaptation and pest resistance.

The expected outputs of Phase III were: sorghum and millet technology transfer; integrated sorghum and millet improvement research programmes; provision of improved sorghum and millet cultivars and genetic material; integrated technologies for pest management and sustainable regional sorghum and millet research programmes; and a plan for the future of SMIP facilities at Matopos station.

An impact study is being conducted against this background and anticipated outputs of the project. A comprehensive framework (Anandajayasekeram *et al.*, 1995) was used to assess the impact of this regional programme. A 'multicriteria' method is used and the analysis looked at the impact in terms of technology development and transfer, human resources development, and assistance to development of an enabling environment. This chapter only deals with the rate of returns estimates for the overall investment of this regional programme.

In an open economy the effects of research in any one country can be divided into a price effect and a supply shift effect. The price effect comes from the world trade of exporting producers after the supply shift leading to a reduction of the world price and consumer surplus gain for importing consumers. As sorghum and millet are not exported by these Southern African countries but essentially substitute for imports, this price effect can be ignored.

The relevant spill-over is the effect that research development in one country has on the supply of the output in another country. Evenson (1989) divides spill-over effects into applied science and basic science. The applied science spill-over here is embodied in new cultivars and the associated improved agronomic practices.

In this chapter, the new cultivars to be introduced into Southern African countries as a result of joint SMIP–NARS activities are described. Then the methodology for estimating these spill-overs is presented and the data collection process described. The benefits of the direct and spill-over impact of sorghum and millet research are calculated next. The final step is the cost calculation. These include the relevant costs for national programmes and SMIP to develop and transfer these technologies to farmers. From these costs and benefits the rate of return for the combined investments of NARS and SMIP was calculated.

New Technologies

SMIP facilitated the development of sorghum and millet research programmes in the SADC region. The national programmes played the central part in sorghum and millet research. As almost all of the new cultivars have been released or pre released recently, their impact is only going to be realized in the future. Although the measurement of their impact in the region is going to be mainly an ex ante analysis, the benefits realized before 1995 are also calculated making this study both ex ante and ex post.

In order to obtain information on sorghum and millet varieties released or pre-released within the region, each country was requested to provide specific information on each of the varieties. This included the year of release or pre-release, the parent material of germplasm and the source of that parent material. This was designed to assess the part played by SMIP in the development of the varieties to be included in the impact assessment exercise. This information will also enable one to estimate the spill-over effects. It is difficult to quantitatively apportion total benefits between SMIP and the national programmes. Therefore the rate of return will measure the return to investments made by both SMIP and NARS in sorghum and millet research and development.

The other information collected on the released or pre-released varieties included the potential target area for the variety, the estimated adoption ceiling and the years it would take to reach that ceiling. This allowed the specification of the adoption profile and the determination of the area that could be under the new varieties on a yearly basis.

Information on yield gains was also collected for each variety released or pre-released in each country. Yield differentials were determined at both improved and farmer management levels in most of the cases. This information is required when estimating the magnitude of the benefits either through a supply shift parameter (for partial equilibrium economic surplus types of impact assessment) or the difference in output for the with-and-without technology cases (simple benefit–cost analysis).

In the next section, a summary of the sorghum and millet information obtained from each member country is provided. This study does not concern itself with the adoption pattern for each of the different varieties released or pre-released in each country. The main assumption is that some combination of the newly released or pre-released varieties will be adopted. The information provided by national programmes on the maximum values of adoption ceilings and yield differentials is considered optimistic. To ensure that the assessment of the impact of the new varieties is a conservative one, the lowest adoption ceiling and yield gains reported were used. The longest estimated number of years to reach the adoption ceiling is the one that is used in this study. Thus, two scenarios are considered in this analysis, that is, an optimistic scenario and a conservative scenario. The differences between the two scenarios were based on the levels of the adoption ceiling and yield gains assumed.

Available sorghum varieties

Eight SADC countries supplied the required information on released and pre-released sorghum varieties as shown in Table 5.1. Although Zambia released several sorghum varieties during the same period, it was claimed that SMIP had very little to do with these releases. Therefore, the Zambian varieties were not included in the analysis. The information on adoption ceiling and yield gains shown in this table is based on conservative estimates. The optimistic estimates of the adoption ceiling and the yield gains for both sorghum and millet are shown in Table 5.1.

Research programmes in Angola were affected by the civil unrest so that even though Angola is a major sorghum producing country, the influence of both the national programme and SMIP can be considered minimal. The omission of Angola can have some effect on the results obtained.

The national research systems had sorghum research programmes already underway when SMIP was established in 1984. These programmes maintained some contact with ICRISAT and had obtained some germplasm for varietal development. National programmes continued their interaction with ICRISAT even after the establishment of SMIP. When SMIP was established, it strengthened these links by making such contacts easier. Thus although the source of the materials used to develop some of the varieties is ICRISAT, SMIP played a part in facilitating germplasm supply and the use of this germplasm to develop appropriate varieties, especially in the 1990s. This was achieved through collaborative research activities, workshops and other formal and informal means.

Most countries released their first varieties in the late 1980s or early 1990s. The early beneficiaries were Tanzania and Zimbabwe followed by Swaziland, Botswana and Mozambique. Malawi and Lesotho released their first varieties in the mid-1990s. It is interesting to note that

Table 5.1. Information on sorghum varieties released by SMIP member countries – conservative scenario.

Country	Number of varieties	Year first released	Target area (000 ha)	Adoption ceiling (%)	Years to ceiling	Yield gains (%)
Botswana	4	1994	30	10	3	15
Lesotho	5	1994	20	40	6	10
Malawi	2	1993	20	35	5	20
Mozambique	3	1989	60	40	10	30
Swaziland	3	1989	10	35	10	15
Tanzania	2	1988	100	30	6	20
Zimbabwe	6	1987	100	30	10	20

Source: SACCAR impact study estimates.

some varieties were released in different countries using different names. For example, SV 1 was released in Lesotho and Zimbabwe but it was called MRS 12 in Swaziland, Chokwe in Mozambique and Pirira 2 in Malawi (Akino and Hayami, 1975).

The conservatively estimated adoption ceiling was highest in Mozambique and Lesotho (40%) and least in Botswana (10%). The highest optimistic estimates given for the adoption ceiling were for Mozambique and Tanzania (60%) and lowest in Botswana (20%) as shown in Table 5.2. The highest conservative yield differentials between the new and the old sorghum varieties were expected in Mozambique (30%) while Lesotho expected the lowest yield gains of only 10%. The highest optimistic yield differentials are expected in Mozambique (65%) and the lowest in Swaziland (35%). The yield differentials/gains were, wherever possible, determined under farmer management conditions.

Millet Varieties

Fewer countries provided information on millet than on sorghum research. Lesotho and Swaziland are not major millet producing countries. The millet research programmes in Malawi and Mozambique have not reached a stage where predictions can be provided when varieties are likely to be released. The two countries are not major millet producers and they only contribute 11,000 and 5000 metric tons (t), respectively, to the total SADC regional production of 617,000 t. The Angolan case is similar to that of sorghum. Thus, while the underestimation of benefits is acknowledged, the underestimation of costs only applies if millet research in Angola benefited from the research programmes in the other five countries. The five countries that provided the required information are shown in Table 5.3.

In all countries millet varieties were released much later than sorghum varieties. However, in almost all cases, the estimated adoption ceiling and yield gains are higher for millet than sorghum. The average adoption ceilings for millet and sorghum are 44% and 31%, respectively. However, the millet varieties are expected to take more time to be adopted as shown by longer estimated years required to reach the adoption ceiling. The average number of years required to reach the adoption ceiling for sorghum is 7 while that for millet is 9. In both sorghum and millet research programmes, it has become apparent that most of the varieties have been released or pre-released recently. Given the problem of seed multiplication and distribution

Table 5.2. Estimated adoption ceilings and yield gains – sorghum and millet (optimistic scenario).

	Sorghum (%)		Millet (%)	
Country	Adoption ceiling	Yield gain	Adoption ceiling	Yield gain
Botswana	20	50	50	30
Lesotho	50	45	–	–
Malawi	45	40	–	–
Mozambique	60	65	–	–
Namibia	–	–	50	30
Swaziland	50	35	–	–
Tanzania	60	50	50	30
Zambia	–	–	40	25
Zimbabwe	50	40	50	40

Source: SACCAR impact study estimates.

Table 5.3. Information on pearl millet variety release and adoption profile – conservative scenario.

Country	Number of varieties	Year first released	Target area (000 ha)	Adoption ceiling (%)	Years to ceiling	Yield gains (%)
Botswana	2	1995	20	50	5	25
Namibia	4	1990	150	50	10	10
Tanzania	2	1994	300	40	10	20
Zambia	2	1987	30	30	10	15
Zimbabwe	2	1992	240	50	10	20

Source: SACCAR impact study estimates.

across the region, very few farmers have been exposed to these new varieties. Therefore, the benefits of these varieties are just beginning to trickle in.

Methodology

The impact of network research programmes such as sorghum and millet is best captured by a framework that takes spill-ins and/or spill-outs into account. Following Davis *et al.* (1987) the spill-over benefits are estimated as the incremental cost savings from the shift of the supply function that results from the imported and adapted technology. The incremental cost saving is equal to a change in output due to imported technology divided by the slope of the supply function. The slope of the supply function is obtained by multiplying the price elasticity of supply by the expected output over the expected price.

However, the Davis *et al.* model, while ideal for a system that involves spill-overs, requires information that is normally hard to find. Foremost among them is the demand and supply elasticities and slopes, quantities produced and consumed, and the whole range of subjectively determined variables. This includes the expected adoption ceiling, the rate of adoption and the probability of research success. There are few or no studies conducted to determine demand and supply functions for sorghum and millet in this region. The difficulty in obtaining production and consumption data stems from the fact that sorghum and millet are not traded on a significant scale within the region, often they are not the major commodities, and are also primarily produced for domestic consumption either in the form of food or beer. Owing to similar weather patterns, a shortage of sorghum or millet in one country means that other countries are also facing shortages. The imports to make up the deficits are usually maize imports. Therefore, the amount produced is often the same as that consumed.

The collection of data on production and consumption by national data collection agencies leaves a lot to be desired as many countries under-invest in this activity. Therefore, many national statistics are unreliable. Currently available data do not permit the use of a partial equilibrium model (Davis *et al.*, 1987).

Assuming that the expected yield differentials include the direct and spill-over effects of research in relatively closed sorghum and millet markets, research benefits can be estimated. The yield differentials and adoption data are used to estimate the benefits of the direct and spill-over impact of sorghum and millet research for each country using both the Akino–Hayami (A&H) method and the simple benefit–cost analysis (BCA) approach. The benefits are measured from the year, varieties were first adopted to 2010. This allows for the inclusion of the benefits that have already been captured by different countries where adoption of the improved sorghum or millet varieties started in the late 1980s.

The second stage of evaluating returns to investments in sorghum and millet research by both national programmes and SMIP is the determination of the research and transfer costs. This, admittedly, is a difficult exercise given the scarcity of data at both national and regional levels. The budgets of national sorghum and millet research programmes are difficult to determine because the accounting systems do not group costs on the basis of commodities, and they are not apportioned by activities. The research and transfer costs include running expenses, salaries and benefits, and overhead and administration expenses. Because the data on national programme and SMIP expenditures are not available in the required form, estimates were made so that the rate of return to investments in sorghum and millet research can be calculated. The next section will give estimates of the net benefits (not incremental gross benefits) of sorghum and millet research in the region.

Ex-ante direct and spill-over benefits

The gross benefits presented in this section are net of adoption costs that farmers incur when they use the new varieties. In this study, only seed costs are considered as the only additional costs that farmers will incur because there is little evidence that fertilizer and other inputs are going to be used on an increased scale when the new varieties have been adopted. In calculating seed costs, the fact that farmers re-use the seeds of new varieties (as the majority of new varieties are not hybrids) was not taken into account. The seed costs are therefore biased upwards. The crucial components in determining seed costs are the area under the new varieties, the seed rates and the seed prices. The seed rates are assumed to be 10 kg/ha and 2 kg/ha for sorghum and millet, respectively. The real seed prices are calculated using the farm level seed prices.

The benefit–cost analysis gross benefit estimates are not significantly different from the A&H estimates and hence only the results from the A&H estimates are reported. The gross benefits initially rise and then level off as the farmers in the respective countries reach the adoption ceiling. The gross benefits using the optimistic scenario are almost two times greater than those based on the conservative scenario reflecting the effect of changing yield gain, and adoption ceiling.

Estimating research and transfer costs

To calculate the rate of return to sorghum and millet research and transfer in the region, the research and transfer expenditures should be determined. For the national programmes, the research costs were estimated by multiplying the full-time equivalent staff in each country's sorghum and millet programme by the expenditure per researcher. The full-time equivalent staff data were compiled by the impact assessment team. The number of full-time equivalent staff working on national sorghum and millet research programmes in 1994 was 38.15. The cost per researcher provided by International Service for National Agricultural Research (ISNAR) is in terms of 1980 PPP (purchasing power party), which need to be converted to conventional exchange rate terms. The data needed to perform this conversion were not available necessitating the use of SMIP estimates. However, the cost per researcher used by SMIP in their proposal for Phases III and IV was considered to be too low in comparison with ISNAR data (Pardey *et al.*, 1995). As a result, the SMIP estimate was therefore doubled from US$25,000 to US$50,000. The transfer costs are estimated at twice the national research costs. SMIP contributes to the transfer of varieties through its on- farm research programmes, some of which are done in collaboration with the national research programme. Transfer costs are normally incurred after the varieties are released.

SMIP costs were also not available in the format that is easily amenable to this analysis. This mainly applies to the cost data for SMIP Phases I and II (from September 1983 to September 1993). The total expenditure of the project for Phases I and II was estimated at US$41.6 million. About US$26 million were spent during this 10-year period for research, and the rest was spent for training and capital items, such as buildings. For annual expenditure estimates for the period 1984 to 1993, the US$26 million is divided by the length of the period (10 years). Annual expenditures for the years 1994–2003 are based on estimates presented in the proposal for Phases III and IV and the Annual Report to the SACCAR Board for 1993 (SADC/ICRISAT, undated). The research and transfer cost data include cost components such as personnel, operational and capital. This was converted to 1995 US dollar equivalents using the US CPI data. The cost estimates for Angola were omitted because there is no information on either sorghum or millet varieties that have been recently released in that country. The development of varieties whose impact is being assessed is assumed to have taken place during the period 1984–1994. Thereafter, there is only maintenance research for these varieties taking about 10% of the total research budget.

Returns to sorghum and millet research and transfer

There are two sets of estimates of returns to sorghum and millet research; those based on conservative estimates of adoption ceiling and yield gains and those based on optimistic estimates of these two variables. After deducting the estimated research and transfer costs from the benefits, incremental net benefits were obtained and shown in Fig. 5.1. The incremental net benefit is used to calculate the net present values (NPV) and the internal rate of return (IRR). To calculate the NPVs an appropriate discount rate for the region as a whole is essential.

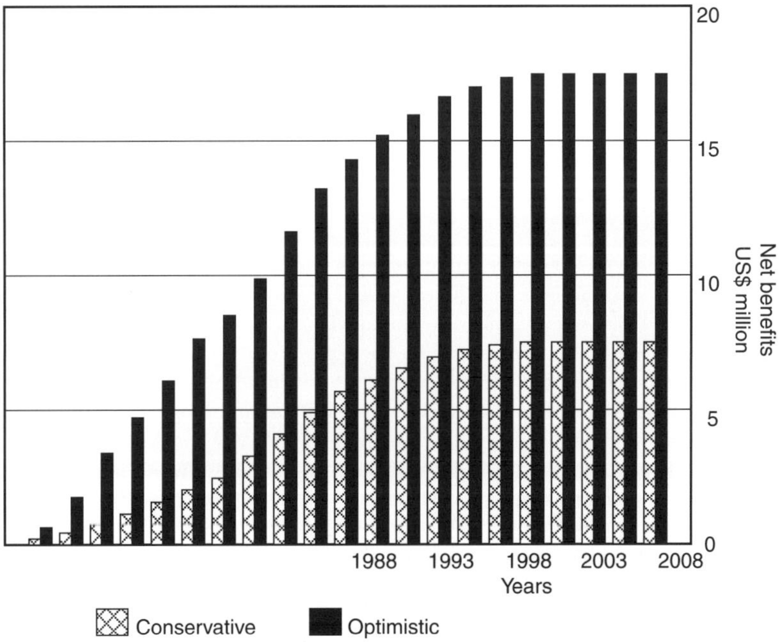

Fig. 5.1. Incremental net benefits of sorghum and millet research. Source: SACAR impact study estimates.

The question of the appropriate discount rate for the region is not easy to answer because the choice of the discount rate in many rates of return studies is often an arbitrary process. The World Bank usually uses between 10 and 12% as the opportunity cost of capital when evaluating the economic worthiness of its projects. However, this is considered by some scholars to be too high a rate resulting in the benefits to research being unnecessarily penalized. However, there is also another school of thought that believes that the opportunity costs for public funds must be set at such high levels to reflect the benefits that accrue to society when appropriate public investments are made, including the dead weight loss.

For Southern Africa, a discount rate of 10% is considered reasonable. If the discount rate used in this analysis is on the high side, then using lower rates will only make the benefits higher. The IRR and NPVs at 10% for the optimistic and conservative scenarios, and the A&H and BCA approaches shows that there are no significant differences between the estimates of returns to sorghum and millet research and transfer obtained using the A&H and the BCA approaches. The estimates of returns from the BCA approach are marginally higher than those obtained using the A&H approach. Thus, using any one of these methods would not result in different outcomes.

If the appropriate discount rate is 10%, then investments in sorghum and millet research and transfer only paid off if the optimistic estimates are a true representation of the adoption of varieties, the yield gains, and the research and transfer costs. The rate of return for this scenario is about 14%. The conservative scenario yields rates of return that are below 10% and thus suggests that if these estimates apply then the region is better off investing its scarce financial and human resources elsewhere and if the competing rates of return on other investments are higher than 5%. The differences in the two estimates suggest that assumptions about adoption ceiling and yield gains are important determinants of the measures of return to the research network.

The picture given by the IRRs is also reflected in the NPVs. The NPVs are positive at 10% only for the optimistic scenario. The region as a whole stands to gain approximately US$16 million as a result of the investment in sorghum and millet research if the optimistic scenario holds true. On the other hand, if the conservative scenario holds true, the region would lose approximately US$18 million as a result of this investment. These estimates are calculated assuming a discount rate of 10%. These are only indicative figures based on the subjective judgement of the breeders and steering committee members. The tentative estimates are also based on the expected values of the key variables. If the underlying assumptions are not realistic, then the estimates may vary. If one could get better estimates of the parameters, the analysis could be repeated. The stability of these measures of return to sorghum and millet research and transfer is further assessed in the next section on sensitivity analysis.

Sensitivity analysis

The effect of changing adoption ceilings and yield gain levels has already been addressed. The next step is to examine the years to reach adoption ceilings and the national research and transfer cost levels. The years to adoption ceiling are reduced to 6 for all cases where they were more than 10 to reflect a more optimistic view on the pace of adoption of the new varieties. The national research and transfer costs have been declining in recent years except for Botswana. The expenditures per researcher are expected to continue falling, or at least remain constant in real terms during the last part of the 20th century and the first decade of the 21st century. Thus, the national research and transfer costs are adjusted to allow for a 3% annual decrease in real budgets for national programmes. Pardey *et al.* (1995) report that 'over the past six years African government expenditures per researcher fell by 20 percent'.

In this study, the cost per researcher is based on the estimates made by ISNAR, i.e. US$50,000 per annum. SMIP has used a figure of US$25,000 in the Phase III feasibility analysis.

As the impact assessment team could not collect data to estimate the annual cost per researcher, it was decided to perform a sensitivity analysis to look at the impact of changing the cost per researcher on the rate of return estimated. Therefore, for the sensitivity analysis four scenarios were selected. First is the case of shorter adoption periods. The second case is that of declining real budgets for national programmes. The third scenario is a combination of the first two and the last evaluates changes in cost per researcher. In this section we used conservative estimates only. Changing both the adoption period and the national programme research and transfer costs has a minimal effect on the IRR. As expected, reducing the number of years to reach the adoption ceiling and adjusting research and transfer costs to reflect declining budgets for the national programmes, increases the IRR marginally.

The results suggest that the estimated rate of return to sorghum and millet research and transfer is stable with respect to a shorter adoption period. There is no evidence of large variability when the number of years to reach the adoption ceiling and the national research and transfer costs are changed. One can be confident that when using conservative estimates the IRR is in the neighbourhood of 5%. Significant deviations from this estimated rate of return are likely to occur when the assumptions and projections used in this study are drastically changed with respect to yield gains and adoption ceiling as happened when optimistic estimates were used.

Conclusions

Like all ex ante assessments the results depend on the quality and accuracy of the projections. The accuracy and quality of the data for such assessments need not be compromised as these studies are used to make decisions on national and regional investments. Ensuring high data quality is made difficult by the nature of agricultural technology development and transfer. The outcomes (research success and adoption) of this process are random events that are almost beyond human control. For instance, the question of what would have happened without the sorghum and millet research network, under the co-ordination of SMIP, is difficult to answer in quantitative terms as is required in a rigorous impact assessment. This is one of the major reasons why this analysis chose to attribute the sorghum and millet research benefits to the NARS and SMIP.

The results suggest that the returns to investments in sorghum and millet research and transfer in the region are relatively low for the conservative scenario of adoption and yield gains. However, when the two variables were changed to the levels assumed under the optimistic scenario, the returns may be comparable with those from other agricultural research investments. The major lesson is that more work is needed to determine the magnitude of key variables such as the adoption ceiling, yield gains and research and transfer costs. Therefore, if the NARS and regional networks are serious about using ex ante impact assessments as a planning tool in the future, then it is vital to establish a formal monitoring and evaluation system that will provide the data needed in a reliable manner.

References

Akino, M. and Hayami, Y. (1975) Efficiency and equity in public research: rice breeding in Japan's economic development. *American Journal of Agricultural Economics* 57, 1–10.

Anandajayasekeram, P., Martella, D.R., Sanders, J. and Kupfuma, B. (1995) *Report on the Impact Assessment of the SADC/ICRISAT Sorghum and Millet Improvement Program*, Vol. 1. SACCAR, Gaborone, Botswana.

Davis, J.S., Oram, P.A. and Ryan, J.G. (1987) *Assessment of Agricultural Research Priorities: an International Perspective.* ACIAR Monograph No. 4. ACIAR, Canberra, Australia.

Evenson, R.E. (1989) Spill-over benefits of agricultural research: evidence from the U.S. experience. *American Journal of Agricultural Economics* 71(2), 447–452.

Pardey, P., Roseboom, H. and Beintema, N. (1995) African Agricultural Research. In: *Africa: Three Decades of Development.* ISNAR Briefing Paper No. 19. ISNAR, The Hague, The Netherlands. pp. 1–15.

6 Impact of Pearl Millet Research and Development in Namibia: the Case of Okashana 1

P. Anandajayasekeram, D.R. Martella, J. Sanders and B. Kupfuma

Abstract: Pearl millet is the most widely grown and utilized cereal in northern Namibia especially in Ovambo and Kavango regions. A millet variety was identified in the late 1980s from the Sorghum and Millet Improvement Program nursery in 1986 and officially released in 1990, under the name Okashana 1. This pearl millet variety is short maturing (85–90 days), and is responsive to moisture and inorganic fertilizer. An impact assessment study was undertaken to calculate the rate of return (ROR) on investment for this technological intervention. The estimated ROR for the R&D investment (development, dissemination and adoption) for Okashana 1 was 13.28%, the estimated ROR was 4.25% when the cost of investment was increased by 10% indicating that the ROR is very sensitive to the R&D costs. The study also revealed Okashana 1 yielded the largest benefits during the dry seasons of 1994 and 1998. During normal rainfall seasons, the net benefits generated by the introduction of Okashana 1 are not likely to replace the current farmer practice of using mixed varieties, a strategy often followed to avoid risk.

Introduction

The principal cereal consumed in Namibia is maize followed closely by millet. Crop production is concentrated in the far north and the main cereal there is pearl millet. About 54% of the population is located in this northern strip along the border with Angola, where the rainfall is higher than in the rest of the country. In a normal rainfall year, 1992/93, millet provided 92% of the cereals produced in Ovambo and 84% in Kavango. This region is predominantly a millet production zone with some other crops and small animals (Keyler, 1994, p. 37).

A millet variety was identified in the late 1980s from a Sorghum and Millet Improvement Program (SMIP) nursery and released in 1990 through the efforts of the Namibian National Agricultural Research System (NARS). The benefits from the introduction of this new cultivar and inorganic fertilizers are estimated. The rate of return (ROR) for the combined research and extension efforts is then estimated. Finally, some constraints on the diffusion of new millet cultivars and inorganic fertilizer are identified.

Cereal production and consumption

In the drought year of 1991/92, when both maize and millet production collapsed in northern Namibia, there was a substantial increase in maize imports. In most of 1992/93, maize meal was cheaper than millet meal; hence it is not surprising that maize was the main cereal consumed. However, rural consumers express a preference for pearl millet (mahango). In fact, only 3% prefer maize in the two predominant agricultural regions of the country (Keyler, 1994). Northern Namibians prefer millet meal, when it is available. The predominant region for pearl millet production is the Ovambo region in northern Namibia. In Ovambo, land is becoming increasingly scarce. Hence, shifting cultivation is no longer possible for many farmers. For additional fields farmers have to walk increasing distances (Keyler, 1994). The farmer response to increasing scarcity of land is to improve land quality through both organic and inorganic fertilizer. There is substantial variability in fertilizer use between regions resulting from differences in soil fertility, length of settlement and the availability of different types of fertilizers.

Predominantly manure has been introduced into the system with only 9% of the farmers in the Ovambo region and 3% in the greater Kavango region utilizing inorganic fertilizers. However, organic fertilizer cannot substitute for inorganic fertilizer unless very large quantities are utilized, i.e. 3–5 t/ha (Sanders *et al.*, 1996). Farmers in northern Namibia do not have sufficient animals to provide these quantities of manure. Therefore, increasing utilization of inorganic fertilizers will be a necessary addition to these crop systems in order to substitute for the increasingly scarce land. Then the organic fertilizers can complement the inorganic fertilizers and improve the water retention capacity and the structure and texture of the soil.

Introduction of Okashana 1

In 1986 the Rossing Foundation, a NGO in Namibia, imported a regional millet variety nursery from SMIP. The farmers were asked to identify the cultivar that they liked. They selected a widely introduced cultivar in India, ICTP 8203. After further testing, this cultivar was pre-released in the 1988/89 season. Then in 1990 Okashana 1 was officially released. By this time it had been substituted with ICMV 88908 by SMIP, a cultivar with the same characteristics as ICTP 8203 but with higher yields in field testing. The characteristics of both varieties are earliness (short season), high yielding ability, large round seeds and resistance to downy mildew. Okashana 1 takes 85–90 days to maturity as compared with 120 days for local varieties.

Diffusion of Okashana 1

In 1989/90, 10.5 t of Okashana 1 seed were produced by SMIP and made available to Namibian farmers for the 1990/91 crop year. Assuming that the seed was planted at a seed rate of 2 kg/ha, there were 5250 ha planted to Okashana 1 in the first year of release.

Unfortunately, after the initial year, seed sales are not very useful for estimating diffusion because farmers begin utilizing their Okashana 1 for seed. However, there was a major drought in 1991/92. Assuming farmers either lost their millet or had to eat all that was left of their Okashana 1, an estimate of diffusion from seed sales can be made for this year. Again, with 90% utilization and 2 kg/ha seed rate, 14,400 ha could be planted in the region. These two point estimates of diffusion from seed sales are complemented with survey estimates.

The data used in the diffusion estimates were obtained from several sources. Seed sale data and area statistics came from the Ministry of Agriculture, Windhoek, Namibia and the rest of the information came from two field surveys conducted during 1992/93 (Keyler, 1994) and 1994/95 crop year by SMIP. The Keyler survey included 200 households from Ovambo and 120 from Kavango and was conducted in the 1992/93 crop year. The SMIP survey was undertaken in the 1994/95 crop year and utilized recall data for the earlier years to obtain the area estimates over time for Okashana 1. The 'estimate' was from subjective judgements based upon the data from the various sources and field interviews. The future estimates for the next 5 years were deliberately conservative and included little further increase from the diffusion levels already attained in 1994/95.

The reported diffusion is more rapid in Kavango reaching 45% of the millet area in 1994/95 as compared with only 17% in the Ovambo region. As the Ovambo region has less rainfall and Okashana 1 is an early maturing cultivar, this is surprising but both surveys report similar results. One explanation is that 1992/93 was also a poor rainfall year for the Kavango but a normal year for the Ovambo region. Hence, the increase in the demand for a short maturing cultivar would have been greater in Kavango than in Ovambo in 1994/95. Another potential explanation could be the differences in soil fertility.

In these higher rainfall conditions of Kavango, Okashana 1 needs to be planted later in the season. According to the SMIP survey, 65% of the farmers in both regions planted Okashana 1 later. Having a combination of cultivars or a portfolio mix then reduces the farmers' risk, especially that due to rainfall variability. Farmers also indicated that they will never replace the landraces completely with Okashana 1.

Yields of Okashana 1 and Local Cultivars

Yield data from trials in northern Namibia vary significantly with the region, seasonal rainfall, soil fertility, planting time and the incidence of bird problems. The 1992/93 crop season was a normal rainfall year. In the Ovambo region there was a strong response to Okashana 1 in two of the three trials. Moreover, there was an excellent response to inorganic fertilizer for both local cultivars and Okashana 1. On these Kalahari sands, the principal constraint on increasing cereal yields is low soil fertility. In 1993/94, the rains stopped early in the Ovambo region and there was a varietal response to Okashana 1 alone and an even larger effect when Okashana 1 was combined with inorganic fertilizer. In West Kavango there was a positive varietal effect for Okashana 1. In contrast in the Eastern Ovambo and East Kavango regions the local landraces performed better and where there was a good response to fertilizer. The poor response of Okashana 1 may be due to early planting in the higher rainfall zone. This later planting requirement is now well known by farmers and SMIP survey data show that almost two-thirds of the farmers planting Okashana 1 also plant it late.

All the yield trials that could be found on Namibia millet production were examined. There were many experiments lost due to birds or other factors. Moreover, the data reported were not very consistent. Nevertheless, a rapid diffusion process is reported for Okashana 1 especially in the Kavango region and to a lesser extent for Ovambo. Especially, if it is planted late in the higher rainfall region of Kavango, appears to be favourable.

The international price of millet (approximate mean value for 1992 and 1993) of $100/t and a freight, handling and transportation charge of $30/t were used in this study. Given the fact that maize is usually imported to offset shortfalls in millet production, the international reference price might alternatively be the maize price. As the international prices for millet and maize are in the same range, this will not affect the analysis. Inorganic fertilizer is already being used in the region. The impact assessment team assumed that fertilizer use would gradually increase. The Akino–Hayami pivotal supply shift was utilized for estimating the economic surplus.

In the two dry years after the introduction of Okashana 1, substantial benefits to the introduction of the early cultivar and fertilization were realized. In these 2 years the benefits were US$341,000 (1992) and US$320,000 (1993, respectively). The key to diffusion will be the availability of seed and inorganic fertilizer. With the frequency of dry years increasing, adoption of Okashana 1 and other similar but higher yielding varieties is going to increase.

The costs of technology development and transfer are made up of adoption costs, research costs and transfer costs. Adoption costs for Okashana 1 are comprised of seed and fertilizer costs. The Zimbabwe prices of seed and fertilizer are used to estimate the adoption costs for Okashana 1 because the price data for Namibia was not available at the time when this analysis was done. The only data on research and transfer costs available were not in the form that allows the determination of the actual costs of the introduction of a specific variety such as Okashana 1. As a result, the costs of developing and transferring Okashana 1 had to be estimated.

To facilitate this estimation, several assumptions were made:

- The national agricultural research budget for 1990/91 and 1991/92 was obtained from the ISNAR report. The budgets for the years before 1990/91 and after 1991/92 were extrapolated assuming a growth rate in the budget of 25% annually. This gave the annual research budgets. The data on staff establishments (total and millet research) allowed the estimation of the total expenditure per researcher that includes salaries (researcher and share of total support staff), capital costs and operation costs.
- The time spent on millet research during the 1991/92 period was 1.3 researcher years out of a total of 21 researcher years. This study assumes that there were 1.3 researcher years for millet research.
- Based on the composition of the millet improvement programme at Omahenene research station during the 1993/94 season, 45% of the millet research budget is spent on the development and testing of non-hybrid varieties such as Okashana 1.
- Not more than a third of the non-hybrid pearl millet research budget was spent on the development and testing of Okashana 1. This was before it was released. After the release, the proportion fell to 10% as only maintenance research was required.
- The proportion of SMIP's material assistance to Namibia's NARS and time spent by SMIP millet scientists in Namibia that can be apportioned to the introduction of Okashana 1 is not more than 10%.
- The cost of cleaning, packing and distributing Okashana 1 seed during the 1992/93 emergency seed production and distribution activity is estimated at US$250/t.

The returns to the introduction of Okashana 1

The subtraction of the cost stream from the net benefit stream yields the incremental net benefits. These are the benefits over and above all the costs. The benefit stream is dependent on the quality of the season while the cost stream steadily increases throughout the period. Okashana 1 yielded the largest benefits during the dry seasons of 1994 and 1998 (the only years with positive incremental net benefits). This assumes that Okashana 1 is planted late and is fertilized. During normal rainfall seasons, the net benefits generated by the introduction of Okashana 1 are modest. During normal seasons the incremental net benefits are negative. This confirms the previous assertion that Okashana 1 is not likely to replace current farmer varieties, but will be a valuable source for variety mixing, a strategy often followed by farmers to avoid risk.

The incremental net benefit stream permits the calculation of the net present value (NPV) and internal rate of return of introducing Okashana 1 in northern Namibia. If the social discount rate (real) is assumed to be below 13.28% then the introduction of Okashana 1 had a positive NPV. The NPVs for discount rates below 13.28% are all positive, and hence if the social discount rate is above 13.28% then investments made in Okashana 1 could have been better elsewhere. In addition the returns to Okashana 1 research and transfer would be different if we change the time period. Clearly, the 12 years used in this study do not sufficiently capture the life span of Okashana 1. Although it is acknowledged that the variety is not going to replace the existing ones completely, the benefits derived from Okashana 1 will not disappear.

Sensitivity analysis

The aim of this sensitivity analysis is twofold. First is the examination of the effect of increasing research and transfer costs by 10% on the ROR to account for any underestimation of research and extension costs. The second is to examine the effect of increasing the time period by 6 more years on the estimated ROR.

When costs are increased by 10%, the ROR falls from 13.2% to 4.2% during a 12-year period of analysis. The increase in research and extension costs would reduce the ROR significantly, even when the period is increased. For the longer time frame, the ROR is 20.13% with the estimated costs. However, when the research and extension costs are increased by 10%, the ROR falls to 13.89%.

Lessons learned

Okashana 1 was rapidly adopted in the Kavango region because of two consecutive dry years of 1991/92 and 1992/93, the higher fertility of the soils and later planting. These factors are encouraging farmers to implement a more rapid diffusion process of Okashana 1 than in the drier and less fertile Ovambo region. The enabling environment to facilitate the adoption of Okashana 1 is good because the extension service is promoting the variety along with inorganic fertilizer. In 1994/95, credit was made available to farmers for fertilizer, seeds and other related activities in northern Namibia. Moreover, processing of millet flour in small (1–2 kg) packets has been undertaken by a local manufacturer.

The development of new cultivars is a continual process in order to respond to farmer requests and to the emerging biotic problems as the agricultural system evolves. Farmers interviewed expressed several complaints about the characteristics of Okashana 1. Poor storability and short stalks are major issues raised by the farmers with respect to this variety. These issues are currently being addressed by the NARS researchers. The eventual substitution of Okashana 1 and successors to Okashana 1 with greater height and longer length of season varieties is expected to accelerate the diffusion rate of the new cultivars.

Conclusions

The introduction of Okashana 1 has largely been an outcome of the efforts of the Namibian NARS. The seed was obtained from SMIP in 1986, and tested by the Rossing Foundation in Namibia. Initial bulking of seed was done by the Foundation and pre-released to farmers during the 1989/90 season. SMIP helped with seed production especially after the drought year of 1991/92. The benefits generated by the introduction of Okashana 1 have proved difficult to estimate because of the lack of appropriate data. Particularly crucial was data to estimate research and transfer costs for both the NARS and SMIP. Underestimation of costs results in an upward bias in the estimation of net benefits. The results of the analysis provide a range within which the actual ROR falls. The estimated ROR is about 13%. Moreover, the return to Okashana 1 is reasonable for a difficult low income farm sector where equity objectives would probably also be very important to the government. More research is required on millet production in Namibia.

References

Keyler, S. (1994) Economics of the Millet sub-sector in Northern Namibia. Report 1: Assessment of the structure and conduct of the Millet Sector, Windhoek: Ministry of Agriculture, Namibia. Unpublished Memo, 165 pp.

Sanders, J.H., Shapiro, B.I. and Ramaswamy, S. (1996) *The Economics of Agricultural Technology in Semiarid Sub-Saharan Africa.* Johns Hopkins University Press, Baltimore, Maryland.

7 Economic Impact of Maize Research in Tanzania

A.J. Moshi, P. Anandajayasekeram, A. Kaliba, D. Martella, W. Mwangi and F.M. Shao

Abstract: Maize was introduced in Tanzania in the 16th century but research on appropriate varieties and management practices did not get underway until the 1940s. In 1974, a National Maize Research Program (NMRP) was established to co-ordinate maize research. During 1974–1994 the NMRP released 15 varieties. This study was conducted to assess the socio-economic impact of maize technology development and transfer investment from 1974 to 1994. Standard pre-tested questionnaire and multi-stage sampling procedures were used for primary data collection. Data were collected from 978 farmers in 53 sites across seven agro-ecological zones. The sample survey revealed that the adoption rate of the improved varieties for the various zones were 28%, 66%, 44%, 24%, 66%, 81% and 36% for the Central, Eastern, Lake, Northern, Southern, Southern Highlands and the Western Zones, respectively. The study demonstrated that farmers adopt the cheapest and low-risk technological components in a stepwise process reflecting the profitability and riskiness of each component. The estimated rate of return for the maize research and development in Tanzania was 19%.

Background

Agriculture is Tanzania's key economic sector. The economic performance of the country is closely related to the growth in the agricultural sector. Agriculture contributes about 52% of Tanzania's gross domestic product (GDP), generates about 75% of the export earnings and employs 84% of the workforce. The agriculture sector is the main source of Tanzania's food supply and raw materials for the industrial sector. Food crops account for 55% of the agricultural GDP, livestock 30% and traditional export crops 8%. Maize is the most important food staple and it is produced on about 1.7 million ha and it accounts for 60% of the total cereals area.

Agricultural research in Tanzania has stagnated for several decades because of funding problems. There are four different sources of funding: government budget allocation, parastal and private enterprises, donor funding and self-financing by research stations. The government allocations are for recurrent and the development budget. The government research allocation is equivalent to about 0.4% of the total research budget. During the 1990/91 financial year only 13% of the original budget request (TSh95 million of the requested TSh708 million) was granted, and the number of research projects (including those of maize) were slashed by almost 50%.

The donor support for agricultural research has fluctuated severely in the past decade. Owing to the funding constraints, there is a need to set research priorities and mobilize additional funds for research.

Maize is the most important staple food in Tanzania. It was introduced in the country from the West Indies via Europe during the 16th century. Currently, it is produced on approximately 1.7 million ha and accounts for about 60% of the total area planted with cereals. Most of the maize in Tanzania is produced by smallholders on 1–3-ha holdings, accounting for about 85% of the total production. A further 10% is produced on medium sized farms (10–100 ha) while large-scale farms (over 100 ha) contribute the remaining 5%. Seventy-five to 80% of the maize produced is consumed at the household level. Surpluses are sold to other farmers, urban dwellers or to maize grain deficit regions. Only about 2% of the maize produced in Tanzania is used as animal feed.

Maize research in Tanzania began in the 1940s. The National Maize Research Program (NMRP) was started in 1974. Since the inception of the NMRP, 15 varieties have been released and several crop management recommendations have been made for various agro-ecological zones (Tables 7.1 and 7.2). Donor and national resources have been invested in maize technology development and transfer (TDT) activities in Tanzania. Thus it was decided to assess the socio-economic impacts of this investment.

Objectives

The overall objective of the study was to assess the socio-economic impacts of the maize research and development (R&D) investment during the period 1974–1994. Specific objectives of the study are: to assess the impact of maize research in Tanzania; to formulate priorities for future maize research in Tanzania; and to derive policy recommendations to improve maize TDT.

Methodology

In assessing economic impacts, R&D is treated as an investment and rates of return (ROR) are then estimated for this investment. ROR summarizes the benefits and costs, and income from the activity in a single number, which can be easily compared with the cost of obtaining funds or ROR obtained from alternative investments. One of the approaches to estimate ROR is the surplus approach that uses a benefit–cost framework to estimate the average rate of return (ARR). The ARR takes the whole expenditure as given and calculates the ROR to the global set of expenditures. The ARR indicates whether or not the entire investment package was successful, but not whether the allocation of resources between investment components was optimal. Owing to lack of consistent time-series data, in this study it was decided to use the surplus approach to estimate the ARR for the R&D investments in the Tanzanian maize research programme.

R&D technology refers to organizational strategies and methods used by research and extension programmes in conducting their work. The intermediate/institutional impact includes the institutional changes that have occurred as a result of a research programme within the research organization, as well as the changes that have occurred in the enabling environment outside the institution that facilitates technology transfer and adoption. The wide applicability of research results over a range of agricultural production conditions or environments, often across commodities are referred to as 'spill-over effects'. These spill-over effects can be intercommodity, intersectoral or intercountry. Spill-over effects of the Tanzanian maize R&D across commodities within the country are assessed in this study.

Table 7.1. Maize varieties released/recommended by the NMRP, 1974–1994.

Variety	Year of release/ recommended	Characteristics	Potential yield (million t/ha)	Target zone/region
Tuxpeno	1976	Open pollinated, white dent, good standability	5.5	0–900 m a.s.l.
H 614	1977	Topcross hybrid, white dent, large ears	10.0	Over 1500 m a.s.l. SH
H 6302	1977	Three-way cross hybrid, white flint, good standability	11.0	Over 1500 m a.s.l. SH
H 614	1979	Topcross hybrid, white dent, good standability	11.5	Over 1500 m a.s.l. SH
Staha	1983	Open pollinated, white flint/dent, streak tolerant	6.5	0–900 m a.s.l. S, E, part of N
Kilima	1983	Open pollinated, white flint/dent, good standability	7.5	E, C, S, N, W, L, SH
Kito	1983	Open pollinated, white flint, early maturity	6.0	900–1500 m a.s.l. C, N, S, N, W, L, SH
TMV-1	1987	Open pollinated, white flint, medium maturity, streak resistant	6.3	E, C, S, N, W, L, SH
TVM-2	1987	Open pollinated, white flint, large ears	9.0	Over 1500 m a.s.l. SH
CH 1	1992	Single cross hybrid, white flint	6.8	0–900 m a.s.l. S, E, part of N
CH 3	1992	Three-way cross hybrid, white flint	6.9	0–900 m a.s.l. S, E, part of N
Kilima-ST	1994	Open pollinated, white flint/dent, good standability, streak tolerant	7.5	900–1500 m a.s.l. C, N, S, N, W, L, SH
UCA-ST	1994	Open pollinated, white dent/flint, streak tolerant	7.5	900–1500 m a.s.l. C, N, S, N, W, L, SH
Kito-ST	1994	Open pollinated, white flint, early maturity, streak tolerant	6.3	E, C, S, N, W, L, SH
Katumani-ST	1994	Open pollinated, white flint, early maturity, streak tolerant	6.3	E, C, S, N, W, L, SH

Key: m a.s.l. = metres above sea level.
Zones: E = Eastern; S = Southern; N = Northern; C = Central; W = Western.
SH = Southern Highlands; L = Lake.

Both primary and secondary data were collected and used in assessing the socio-economic impact of maize TDT activities in Tanzania. All available secondary information regarding maize TDT such as hectarage, production, consumer price index, elasticities, research and transfer costs, input and output prices and other relevant information were collected from various published and unpublished reports. Primary data were collected through a farm level survey of

Table 7.2. Sites and farmers interviewed by zone in the farm survey.

Zone	Percentage contribution to national production	Number of sites	Number of farmers
Southern Highlands	46	21	396
Lake	17	9	162
Northern	11	7	126
Western	11	6	114
Eastern	9	5	90
Southern	2	2	36
Central	3	3	54
Total	100	53	978

Source: Tanzania farm survey, 1994.

978 farmers distributed across the various agro-ecological zones using a pre-tested standard questionnaire. The contribution of the various zones to the national production was used as a basis for allocating the sample farmers across the zones. The contribution of each of the zones, the number of sites and the number of farmers interviewed are presented in Table 7.2. Within each zone, production statistics at district level were collected. These were used as a basis for selecting the districts within the zone. Villages within each district were chosen based on the knowledge of maize production, as well as logistical feasibility, i.e. accessibility and cost. Eventually a cluster of farmers was interviewed within each village. Thus, a multistage sampling procedure was used in choosing farmers for interviews. A total of 978 farmers from 53 sites across the zones was interviewed.

Research and transfer costs were collected from the various collaborators. Survey information on the diffusion of technologies was used to estimate the gross benefits of the maize TDT in Tanzania. Using the surplus approach (Akino–Hayami method and simple benefit–cost method) the ROR for investments in maize TDT for the period 1974–1994 was computed.

Survey Results

Maize is produced in all seven agro-ecological zones. However, the high-potential area (>1500 m above sea-level and >1000 mm of rainfall) produced 55%, the medium-potential area (1000–1500 m above sea-level and 800–1000 mm of rainfall) produced 21%, and the low-potential area (<800 mm rainfall) produced 14% of the national maize production during the 1991–1994 period. The sample survey revealed that the adoption rates of the improved varieties for the various zones are 28%, 6%, 44%, 66%, 24%, 81% and 36% for Central, Eastern, Lake, Northern, Southern, Southern Highlands and Western zones, respectively.

Farmers demonstrated stepwise adoption behaviour, where they adopted technological components according to their relative profitability and riskiness (Byerlee and Polanco, 1986). The adoption of technology is also related to returns on investment at the farm level. In zones where there are high returns to investment (high potential areas), and the risk of losing a crop is low, even relatively expensive technologies (high fertilizer use in the Northern and Southern Highlands zones) have been adopted. This demonstrates the stepwise adoption behaviour of the small-scale farmers, and their rational decision making process.

Both local and improved varieties are grown by the sample farmers. Identification of current varieties is difficult because of recycling of seed, a common phenomenon in all zones. Farmers recycle both hybrid and composite seeds. The number of years of recycling varied between 5 and 15 years.

Public extension is the major source of technological information for maize farmers. However, lack of knowledge has been often mentioned by the sample farmers as one of the constraints on the adoption of new technology.

There is a need to strengthen the linkages between researchers, extension workers, and farmers in order to promote the smooth flow of technologies to farmers and feedback from farmers to researchers in their problems and needed research. Formation of farmer groups should be encouraged and strengthened. This will not only facilitate the speedy diffusion of technologies, but will also assist in obtaining credit from the formal sources.

Impact results

As Tanzania is a net importer of maize during most years, the study focused on the economic benefit accrued to the domestic producers and consumers, and used the import parity price to measure the benefits.

Returns to investment

The estimated internal rate of return (IRR) for the investments in maize TDT was 23% for the Akino–Hayami method and 19% for the simple benefit–cost approach (Table 7.3). The latter approach is the most conservative estimate. Given the opportunity cost of capital in Tanzania, this appears to be a very attractive investment. Using the benefit–cost approach the net present value (NPV) of this investment at 5% discount rate was US$1.1 billion. Even at the 15% discount rate, the estimated NPVs are US$255 million and US$108 million for the Akino–Hayami and benefit–cost approach, respectively.

Effects of increasing the research and extension cost

When the Government of Tanzania's research and extension costs were increased by 20%, the estimated IRR for the Akino–Hayami method was 20.1%. The NPV for 5, 10 and 15% discount rates were US$1.6 billion, US$608 million and US$188 million respectively (Table 7.4). However, when the research and extension costs were increased to 50%, the estimated IRR

Table 7.3. Estimated net present value and internal rate of return for maize R&D investment in Tanzania, 1965–1994.

Discount rate used (%)	Net present value		Internal rate of return	
	Akino–Hayami	Benefit–cost	Akino–Hayami	Benefit–cost
5	1,707,505,917	1,107,858,232	23	19
10	687,739,662	299,913,930	23	19
15	255,223,183	107,791,365	23	19

Source: Impact Assessment Team Estimates.

Table 7.4. Estimated net present value and internal rate of return for maize R&D investment in Tanzania: increased research and extension costs 1965–1994.

	Net present value				Internal rate of return			
	Akino–Hayami		Benefit–cost		Akino–Hayami		Benefit–cost	
Discount rate used (%)	20% increase	50% increase	20% increase	50% increase	20% increase	50% increase	20% increase	50% increase
5	1,609,035	1,009,387	1,439,236	839,588	20.1	16.3	16.6	13
10	607,750	319,924	469,819	181,994	20.1	16.3	16.6	13
15	187,745	40,314	71,385	76,042	20.1	16.3	16.6	13

Source: Impact Assessment Team Estimates.

declined to 16.6%. The NPV for the three discount rates changed to US$1.4 billion, US$470 million and US$71 million, respectively. Despite the fact that the IRR and NPV declined with the rise in cost, given the opportunity cost of capital in Tanzania, the maize research and extension investments were still profitable to the society.

Effect of parity price

In the base estimation of economic benefits, the parity price of maize at Dar es Salaam was used. This will not truly reflect the real value of maize, as maize is produced and consumed at zonal levels. It was therefore decided to repeat the analysis using parity price at the producing centres. When the 1994 road transport and handling costs were used to estimate the parity price at the production centres, and analysis was repeated, the IRR increased to 23.6% for the Akino–Hayami method and to 19.6% for the benefit–cost approach (Table 7.5). Using the import parity price at the production centre, increases the surplus generated in terms of NPV, but the IRR was not very sensitive to changes in domestic transport and handling costs.

Effect of increased overhead and administrative cost

The World Bank estimated 20% of the sum of the recurrent expenditure and salaries and benefits to be the overhead and administrative cost. In the base analysis, this estimated percentage was used to calculate the overhead and administrative cost of the Division of Research and Development headquarters. The overhead and administrative costs of the zonal research institutes were not included. Thus, it was decided to raise this to 40% in order to accommodate the zonal overhead and administrative costs. Analysis indicated that the estimated IRR declined by 1%. At the 5% discount rate, the estimated NPV declined by US$28 million. Although the increase in administrative and overhead cost reduces the overall benefit to the society, the estimated IRR and NPV are not very sensitive to this parameter.

Impact on food security

The study indicated that the increase in maize yield and marketable surplus between 1974 and 1994 could largely be attributed to the improved varieties and recommended maize

Table 7.5. Estimated net present value and internal rate of return for maize R&D investment assuming parity price at producing area 1965–1994.

	Net present value (millions)				Internal rate of return			
	Akino–Hayami		Benefit–cost		Akino–Hayami		Internal rate of return	
Assumption	5%	10%	15%	5%	10%	15%	Akino–Hayami	Benefit–cost
1994 transport cost	1896	762	288	1247	454	131	23.6	19.6
Transport cost reduced by 50%	1802	725	271	1177	427	118	23.6	19.3

management practices. The maize TDT programme has contributed to both the household and national food security.

Environmental impact

The environmental impacts of the maize technologies are externalities to the advocated technical innovation aimed at increasing agricultural productivity. The row planting of maize adopted by over 90% of sample farmers across all zones is a good practice that controls soil erosion. Breeding for pest and disease resistance (Kilima, TMV-1, Kito) with flinty hard pericarp, reduce the need for chemical treatment, both in the field and storage structures. These are positive environmental impacts. Continuous use of inorganic fertilizer, residual effects of insecticides (such as DDT), and the slash and burn cultivation systems do have negative impact on the environment. However, the percentage of farmers following these practices is relatively small.

Human resources development

Over the period 1974–1994, six NMRP staff were trained to PhD level, 16 to MSc level and 12 to BSc level. Over 60 staff (research officers, field officers) underwent short-term training inside and outside Tanzania.

The available information clearly demonstrates that the ongoing maize R&D programme has had a substantial socio-economic impact on society, and the investment is socially attractive and very profitable.

Conclusions and Recommendations

Between 1974 and 1994, the NMRP released 15 varieties and several site-specific crop management recommendations for the various agro-ecological zones. The estimated ROR for the maize R&D investment in Tanzania, even under the most conservative scenario was 19%. This means that for every dollar invested in the maize R and D, the society received a return of 19 US cents. The estimated NPV at the 5% discount rate for this investment was US$1.1 billion (1997 constant dollar). For the 10 and 15% discount rates, the estimated NPVs were US$300 million and US$105 million respectively. This clearly demonstrates that the past investments in maize R&D activities in Tanzania are profitable, and generated a competitive ROR. The estimated ROR for the various scenarios ranged between 13 and 23.6%.

In addition to the 19% return, the investment also contributed to capacity building, food security, and establishing linkages with other regional and international organizations. These elements will eventually contribute to the overall agricultural and economic growth and development of Tanzania. However, these benefits are not included in the ROR estimates.

The NMRP has developed several varieties. Currently there are three components of the program addressing the three major agro-ecological zones. The future breeding initiatives should be targeted to the specific problems identified in the adoption surveys. With respect to crop management practices, there is a need for site-specific recommendations with heavy emphasis on on-farm research and testing.

The public extension service is the major source of technological information for the maize farmers in Tanzania. However, lack of knowledge was often mentioned by the sample farmers as one of the constraints to adoption of technologies. Thus, there is a need to strengthen the linkages between researchers, extension workers and the farmers in order to promote the smooth flow of technologies.

Farmers recycle both hybrid and composite seeds. There is a need to undertake research to assess the losses due to recycling of hybrids and composites and the information obtained should be effectively communicated to the farmers through the extension services. Credit availability and use by small-scale farmers remained low in all zones. The major source of credit was the informal sector. All farmers indicated that it was difficult to get credit due to cumbersome procedures, lack of collateral and ignorance. Thus, the issue of credit should be addressed in order to increase the use of purchased inputs. The current agricultural input supply fund initiated by the Ministry of Agriculture and Food Security is a step in the right direction. Unfortunately most input supply stores are located in the town centres. There is a need to improve the input delivery and product marketing system. Participation of the private sector and NGOs in this arena should be encouraged.

Currently there are a large number of recommendations for each zone. Some are blanket and some are site-specific. Thus, there is a need to review and update the recommendations, and where appropriate, verify the existing recommendations under farmer conditions.

Acknowledgements

The authors would like to acknowledge the technical and financial support provided by SACCAR under a USAID Grant, and CIMMYT through its projects for Strengthening Economics and Policy Research for NARS in Eastern Africa and the CIDA funded East African Cereals Project. Thanks to E. Nkonya, P. Xavery, N. Katinila, J. Mafuru, B. Bisanda, H. Ulotu and A. Mwilawa for administering the questionnaires and preparing the zonal adoption reports.

References

Akino, M. and Hayami Y. (1975) Efficiency and equity in public research: rice breeding in Japan's economic development. *American Journal of Agricultural Economics* 57, 1–10.

Anandajayasekeram, P. and Martella, D. (1995) Institutionalization of Impact Assessment: SACCAR's Experience in Southern Africa. A paper presented at the USAID Collaborative Workshop on Agricultural Technology Development and Transfer in sub-Saharan African, Harare, Zimbabwe, January 24–27.

Anandajayasekeram, P., Martella, D.R., Sanders, J. and Kupfuma, B. (1995) *Report on the Impact Assessment of the SADC/ICRISAT Sorghum and Millet Improvement Program*, Vol. 1. SACCAR, Gaborone.

Anandajayasekeram, P., Martella, D.R. and Rukuni, M. (1996) *A Training Manual on Research and Development Evaluation and Impact Assessment of Investments in Agricultural and Natural Resources Research.* SADC-SACCAR, Gaborone.

Anderson, J.R. and Herdt, R.W. (1996) Reflections on impact assessment. *Proceedings of the ISNAR/Rutgers Agricultural Technology Management Workshop, 6–8 July 1988, Rutgers University, New Jersey*, Vol. II: *Assessing the Impact of Agricultural Research.* ISNAR, The Hague, The Netherlands.

Bolton, A. (1971) Territorial maize variety trials in Tanzania 1966–70. *East African Agriculture and Forestry Journal* 37, 109–124.

Byerlee, D. and de Polanco, E.H. (1986) Farmers' stepwise adoption of technological packages: evidence from the Mexican Altiplano. *American Journal of Agricultural Economics* 68, 519–527.

CIMMYT (1992) *CIMMYT World Maize Facts and Trends: Maize Research Investments and Impacts in Developing Countries.* CIMMYT, Mexico, D.F.

Darrah, L.L. (1976) Altitudinal and environmental responses of entries in the 1972–73 Eastern African Maize Variety Trials. *East African Agriculture and Forestry Journal* 41, 273–288.

Department of Research and Training (DRT), Tanzania (1991) *National Agricultural and Livestock Research Master Plan – Annex I, Part A.* ISNAR, The Hague, The Netherlands.

Ellis, F. (1982) Agricultural price policy in Tanzania. *World Development* 10(4), 263–283.

Griliches, Z. (1958) Research costs and social returns: hybrid corn and related innovations. *Journal of Political Economy* 66, 419–431.

Harrison, M.N. (1970) Maize improvement in East Africa. In: Leakey, C.L.A. (ed.) *Crop Improvement in East Africa,* CABI, Farnham, UK, pp. 22–59.

Horton, D.E. (1990) Assessing the impact of international research: concepts and challenges. In: Echeverria, R.G. (ed.) *Methods for Diagnosing Research System Constraints and Assessing the Impact of Agricultural Research*, Vol. II. ISNAR, The Hague, The Netherlands.

Isinika, A.C. (1995) Assessing the effect of agricultural research expenditures on agricultural productivity in Tanzania. PhD thesis, University of Kentucky, Lexington, Kentucky.

Norton, G.W. and Davis, S. (1981) Evaluating returns to agricultural research: a review. *American Journal of Agricultural Economics* 63(4), 685–699.

Oehmke, J.F. (1992) *Technology, Impact and Agricultural Transformation: Lessons Learned from Impact Studies.* Paper presented at the Symposium on the Impact of Technology on Agricultural Transformation in Africa, October 14–16, 1992, Washington, DC.

Pardey, P.G., Roseboom, J. and Beintema, N.M. (1995) *Agricultural Research in Africa: Three Decades of Development.* Briefing Paper No. 19. ISNAR, The Hague, The Netherlands.

Schuh, G.E. and Tollini, H. (1979) *Costs and Benefits of Agricultural Research: State of The Arts.* World Bank Staff Paper No. 360. World Bank, Washington, DC.

8 Economic Returns from Livestock Research and Development in Tanzania: 1966–1995

ALOYCE KALIBA, SEAN FOX AND DAVID NORMAN

Abstract: The objective of this study was to estimate the economic return to public investment in livestock research and development in Tanzania from 1966 to 1995. The supply response model was used to estimate supply elasticities for cattle and small ruminant meat and milk products. The results showed that on average, all commodities were equally responsive to investment in research and development as well as to market prices. All economic profitability measures were positive suggesting that the society gained through investing in livestock research and development. Policies to improve market incentives may increase productivity, reduce risk and encourage farmers to engage in livestock marketing activities.

Introduction

While there are few research reports related to estimating the economic returns from livestock research and development, there is an extensive literature concerning the estimation of economic returns from public investment in research and development (R&D). Economic surplus is the major technique used in measuring the benefits from R&D. The basic assumption is that there is an induced shift in supply due to investment into R&D. The supply shift allows the estimation of gains by producers and consumers (society welfare gain). Net economic gain is calculated by deducting the investment cost from the society welfare gains. Then economic profitability measures are estimated. While induced supply shift can be parallel (Schultz, 1953), pivotal (Linder and Jarrett, 1987) or kinked (Rose, 1980), the original supply curve can be completely elastic, non- linear or linear with constant elasticity (Peterson, 1967; Akino and Hayami, 1975; Voon and Edwards, 1991). Thus, the estimated social benefits depend on the assumption attached to the type of supply shift and estimated supply and demand elasticities. In return, it influences the estimated magnitude and distribution of research returns. However, little empirical evidence exists on the nature of both demand and supply shifts. Also, the technique cannot estimate the returns with respect to individual agencies involved in R&D. The use of demand and supply curves requires knowledge on the extent of adoption of technologies, and both the demand and supply elasticities of the commodity concerned. This knowledge is currently unavailable for the Tanzanian livestock sector at the national level.

 Impact of Science on African Agriculture and Food Security (eds P. Anandajayasekeram, M. Rukuni, S. Babu, F. Liebenberg and C.L. Keswani)

An alternative approach is to use the Supply Response Model to estimate returns from R&D. The model is based on the price adaptive expectation theory as proposed by Nerlove (1956) and developed by Houck and Ryan (1972). The Supply Response Model estimates production as a function of expected market conditions or price. Production or hectarage is assumed to be a function of relative expected price and other important factors such as expenditure on R&D and weather conditions. Estimation of a supply response model has two advantages, both the short-term and long-term elasticities of commodity supply can be estimated, and returns from specific investment projects can be estimated directly from the model. The procedure is more direct, and avoids researchers' a priori assumptions on the nature of supply and demand shifts.

Objectives

The objectives of this chapter are twofold: (i) to develop an empirical framework, which estimates returns from R&D using a Supply Response Model; and (ii) to estimate short-term economic returns from investment in livestock research and development by the Government of Tanzania (GOT) and social welfare gain at the national level. The study focuses on cattle and small ruminant meat and milk production, as livestock research resources are traditionally concentrated on increased production of the three commodities. The empirical model is applied on time-series data, for livestock products supply and GOT investment in livestock research and development for the 1966/95 period. Data were obtained from the Food and Agricultural Organization (FAO) online statistical database and the Department of Research and Development in the Ministry of Agriculture and Cooperatives, Tanzania.

Livestock Research

In Tanzania, arable land represents 5% of the total land area. Permanent crops represent 1%, meadows and pastures 40%, forest and woodlands 47% and others 7%. The climate varies from tropical along the coast, to temperate in the southern highlands. The terrain changes from plain plateau along the coast, to undulating plateau in central regions and highlands in northern and southern regions. Tanzania has vast pastoral resources estimated at 51% of the total land area. It has a cattle stock of 17.7 million, plus 455,000 pigs, 3.5 million sheep, 12.5 million goats and 30 million chickens. There are 4.8 million dairy cattle that produce 835,000 t of milk (FAO, 2003). Most of the meat and milk consumed are produced by zebu cattle using traditional livestock production methods.

Red meat (beef, veal, mutton and lamb) from cattle and small ruminants is the single largest commodity produced by the livestock sector. The sector produces about 256,000 t of red meat and 835,000 t of milk per year (FAO, 2003). The per capita consumption of red meat and milk is low and the national livestock research system is aimed at increasing the per capita consumption, through the introduction of genetically superior beef and dairy cattle in the traditional production system, improving feeding and husbandry practices, encouraging efficient use of crop residues and effective routine disease control measures.

The Tanzanian livestock research system is divided into three major research programmes, i.e. beef cattle, dairy cattle and small ruminants. These programmes are co-ordinated through research institutions and centres in the Eastern, Western, Southern, Southern Highlands, Northern, Lake and Central zones. Historically and despite frequent changes of agricultural research management structures, the Livestock Production Research Institute (LPRI), located in Central Zone, has remained a primary centre of livestock research. The centre was established in 1932, as a veterinary research centre for developing animal vaccines. Achievements of the centre include the development of the Mpwapwa breed of cattle and on-farm evaluation of some exotic breeds under small-scale farmers' conditions. Other livestock research institutions include: Tanga Livestock Research Center, located in the Eastern Zone, dealing with small-scale dairy development for coastal areas; Kongwa Pasture Research Center, located in the Central Zone dealing with pasture research; West Kilimanjaro Farm, which acts as a livestock gene bank and is located in the Northern Zone; and Uyole Agricultural Research Center, which concentrates on dairy cattle research and development in the Southern Highlands.

Data and Estimation Procedure

Production data were collected from the FAO online statistical database and the Department of Research and Development of the Ministry of Agriculture and Cooperative (MOAC), Tanzania. Complete data required for the analysis were available for the 1966/95 period. The average producer prices of cattle, sheep and goat meat, and milk in Tanzania shillings were deflated using the Consumer Price Index (CPI) to get the real price, while expenditure on research and development were deflated using the nominal inflation rate.

The quantities of goat and sheep meat were aggregated together to form quantities of small ruminant meat, and the weighted average price for small ruminant meat was calculated using individual price and quantities. Research and development costs for each year included the government budgetary allocation to recurrent and development expenditure, salaries and benefits, overhead and administration expenditures. Rainfall is the average annual rainfall in millimetres from four meteorological stations located at LPRI Mpwapwa, Dodoma Municipal airport, Kondoa and Iramba districts.

The real producer price for 1 kg of cattle meat was TSh878 in the 1966–1979 period and TSh382 in the 1980–1995 period. The average real price for the two periods was TSh551/kg of cattle meat. In the same periods, price of small ruminant meats decreased from TSh904 to TSh406/kg. Real price of milk declined from TSh209 to 89/kg. While quantity of cattle meat, small ruminant meat and milk produced fell by 36, 28 and 32%, in the same two periods, the herd size of both cattle and small ruminants decreased by 20%. Falling real price and thus herd size and quantity produced is a reflection of the worsening economic situation of smallholder farmers in Tanzania for the 1975–1995 period.

The estimated supply response models for each commodity (cattle meat, small ruminant meat and milk) were specified as follows:

$$\Delta \ln Q_t = \alpha_0 + \alpha_1 \Delta \ln LU_t + \alpha_2 \Delta \ln Q_{t-1} + \alpha_3 \Delta \ln P_{t-1} + \alpha_4 W_t + \alpha_5 \Delta \ln RD_{t-1} + \alpha_6 \Delta \ln RD_{t-2} + \lambda \mu_t + \eta \quad (1)$$

Where ln is the natural logarithm, LU is the national cattle or small ruminant herd size, P is average price, W is annual rainfall, RD is investment in R&D, α's are parameters of the model, λ is the adjustment parameter or error correction coefficient and μ and η are, respectively, truncated and normally distributed error terms of the supply response model. Livestock herd size was included in the model, as both quantities of meat and milk produced depend on the number of available livestock. The lag on the R&D variable is intended to capture the delayed effects of R&D. The rainfall variable was not transformed into logarithm due to the presence of substantial negative values. Owing to the small sample size, a limited number of lagged variables were included in the model, to have enough degrees of freedom.

Before estimating the supply response mode, testing for existence of unit roots in all time series variables was performed by the augmented Dickey and Fuller (1981) test using the following equation, where Y_t is the variable of concern and t is a time trend.

$$\Delta Y_t = \alpha + \beta Y_{t-1} + \sum_{j=1}^{m} \gamma Y_{t-1} + \rho_t + \varepsilon_1 \quad (2)$$

Table 8.1. Test results for presence of unit roots.*

Variable	Dickey–Fuller	Perron–Phillips
Log quantity of cattle meat	2.826	3.133
Log quantity of small ruminant meat	3.653	3.189
Log quantity of poultry products	1.615	1.824
Log quantity of milk	3.467	3.961
Log price of poultry	0.961	0.994
Log price of milk	0.818	1.487
Log price of cattle meat	1.700	1.529
Log price of small ruminant meat	1.031	1.113
Log RAIN	9.134†	9.204†
Log R&D	6.576	4.154

*The calculated asymptotic critical value was 6.25.
†Statistically significant at 5% probability level.

The null hypothesis is ($\beta = 0$) while the alternative hypothesis is ($\beta \neq 0$), (i.e. Y_t has unit root against the alternative that it integrated of order zero). The SHAZAM program used in the analysis reports both ADF test statistics and the critical value. If a test statistic is smaller than the estimated critical value, then there is evidence of co-integration/presence of unit root, indicating that the variable has a long-term relationship. At the 95% significance level, the null hypothesis was accepted in all cases, except for rain variables. These results indicate that all time series are non-stationary and have a unit root. To confirm these results, the Phillips and Perron (1988) test for the presence of unit root was also performed. At the 95% significance level, the null hypothesis was rejected again except for the rainfall variable. The ADF and Phillips and Perron (1988) results suggested that all the time-series variables, except the rainfall variable, have a unit root and hence the first difference of the supply response model was appropriate (see Table 8.1 for detailed results).

The three equations (i.e. cattle meat, small ruminant meat and milk production) were jointly estimated as a seemingly unrelated regression system with vector autoregressive errors. Autocorrelation was admitted both across as well as within equations. The procedure allowed for contemporaneous correlation across equations and serial correlation within equations (Savin and White, 1978). Contemporaneous correlation arises, because quantities and price of commodities were jointly determined at the household and market level, respectively. Serial correlation arises due to the presence of lagged variables.

Results and Discussion

Supply response models

Table 8.2 presents the results of the supply response model of cattle meat. The estimated R^2 between observed and predicted values was 40%, indicating that the exogenous variables included in the model at least explain 40% of the variation in cattle meat supply. Statistically significant variables include the real price of cattle meat and the lagged variable of investment in R&D. The estimated coefficients on real price of cattle indicated that, in the short run, increase in the price of cattle meat will increase supply by about 0.12%. A similar interpretation can be made to the coefficients of R&D and rainfall variables. A positive and significant relationship existed between investment in livestock R&D and cattle meat supply when the

Table 8.2. Results of the supply response model for cattle meat.

Variable	Coefficient	*T*-ratio
Constant	0.059	0.333
Herd size (cattle)	5.847	0.517
Lag of quantity of cattle meat	–0.049	–0.212
Real price of cattle meat	0.121	2.432**
Investment in livestock R&D: *t*	0.009	0.350
(in constant shillings terms) *t*–1	0.042	1.517*
Rainfall	–0.063	–2.011**
R^2 between observed and predicted values	40.24	
Residual variance	0.001	
λ	0.033	

**Significant at 5% probability level, *Significant at 10% probability level.

Table 8.3. Results of supply response model for small ruminant meat.

Variable	Coefficient	*T*-ratio
Constant	–4.084	–2.098**
Herd size (small ruminants)	–268.222	–2.088**
Lag of quantity of small ruminant meat	–0.052	–0.281
Real price of small ruminant meat	0.046	1.575*
Investment in livestock R&D: *t*	0.016	1.107
(in constant shillings term) *t*–1	0.031	2.382**
Rainfall	0.0586	2.986**
R^2 between observed and predicted values	37.85	
Residual variance	0.001	
λ	0.18	

**Significant at 5% probability level, *Significant at 10% probability level.

former is lagged. Estimated long-run price and R&D elasticities of cattle meat were 0.12 and 0.05, respectively. Hence, an increase in the real price of cattle meat or expenditure on R&D will increase long-run supply by 0.12 and 0.05, respectively. The estimated error correction coefficient (λ) was 0.033 and was statistically not significant. Non-significance of the coefficient implied that causality from the independent variables to dependent variable is weak. Its small magnitude indicated that the speed of adjustment toward a long-run equilibrium after changes in price or R&D investment was slow (i.e. non-significance of the error correction coefficient implies that production does not adjust quickly to shocks caused by variability of variables included in the model and causality of independent variables to dependent variables is relatively weak).

Table 8.3 presents a supply response model for small ruminant meat. The estimated R^2 between observed and predicted values was 38%, indicating that the variables included in the model at least explain 38% of the variations in small ruminant meat supply. Statistically significant variables were herd size, real price of small ruminant meat, lagged investment in R&D and rainfall. From Table 8.3, the short run price and R&D elasticities of supply for small ruminant meat were 0.05 and 0.02, which is very inelastic. The corresponding long-run elasticities were 0.04 and 0.05, respectively. Again, the estimated error correction coefficient for the small ruminant meat model was non-statistically significant and small in magnitude (0.18), indicating

Table 8.4. Results of supply response model for milk variable.

Variable	Coefficient	*T*-ratio
Constant	0.006	0.046
Herd size (cattle)	1.347	0.150
Lag of quantity of milk	0.272	1.221
Real price of milk	0.001	0.056
Investment in livestock R&D: *t*	0.012	0.728
(in constant shillings terms) *t*–1	–0.007	–0.357
Rainfall:	0.0197	0.717
R^2 between observed and predicted values	19.05	
Residual variance	0.001	
λ	0.052	

that supply was very vulnerable to production shock. Any production shock has a long-term residual impact, as production does not adjust quickly to its long-run equilibrium path. Note a relatively large and negative coefficient on herd size. This may represent an overemphasis of the negative production relationship of increasing number of small ruminants. Small ruminants are important agents of land degradation in most of sub-Saharan Africa.

Table 8.4 presents the results of the milk supply response model. The estimated R^2 for the milk model was 19%. Although there were positive relationships between dependent and all independent variables, except lagged R&D variable, the relationship was not statistically significant. The estimated coefficients were interpreted as in the cattle meat or small ruminant supply response models.

Economic returns and profitability of R&D

The net social welfare gain was estimated by aggregating the gross returns of the different commodities, (i.e. cattle meat, small ruminant meat and milk). Net returns were estimated as gross returns minus the total GOT real budgetary cost of R&D. Because the models were estimated using real prices and costs, the estimated gross returns are also in real terms. The data on R&D costs could not be disaggregated, because the budget is allocated to the entire livestock research programme, rather than individual commodities. Economic profitability measures are presented in Table 8.5. In the table, Marginal Internal Rate of Return of Research (MIRR) represents the modified internal rate of return estimated by the following formula:

$$MIRR = \left[\frac{-NPV\,(r, R_{RD}) \times (1+r)^n}{NPV\,(f, RD) \times (1+f)} \right]^{\frac{1}{n-1}} - 1 \qquad (3)$$

Where NPV (r, GR) is the net present value of gross return (R_{RD}) from R&D discounted by r (reinvestment rate), NPV (f, RD) is the net present value of R&D costs (RD) discounted by f (finance rate) and n is the number of cash flows. In the formula, f, accounts for the opportunity cost of money or capital used for R&D rather than being invested elsewhere. Also, r accounts for money or capital taken from R&D for reinvesting in the production process.

For the 1990–1995 period, the inflation and average interest rate of long-term bonds have averaged 11 and 15%, respectively. The calculated social cost of capital was 4.5%. Therefore, returns and costs were discounted by 5, 10 and 15% to get the MIRR, the net present value and the net benefit–cost ratio. The results are presented in Table 8.5. From the table,

Table 8.5. Estimated short-run economic profitability measures.

Discount factor (%)	NPVs (billion TSh)	Benefit–cost ratio	MIRR (%)
5	216.222	456.78	29.78
10	85.648	362.14	33.08
15	39.886	268.74	36.74

at 5% finance and reinvestment rates, the calculated internal rate of return was 29.8%. These results imply that at a 5% opportunity cost of capita and reinvestment rates, every shilling invested in livestock R&D returned 30 cents. At 10 and 15% finance and reinvestment rates, the returns were 33 and 37 cents per shilling invested, respectively. When the reinvestment rate was set to zero, the MIRR figures were 27.5, 28.0 and 28.5%, respectively. These results fall within the range of internal rates of return reported by other studies. For example, Kaliba (1995) estimated an internal rate of return of 16% for traditional free range grazing zebu cattle and a 23–41% internal rate of return for improved zero grazed dairy cattle in Central Tanzania. Despite the many problems associated with traditional livestock production, under these scenarios the investment in livestock R&D in Tanzania was economically profitable.

In general, other estimated economic profitability measures (NPV and benefit–cost ratio) were all positive, suggesting that investment in livestock R&D was economically viable and socially justified. At a 5% discount rate, for the 1966–1995 period, net welfare gain from livestock R&D was 216 billion Tanzania shillings. At 10 and 15%, the net welfare gain was 86 and 40 billion Tanzania shillings, respectively. The estimated profitability measure was relatively large, as compared with those estimated for maize production. Moshi *et al.* (1997) estimated an internal rate of return of 23% for maize research and development in Tanzania. Therefore, the tendency of favouring food crop research programmes over livestock research, does not appear to be economically justified. High return associated with inelastic supply responsiveness is an indication of potential underinvestment in livestock research and development in Tanzania.

Conclusions

This study used the supply response approach to estimate the returns from livestock R&D in Tanzania. The results imply that market incentives alone are not enough to increase the supply of livestock products in Tanzania. Market incentives and R&D were found to be of equal importance in influencing the supply of cattle meat, small ruminant meat and milk. Investment in livestock R&D has had a notable impact on cattle and small ruminant meat supply, indicating that research and development can increase the production of meat. In all models, the estimated coefficients that measure the speed of adjustment towards long-term equilibrium were statistically insignificant. The results reinforce the idea that the adjustment towards long-term equilibrium is not instantaneous and a significant time lag exists, before supply adjusts to its long-term trend after a production shock. Livestock production is affected by many biological and management factors. Production shocks will interfere with interwoven factors that require a considerable time to adjust towards a long-term growth path, after the initial shock. The slow adjustment in the production process and vulnerability to external shocks, emphasizes the importance of developing cushioning technologies to reduce the gravity of production risks.

In subsistence agriculture it is difficult to quantify all the benefits of livestock production. Households, as economic units, are major producers as well as consumers. What is taken to the market is the household surplus. Also, traditional livestock producers do not regard their livestock as an end product in itself, but as a store of wealth and economic units capable of generating a stream of benefits or use value over their economic lifetime. Livestock are raised not only to produce meat and milk, but also serve other important socio-economic functions. Livestock are sold when they no longer generate a stream of benefits, when they are non-essential to the herd or when there are heightened risks of mortality, such as droughts (Benhke, 1992; Hortzman *et al.*, 1992). Thus, the decision to sell (supply) is normally determined by factors outside the market. To stimulate farmer participation in marketing, incentives need to be higher than the socio-cultural value attached to livestock. Policies aimed at creating livestock marketing incentives through increased efficiency and performance of existing marketing systems would be highly desirable.

References

Akino, M. and Hayami, Y. (1975) Efficiency and equity in public research: rice breeding in Japan's economic development. *American Journal of Agricultural Economics* 57, 1–10.

Araji, A.A., White, F.C. and Guenthner, J.F. (1995) Spillovers and returns to agricultural research for potatoes. *Journal of Agricultural and Resource Economics* 20(2), 263–276.

Benhke, R.H. (1992) *New Directions in African Range Management Policy*. Pastoral Development Network Paper 32C. Overseas Development Institute (ODI), London.

Buhr, B.L. (1993) *A Quarterly Econometric Simulation Model of the US Livestock and Meat Sector*. Staff Paper P93–12. Department of Agricultural and Applied Economics, University of Minnesota, St Paul, Minnesota.

Dickey, D.A. and Fuller, W.A. (1981) Likelihood ratio statistics for autoregressive time series with a unit root. *Econometrica* 49, 1057–1072.

FAO (2003) *FAO Year Book*. FAO, Rome.

Griffiths, W.E., Carter Hill, R. and Judge, G.G. (1993) *Learning and Practicing Econometrics*. John Wiley & Sons, New York.

Hallami, D. and Zanoli, R. (1993) Error correction models and agricultural supply response. *European Review of Agricultural Economics* 20, 151–166.

Hortzman, J.S., Kulibaba, N.P. and Stathacos, C.J.D. (1992) *Livestock Marketing and Trade in Central Corridor of West Africa. Agricultural Marketing Improvement Strategies Project, Sahel West Africa Office*. USAID, Washington, DC.

Houck, J.P. and Ryan, M.E. (1972) Supply analysis for corn in the United States: the impact of changing government programs. *American Journal of Agricultural Economics* 54, 184–191.

Kaliba, A.R.M. (1995) The socio-economics of improved cattle raising under the stall feeding management system in semi-arid central Tanzania. MSc thesis, Department of Agricultural Economics, Kansas State University, Kansas.

Linder, R.K. and Jarret, F.G. (1978) Supply shifts and the size of research benefits. *American Journal of Agricultural Economics* 60, 48–56.

Moshi, A., Anandajayasekeram, P., Kaliba, A.R.M., Martella, D., Mwangi, W. and Shao, F. (1997) *Economic Impact of Maize Research in Tanzania*. SACCAR, Gaborone.

National Sample Census of Agriculture (1994/95) *Tanzania Mainland Report*, Vol. II. Planning Commission. Dar-es-Salaam, Tanzania.

Nerlove, M. (1956) Estimation of supply of selected agricultural commodities. *Journal of Farm Economics* 38, 495–509.

Peterson, W.L. (1967) Returns to poultry research in the United States. *Journal of Farm Economics* 49, 656–669.

Phillips, P.C.M. and Perron, P. (1988) Testing for unit root in time series regression. *Biometrika* 75, 335–346.

Rose, R.N. (1980) Supply shifts and research benefits: comments. *American Journal of Agricultural Economics* 62, 843–837.

Savin, N.E. and White, K.J. (1978) Estimation and testing for functional form and autocorrelation: a simulation approach. *Journal of Econometrics* 44, 1–2.

Schultz, T.W. (1953) *The Economic Organization of Agriculture*. McGraw-Hill, New York.

Voon, J.P. and Edwards, G.W. (1991) The calculation of research benefits with linear and nonlinear specification of demand and supply functions. *American Journal of Agricultural Economics* 72, 415–420.

Weliwita, A. and Govindasamy, R. (1997) Supply response in the northeastern fresh tomato market: cointegration and error correction analysis. *Agricultural and Resource Economic Review*, 248–255.

9 Impact of the Russian Wheat Aphid Integrated Control Programme in South Africa

CARISSA N. MARASAS

Abstract: This study was undertaken to describe the adoption and impact of the Russian wheat aphid (RWA) integrated control programme in the Central and Eastern Free State of South Africa. Rapid adoption of RWA-resistant cultivars has been demonstrated since their release in 1993. The area sown to the cultivars increased from 3% in 1993 to 46% in 1997. Correlation and multiple regression analyses suggested that adoption of the cultivars was influenced by the farmer's educational level, wheat area and experience with RWA, the quality grades realized by resistant cultivars and the combined weighted average yield of resistant and susceptible cultivars. The Akino and Hayami, cost–benefit and unit cost saving variations of the economic surplus approach were used to estimate the macro-level returns on the public sector investment from 1980 to 2005. The internal rate of return ranged between 21.6 and 26.8%, and the net present value at the 5% discount rate ranged between 37.5 and 73.5 million South African Rand. The unit cost saving approach yielded the most conservative and the cost–benefit method the most optimistic results. Partial budget analyses indicated that incremental net benefits ranging between 140 and 329 Rand/ha in 1997 prices were generated at the farm level.

Introduction

The Russian wheat aphid (*Diuraphis noxia*) (RWA) was first reported in South Africa in 1978 (Walters *et al.*, 1980) and has been reported to cause yield losses up to 92% (Walters, 1984; Tolmay and Wessels, 1996). Public sector research to address this problem was initiated at the ARC-Small Grain Institute (ARC-SGI) of the South African Agricultural Research Council. The Russian wheat aphid integrated control programme (RWA-ICP) involved three stages of technology development, including chemical control, resistant cultivars and biological control. This chapter describes the adoption and impact of the research programme in South Africa.

Background

Public sector research on chemical control started at the ARC-SGI in 1980 and culminated in appropriate control recommendations. However, apart from these recommendations,

 Impact of Science on African Agriculture and Food Security (eds P. Anandajayasekeram, M. Rukuni, S. Babu, F. Liebenberg and C.L. Keswani)

most insecticides and chemical control technologies were developed by the private sector. Insecticides for RWA can be sprayed with tractors or airplanes, or applied as a seed treatment. The seed treatment was developed at a later stage and became available to farmers in 1993. Though chemical control provided a partial solution to the RWA problem, it was considered expensive and harmful to humans and the environment.

Host plant resistance breeding for RWA commenced in 1985 and the ARC-SGI released the first resistant cultivar in the world in 1993, known as Tugela-DN. Private seed companies have also developed resistant cultivars since then. The ARC-SGI had released five and private seed companies eight of the 13 resistant cultivars available to producers at the time of this study. Research on biological control of RWA started in 1989 with the introduction of a natural enemy from Russia. This parasitoid (*Aphidius matricariae*) was released on a limited scale during 1996 and 1997, but biological control was still in its developmental stages at the time of this study.

The analyses presented in this chapter focused on the technologies developed by the public sector and focused on the RWA-resistant cultivars released by the ARC-SGI. Appropriate data were not available to assess the impact of chemical and biological control; most chemical control technologies were developed by the private sector; and biological control was still under development.

Objectives

The objectives of this study are to describe the adoption of the RWA-ICP in the study area; assess the economic impact of the RWA-resistant cultivars at the macro level, to South African society; and to assess the economic impact of the RWA-resistant cultivars at the micro level, to farmers.

Methodology

The analyses presented in this chapter employed the results of farm-level surveys undertaken in 1997, in which data were collected on the farmers' wheat management practices in 1996. Data were collected through personal interviews using standard pre-tested questionnaires (Marasas, 1999). Information from a preliminary mail survey conducted in 1996 (Marasas *et al.*, 1997) was applied in compiling the questionnaires. The study focused on the dryland summer wheat production areas of the Central and Eastern Free State of South Africa, where RWA occurred the most often. These two regions are separated according to climate and production considerations.

A stratified random sample of 90 Central and Eastern Free State farmers was selected from the 14 major relatively homogeneous farming areas in the two regions. A 'relatively homogeneous farming area' has a fair degree of uniformity in agricultural use, attainable yields and production practices. It consists of one or more land types, grouped together by taking into account the macro-climate, topography, geology, soil pattern, yield potential of resources, adapted crops and vulnerability to wind and water erosion (Scheepers *et al.*, 1984). The number of farmers from each major homogeneous farming area was selected by the respective areas under ploughable soil types suitable for wheat production. This aimed to represent the variability in wheat production conditions in the regions. The sampling frame consisted of producers' address lists.

Farmers in the regions formerly known as Thaba Nchu and Qwa Qwa were also included in the study. Some climate and production considerations are similar in the Central Free State and Thaba Nchu, and the Eastern Free State and Qwa Qwa. However, Thaba Nchu and Qwa Qwa were not classified into homogeneous farming areas at the time. As less information was available, and few farmers in these two regions had planted wheat in 1996, all farmers who could be contacted through the institutions working with them were interviewed. This included 20 producers, but the combined wheat area for the two regions comprised a relatively small proportion of the study area.

The survey results were separated for the Central Free State, Eastern Free State, Thaba Nchu and Qwa Qwa, as the farmers' production conditions and wheat management practices differed. The factors affecting the producers' adoption of RWA-resistant cultivars were assessed by correlation and multiple regression analyses. The impacts associated with the RWA-ICP were identified according to the framework suggested by Anandajayasekeram *et al.* (1996). The macro-level economic impact of the cultivars was estimated using three variations of the economic surplus approach, and the micro-level impact was assessed with partial budgets. Primary and secondary data were used. Environmental, institutional and spill-over impacts were qualitatively described.

Results and Discussion

Although regional differences were observed, the farmers' wheat management practices such as planting, fertilizer and herbicide application, and chemical control for pests and diseases other than RWA, did not differ significantly between RWA-resistant and susceptible cultivars. The major differences between the two types of cultivars included the respective yield levels and the farmers' RWA control practices. These are outlined in the following two sections, and are pertinent to the adoption and impact analyses presented in subsequent sections. A detailed description of the farmer and farm characteristics, wheat management practices and RWA control in the study area is provided in Marasas (1999).

Wheat yields of Russian wheat aphid resistant and susceptible cultivars

Yields varied between the numerous wheat cultivars planted in 1996. Pooled weighted average yields were therefore calculated for resistant and susceptible cultivars, respectively (Table 9.1). Resistant cultivars demonstrated a weighted average yield advantage over the susceptible types, amounting to 0.15 t/ha in the Eastern Free State, 0.25 t/ha in the Central Free State and 0.36 t/ha in Thaba Nchu. Qwa Qwa was the exception where the weighted average yield for susceptible cultivars was higher. However, difficulties were experienced in obtaining consistent yield estimates in this region due to problems with flooding and record-keeping by the farmers. Yield gain was therefore not considered as a realized benefit of the resistant cultivars in Qwa Qwa. This nevertheless comprised a minor proportion of the overall study area.

Table 9.1. Weighted average wheat yield in the study area (1996).

	Weighted average yield (t/ha)	
Region	Resistant cultivars	Susceptible cultivars
Central Free State	2.20	1.95
Eastern Free State	2.99	2.84
Thaba Nchu	1.47	1.11
Qwa Qwa	2.15	2.29

Russian wheat aphid control practices

RWA has been perceived as a problem at any given time by 90% of Eastern Free State, 68% of Central Free State, 46% of Thaba Nchu and 43% of Qwa Qwa farmers. The ARC-SGI recommendations for RWA control comprised the use of resistant cultivars, or chemical control by either seed treatment or spraying for susceptible cultivars. However, the survey results demonstrated that the farmers used either no control, spraying, seed treatment, or both spraying and seed treatment, and that a proportion of resistant cultivars was also treated (Table 9.2). No Thaba Nchu farmers applied chemical control in 1996. Although spraying was continued on resistant cultivars in the remaining regions, only Eastern Free State producers continued seed treatment on these cultivars.

An imidacloprid insecticide was the only RWA seed treatment available at the time, and was applied once per season by mixing the insecticide with the wheat seed before planting. The farmers sprayed various mixtures of RWA insecticides including monocrotofos, parathion, demeton-S-methyl, chlorpyrifos and mevinphos. Most producers sprayed only once in 1996. Tractors, airplanes or both methods were used to apply the RWA insecticides. Producers mostly combined the RWA insecticides with herbicides, fungicides and other insecticides in the same spray application, which reduced the application cost of spraying RWA insecticides. Cost savings in the economic analysis were therefore accrued to RWA insecticide costs only, as spray application costs were assumed to continue for the other chemicals even if all RWA spraying was eliminated. Although chemical control methods for RWA were still used in 1996, the farmers indicated a decline in their use since previous years, and projected further declining trends into the future.

Adoption of the Russian wheat aphid integrated control programme in the study area

The sample farmers rapidly adopted the RWA-resistant cultivars developed by the ARC-SGI since their first release in 1993. The area sown to the cultivars increased from 3% in 1993 to

Table 9.2. The percentage study area planted to resistant and susceptible cultivars and treated with different methods of Russian wheat aphid control (1996).

	Percentage area treated			
	No control	Spraying	Seed treatment	Spraying and seed treatment
Central Free State:				
Resistant cultivars	85	15	–	–
Susceptible cultivars	53	12	4	31
Eastern Free State:				
Resistant cultivars	45	5	28	22
Susceptible cultivars	5	14	55	26
Thaba Nchu:				
Resistant cultivars	100	–	–	–
Susceptible cultivars	100	–	–	–
Qwa Qwa:				
Resistant cultivars	70	30	–	–
Susceptible cultivars	43	23	34	–

Notes: (–) indicates that the treatment was not used in the region. Estimates add to 100% across the rows of Table 9.2.

46% in 1997, and the farmers projected this area to increase further to 52% by the year 2000. In 1997, the area planted to the cultivars in each region was 68% in Qwa Qwa, 49% in the Eastern Free State, 41% in the Central Free State and 35% in Thaba Nchu. The estimates exclude RWA-resistant cultivars released by seed companies, as this study focused on the public sector investment in these cultivars. However, the survey results indicated that the ARC-SGI accounted for the major market share in RWA-resistant cultivars over the period 1993–1997 (Marasas, 1999).

The factors affecting the farmers' adoption of RWA-resistant cultivars were assessed by correlation and multiple regression analyses. The percentage area sown to the cultivars in 1996 was assumed as the dependent variable, and the analyses therefore described the behaviour of adopters only. This included 73 of the 110 producers interviewed. All variables were converted to log values to improve the fit of the model. Pearson's correlation coefficients were used for continuous and Spearman's correlation coefficients for qualitative values. Several explanatory variables were initially considered, but not all of them significantly explained the farmers' adoption behaviour (Marasas, 1999). Correlation was considered significant at the 5% level. The analyses were initially separated for the four farmer groups, but the regions as a variable did not significantly explain the producers' adoption decisions.

The farmer's educational level, wheat area and the quality grades realized by resistant cultivars significantly explained adoption individually and in combination with other factors (Tables 9.3 and 9.4). The farmer's experience with RWA was significant as an individual factor, while the combined weighted average yield of resistant and susceptible cultivars was significant in association with other factors. The percentage area sown to resistant cultivars increased when the farmer has studied at the tertiary level, experienced RWA as a problem at any given time,

Table 9.3. Correlation analysis of the factors significantly* explaining the adoption of Russian wheat aphid resistant cultivars in the study area (1996).†

	Dependent variable (percentage area sown to resistant cultivars)		
Explanatory variables	Correlation coefficients	Significance‡	Explanation
Farmer's educational level	–0.23545	0.0449 (S)	Qualitative data coded for confirmation or denial that the farmer has studied at the tertiary level. Confirmation corresponds with a lower code value
Farmer's wheat area	–0.24994	0.0330 (P)	Continuous variable
Quality grades realized by resistant cultivars	–0.25722	0.0280 (S)	Qualitative data coded by the wheat quality grade realized. A higher wheat quality grade corresponds with a lower code value§
Farmer's experience with the Russian wheat aphid	–0.27339	0.0193 (S)	Qualitative data coded for confirmation or denial that the farmer has experienced Russian wheat aphid as a problem at any given time. Confirmation corresponds with a lower code value

*Indicates statistical significance at the 5% level.
†Sample size (*n*) = 73 farmers.
‡(P) denotes the Pearson's coefficients used for continuous data and (S) denotes the Spearman's coefficients used for qualitative data.
§The wheat quality grade classification effective in South Africa at the time is summarized in ARC-Small Grain Institute (1998) and Marasas (1999). Wheat quality is one of the major factors determining the wheat price received by farmers.

Table 9.4. Multiple regression analysis of the factors significantly* explaining the adoption of Russian wheat aphid resistant cultivars in the study area (1996).†

Variables‡	T for H_0: Parameter = 0	Prob > |T|
Intercept	8.571	0.0001
Farmer's educational level	–2.392	0.0195
Farmer's wheat area	–3.113	0.0027
Quality grades realized by resistant cultivars	–2.251	0.0277
Combined weighted average yield of resistant and susceptible cultivars	2.268	0.0265

F value = 5.270; Prob > *F* = 0.0009; R^2 = 0.2366§; Adjusted R^2 = 0.1917; and d.f. = 68.
*Indicates statistical significance at the 5% level.
†Sample size (*n*) = 73 farmers.
‡The farmer's educational level and the quality grades realized by resistant cultivars are qualitative data, and the interpretation of their correlation is explained in Table 9.3.
§Though the relatively low R^2 value initially caused some concern, it should be reasonable for cross-sectional farm-level data (Intriligator, 1978; CIMMYT, 1993; Bua, 1998; Marasas, 1999). Moreover, all variables included in the analysis were significant, and the signs of the estimated coefficients were either as expected, or could be explained by observations from the survey results.

with higher quality grades realized by resistant cultivars, and with higher combined weighted average yields of resistant and susceptible cultivars.

However, the dependent variable correlated negatively with the farmer's wheat area. Though this finding initially seemed surprising, the survey results provided possible explanations. The producers in the study area planted various cultivars to spread their risk. Although the proportion of resistant to susceptible cultivars has increased substantially since 1993, this process of replacement may take longer over large wheat areas involving numerous cultivars. Large-scale chemical control could also be difficult to restrict to susceptible cultivars alone, which may render resistant cultivars less attractive over large areas. If land could furthermore be assumed as a proxy for wealth, larger-scale farmers might be less concerned about the costs of chemical control, and might continue sowing larger proportions of susceptible cultivars for other desirable characteristics. For these reasons, smaller-scale farmers might have planted larger proportions of resistant cultivars sooner than larger-scale farmers.

The economic impact of Russian wheat aphid resistant cultivars at the macro level

The Akino and Hayami (1975), cost–benefit and unit cost saving variations of the economic surplus approach were used to estimate the benefits of the RWA-resistant cultivars. The analysis followed a partial equilibrium setting and focused on South African producers and consumers. As the concern was with the public sector investment in the cultivars, social prices were used as applicable.

Weighted average yields for resistant and susceptible cultivars were obtained from the survey results (Table 9.1). The yield advantage of resistant cultivars was calculated as dictated by the respective methods. Weighted average cost savings due to reduced chemical control were accrued to the RWA-resistant cultivars. These were represented by the costs of chemical control on susceptible cultivars, calculated for the various options used by the farmers and weighted by area subjected to each regime (Table 9.2). Savings were only accrued to the proportion of the area planted to resistant cultivars, which was not chemically treated for RWA in 1996. For the Akino and Hayami and cost–benefit approaches, cost savings were added to the yield gain of resistant cultivars and multiplied with the remaining parameters to calculate the gross benefits.

Information on the wheat area and production for the Central Free State, Eastern Free State, Thaba Nchu and Qwa Qwa was obtained from: per silo estimates from the former South African Wheat Board; published reports and documents; and development corporations, co-operatives and other organizations operating in the study area. Adoption data for the RWA-resistant cultivars were obtained from the survey results. Wheat import parity prices for the study area were obtained from the South African Grain Information Service and were adjusted to social prices using existing social exchange rates for South Africa (Jooste and Van Zyl, 1997). In the Akino and Hayami approach, South African estimates of the elasticity of wheat supply (Niebuhr and Van Zyl, 1990) and of demand for bread (Niebuhr and Van Zyl, 1992) were applied. The latter was assumed to approximate the demand elasticity for wheat, as most of the demand for wheat is for human consumption, and largely in the form of bread products.

Enterprise budgets from the Free State Department of Agriculture were used to estimate the total production costs of resistant and susceptible cultivars for the unit cost saving approach. The survey results were used to adjust the budgets for the differences in yield, seed cost and RWA control practices between resistant and susceptible cultivars. Estimates of regular labour, depreciation of machinery and the rental value of land were included as approximations of the fixed costs of production. The weighted average yields of resistant and susceptible cultivars were used to calculate the respective average total production costs. The market prices in the budgets were adjusted to social prices using existing shadow exchange rates for South Africa (Jooste and Van Zyl, 1997).

Technology development, transfer and adoption costs of the RWA-ICP were subtracted from the gross benefits in the Akino and Hayami and cost–benefit approaches. An additional seed cost of 10 Rand/ha was the only adoption cost associated with resistant cultivars. The unit cost saving approach included adoption cost in the enterprise budgets used to calculate the difference in average total costs between resistant and susceptible cultivars. The RWA-ICP comprised an integrated strategy involving infrastructure, knowledge and support extending across the various stages of chemical control, resistant cultivars and biological control. We therefore applied the full research cost of the three phases of the programme in the analysis, since 1980 and projected to 2005. Given that the benefits of chemical and biological control were not included in the analysis, the benefits were therefore underestimated. However, aspects of chemical and biological control were addressed in the sensitivity analysis. Guidelines were used to collect research development and transfer cost information from the various institutions involved in the RWA-ICP since 1980.

The internal rate of return and net present value were used to estimate the returns on the research investment, using a standard cost–benefit framework of analysis (Gittinger, 1982). Annual nominal values were adjusted to real terms using the South African Consumer Price Index. Discount rates of 3%, 5% and 10% were used to calculate the net present value, given the debate in the economics literature on choosing appropriate interest rates (Alston *et al.*, 1995). The economic real interest rate was assumed as a measure of the opportunity cost of capital. This was derived from the deflated long-term government bond rate in South Africa and was estimated at 5.5% at the time of this study (*Quarterly Bulletin of the South African Reserve Bank*, 1998).

Table 9.5 shows the internal rate of return and net present value of the research investment, as estimated with the three surplus methods over the period 1980–2005. The internal rate of return was 21.6% for both the unit cost saving and Akino and Hayami methods; and was 26.8% for the cost–benefit approach. The internal rates of return of over 20% were

Table 9.5. Internal rate of return and net present value of the Russian wheat aphid integrated control programme using three economic surplus methods (1980–2005).

Measures of project value	Unit cost saving	Akino–Hayami	Cost–benefit
Internal rate of return (%)	21.6	21.6	26.8
Net present value (million Rand)			
3% discount rate	57.4	60.0	110.6
5% discount rate	37.5	39.1	73.5
10% discount rate	12.8	13.2	26.9

substantial compared with the social opportunity cost of capital of 5.5%. The results imply that society receives a return of over 20 cents for every South African Rand invested in the RWA-ICP, after paying the full cost of the programme. The net present value at the 5% discount rate was 37.5 million Rand for the unit cost saving; 39.1 million Rand for the Akino and Hayami; and 73.5 million Rand for the cost–benefit method. The net present value of the investment was sensitive to the discount rate, but remained substantial even at the 10% discount rate. RWA-resistant cultivars thus proved to be a profitable public sector investment, even when stringent investment criteria were applied.

The cost–benefit approach yielded the most optimistic results, while the unit cost saving method was the most conservative. The returns estimated with the unit cost saving and Akino and Hayami approaches were within a similar range and therefore comparable. The internal rates of return were the same, and the net present values at the 5% discount rate differed by only 1.6 million Rand. However, the returns increased when the cost-benefit method was used instead.

The economic impact of Russian wheat aphid resistant cultivars at the micro level

The micro-level economic impact of the RWA-resistant cultivars was estimated using partial budget methodology (CIMMYT, 1988; Anandajayasekeram *et al.*, 1996). The investment was viewed from a producer's perspective and costs and benefits were calculated at the farm level (Marasas, 1999). As partial budgets involve only the production factors relevant to a specific investment decision, the analysis focused on the costs and prices differing between RWA-resistant and susceptible cultivars. The major agronomic differences between the two types of cultivars were the respective yields and the farmers' RWA control practices. Differences in wheat prices and seed costs were also included in the budgets.

Using producers' estimates obtained from the survey results, we calculated weighted average costs and prices, by area subjected to the various applicable management options used by the farmers, for resistant and susceptible cultivars, respectively. We then calculated the respective net benefits of resistant and susceptible cultivars, and the incremental net benefits of resistant cultivars. Separate budgets were developed for the RWA control options, shown in Table 9.2. The analysis focused on the Central and Eastern Free State, because Thaba Nchu producers did not use chemical control in 1996, and the lack of a demonstrable yield advantage precluded conclusive estimation of the incremental net benefits of RWA-resistant cultivars in Qwa Qwa.

Under the assumptions employed in this study, the incremental net benefits of RWA-resistant cultivars at the micro level ranged between 140 and 329 Rand/ha in 1997 prices

(Table 9.6). The benefits remained substantial for all alternatives considered in the analysis, even though chemical control was continued on a proportion of the resistant cultivars. This could be ascribed to the yield advantage of the cultivars. The findings suggest that the producers' continued use of chemical control on resistant cultivars was therefore not irrational, but might be explained by other considerations.

First, farmers indicated that they continued the practice as a perceived 'extra precaution' to avoid RWA. This could be especially applicable to areas where farmers have experienced more problems with RWA. For example, 90% of Eastern Free State producers have perceived RWA as a problem at any given time, and these farmers also continued chemical control on RWA-resistant cultivars to the largest extent. Second, as the producers planted various resistant and susceptible cultivars, the logistics of large-scale chemical control could be difficult to restrict to susceptible cultivars.

Nevertheless, the economic benefits of the resistant cultivars could improve if used without chemical control. Potential incremental net benefits ranging between 42 and 91 Rand/ha

Table 9.6. The farm-level incremental net benefits of changing from Russian wheat aphid susceptible to resistant cultivars (1997).

Options		Incremental net benefits (Rand/ha)	
Changing from	Changing to	Central Free State	Eastern Free State
Susceptible cultivars with no chemical control	Resistant cultivars with no chemical control	211	232
	Resistant cultivars with spraying	144	190
	Resistant cultivars with seed treatment	–	182
	Resistant cultivars with spraying and seed treatment	–	140
Susceptible cultivars with spraying	Resistant cultivars with no chemical control	267	280
	Resistant cultivars with spraying	200	238
	Resistant cultivars with seed treatment	–	230
	Resistant cultivars with spraying and seed treatment	–	188
Susceptible cultivars with seed treatment	Resistant cultivars with no chemical control	273	275
	Resistant cultivars with spraying	206	233
	Resistant cultivars with seed treatment	–	225
	Resistant cultivars with spraying and seed treatment	–	184
Susceptible cultivars with spraying and seed treatment	Resistant cultivars with no chemical control	329	323
	Resistant cultivars with spraying	262	282
	Resistant cultivars with seed treatment	–	274
	Resistant cultivars with spraying and seed treatment	–	232

Note: (–) indicates that the treatment was not used in the region.

Table 9.7. The farm-level incremental net benefits of changing from Russian wheat aphid resistant cultivars with chemical control to resistant cultivars without chemical control (1997).

Options		Incremental net benefits (Rand/ha)	
Changing from	Changing to	Central Free State	Eastern Free State
Resistant cultivars with spraying	Resistant cultivars with no chemical control	67	42
Resistant cultivars with seed treatment	Resistant cultivars with no chemical control	–	50
Resistant cultivars with spraying and seed treatment	Resistant cultivars with no chemical control	–	91

Note: (–) indicates that the treatment was not used in the region.

could be realized (Table 9.7). Technology transfer efforts should therefore encourage producers to discontinue chemical control on resistant cultivars in order to reap their full economic benefits.

Conclusions

This study was undertaken to describe the adoption and impact of the RWA-ICP in the Central and Eastern Free State of South Africa. The analyses focused on the resistant cultivar phase of the programme, because these impacts could be most clearly measured and demonstrated. The area planted to resistant cultivars increased from 3% in 1993 to 46% in 1997, and producers projected this area to increase further to 52% by the year 2000. Correlation and multiple regression analyses suggested that adoption of the cultivars was influenced by: the farmer's educational level, wheat area and experience with RWA; the quality grades realized by resistant cultivars; and the combined weighted average yield of resistant and susceptible cultivars.

Substantial economic benefits were demonstrated for the RWA-ICP at the macro and the micro level. At the macro level, the Akino and Hayami, cost–benefit and unit cost saving variations of the economic surplus approach were used to estimate the returns on the research investment over the period 1980–2005. The internal rate of return ranged between 21.6 and 26.8%, and the net present value at the 5% discount rate ranged between 37.5 and 73.5 million South African Rand. The sensitivity of the results was analysed by alternative assumptions on various parameters, but the returns remained substantial for all scenarios considered. At the micro level, partial budget analyses indicated that the resistant cultivars generated incremental net benefits ranging between 140 and 329 Rand/ha in 1997 prices. Positive environmental, institutional and spill-over impacts were also associated with the research programme.

Even though chemical control was continued on some of the resistant cultivars, their economic benefits remained substantial. This could be ascribed to the yield advantage of the cultivars. However, the benefits of the cultivars could be improved if they were used without chemical control. The analyses showed that various factors apart from pest resistance also influenced the adoption and impact of the RWA-resistant cultivars. This emphasizes that agricultural technologies are often of an integrated nature, and innovations such as germplasm improvement for pest resistance can therefore not be developed in isolation from other considerations.

This study is the first in Southern Africa to compare three variations of the economic surplus approach, using the same set of empirical data. The internal rates of return estimated with the unit cost saving and Akino and Hayami methods were the same and the net present values at the 5% discount rate differed by only 1.6 million Rand. However, the returns increased when the cost–benefit method was used instead. The unit cost saving method thus yielded the most conservative and the cost–benefit method the most optimistic results. Nevertheless, each method has benefits and may be preferred for specific reasons. Any one of the three methods could therefore be used to evaluate a research programme. However, once an approach has been chosen, it should be consistently applied to compare alternative initiatives. Even among methods classified as surplus approaches, the estimated benefits may differ depending on assumptions regarding the nature of the supply and demand curves and the research-induced shifts in the supply curves. The results of this study underline the importance of adoption and impact studies as tools to assess the benefits of agricultural technologies, and to provide feedback and insight into producers' decision-making behaviour.

References

Akino, M. and Hayami, Y. (1975) Efficiency and equity in public research: rice breeding in Japan's economic development. *American Journal of Agricultural Economics* 57, 1–10.

Alston, J.M., Norton, G.W. and Pardey, P.G. (1995) *Science Under Scarcity: Principles and Practice for Agricultural Research Evaluation and Priority Setting.* Cornell University Press, Ithaca, New York.

Anandajayasekeram, P., Martella, D. and Rukuni, M. (1996) *A Training Manual on R&D Evaluation and Impact Assessment of Investments in Agricultural and Natural Resources Research.* SACCAR, Gaborone, Botswana.

ARC-Small Grain Institute (1998) *Guidelines for Wheat Production in the Summer Rainfall Region.* ARC-Small Grain Institute, Bethlehem, South Africa.

Bua, A. (1998) Evaluation of participatory research approaches in the development, transfer and adoption of cassava technologies in Uganda. PhD thesis, Wye College, University of London.

Gittinger, J.P. (1982) *Economic Analysis of Agricultural Projects*, 2nd edn. John Hopkins University Press, Baltimore, Maryland.

International Maize and Wheat Improvement Center (CIMMYT) (1988) *From Agronomic Data to Farmer Recommendations: an Economics Training Manual.* Revised Edition. CIMMYT, Mexico.

International Maize and Wheat Improvement Center (CIMMYT) (1993) *The Adoption of Agricultural Technology: a Guide For Survey Design.* CIMMYT, Mexico.

Intriligator, M.D. (1978) *Econometric Models, Techniques and Applications.* Prentice-Hall, Englewood Cliffs, New Jersey.

Jooste, A. and Van Zyl, J. (1997) *Regional Agricultural Trade and Changing Comparative Advantage in Southern Africa.* University of Pretoria, South Africa.

Marasas, C.N. (1999) Socio-economic impact of the Russian Wheat Aphid Integrated Control Program. PhD thesis, University of Pretoria, South Africa.

Marasas, C.N., Anandajayasekeram, P., Tolmay, V., Martella, D., Purchase, J. and Prinsloo, G. (1997) *Socio-Economic Impact of the Russian Wheat Aphid Control Research Program.* Report by the Agricultural Research Council and SACCAR. SACCAR, Gaborone, Botswana.

Niebuhr, H.G. and Van Zyl, J. (1990) Ekonomiese ontleding van aanbod van koring in Suid-Afrika: 'n Streeksbenadering. [An economic analysis of wheat supply in South Africa: A regional approach]. *Agrekon* 29(4), 324–334.

Niebuhr, H.G. and Van Zyl, J. (1992) Die vraag na brood in Suid-Afrika, 1984–90: 'n Streeksbenadering. [The demand for bread in South Africa, 1984–90: A regional approach]. *Agrekon* 31(3), 128–135.

Quarterly Bulletin of the South African Reserve Bank (1998) Number 209, September 1998.

Scheepers, J.J., Smit, J.A. and Ludick, B.P. (1984) *An Evaluation of the Agricultural Potential of the Highveld Region in Terms of Dryland Cropping and Livestock Production.* Technical Communication Number 185, Department of Agriculture, South Africa.

Tolmay, V. and Wessels, C.V. (1996) *Evaluation of Wheat Breeding Material for Resistance to the Russian Wheat Aphid,* Diuraphis noxia. Project progress report, ARC-Small Grain Institute, Bethlehem, South Africa.

Walters, M. C. (1984) Introduction. In: Waters, M.D. (ed.) *Progress in Russian Wheat Aphid (*Diuraphis noxia *Mordw.) Research in the Republic of South Africa.* Proceedings of a Meeting of the Russian Aphid Task Team held at the University of the Orange Free State, Bloemfontein, 5–6 May 1982. Technical Communication No. 191, Department of Agriculture, South Africa, pp. 1–2.

Walters, M.C., Penn, F., Du Toit, F., Botha, T.C., Aalbersberg, Y.K., Hewitt, P.H. and Broodryk, S.W. (1980) *The Russian Wheat Aphid.* Farming in South Africa. Leaflet Series, Wheat G.3, South Africa.

10 The Impact of Public Investment in Maize Research in Kenya

DANIEL D. KARANJA*

Abstract: Kenya has been transformed into a net importer of maize, which is its major staple. As maize accounts for a major portion of calorific and protein intake for more than 90% of Kenyans, the fact that Kenya must increase farm productivity and income is no longer debatable. Past successes in maize production in Kenya were achieved by exploiting a synergy between technology development, dissemination, and seed multiplication and distribution programmes. However, lack of funding, poor linkages between research and farmers, lack of private investments in maize research, and human capital turnover are problems that need serious attention. Two stage linear regression analysis is used to calculate the rate of return to investment in Kenyan maize research for the 1955–1988 period. The result indicated that improvement in maize yield and expansion of maize area are explained by, *inter alia*, increase in research and extension expenditures and the spread of hybrid seed; yield is also positively influenced by use of fertilizer while greater area expansion is stimulated by higher maize producer prices. The results also indicate that maize research, extension and seed programmes contributed to attainment of higher maize yields, expansion of maize area and to growth in output.

Introduction

Many African nations were self-sufficient in food production at independence about 40–45 years ago. However, the combination of major oil crises in the 1970s, adverse weather, poor macroeconomic and sectoral performance in the 1980s and 1990s, and declining public investment in agricultural research, development and infrastructure, undermined the capacity of these economies to supply sufficient food to their populations. Further, rapid population growth and persistent decline in the natural resource base, resulted in a decline in per capita food production, leading to unmet food demand. The ultimate effect was a growing reliance on food imports, and food aid, increased poverty and civil strife. However, Kenya was an exception and it marshalled resources and talent to generate a mini-green revolution in maize

* This chapter is based on the author's Masters thesis research of 1989. Since then, further maize institutional reform and research has occurred.

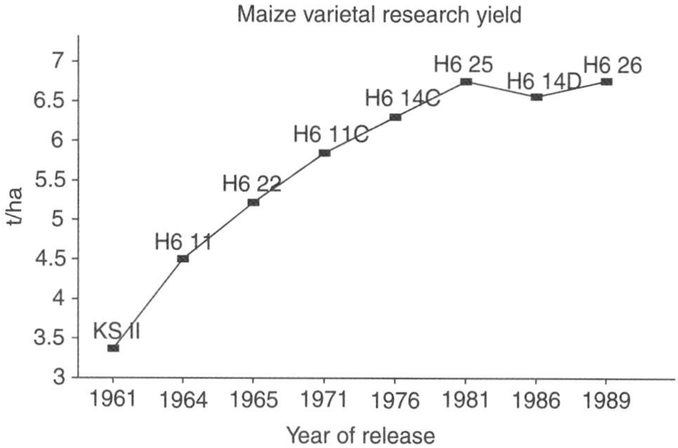

Fig. 10.1. Research yields for highland maize varieties in Kenya, 1961–1989.

from 1955 to 1979. However, since the late 1970s real research expenditures on maize have declined in Kenya.

As maize accounts for a major proportion of calorific and protein intake for more than 90% of Kenyans, it is imperative that ways and means of improving its productivity be sought. During the 1961–2000 period Kenya developed 25 new maize varieties (17 for high altitude, six for medium altitude) with a potential yield ranging between 2 and 6.8 t/ha. Recent evidence indicates that average maize output and area have stagnated below 2.5 million t and 1.5 million ha, respectively (Fig. 10.1). Future growth in maize productivity will rely more on improvements in yield, than area expansion due to land constraints (Hassan and Karanja, 1997).

Economic Analysis: the Rate of Return to Maize Research 1955–1988

Karanja (1990) used a production function approach to calculate the rate of return (ROR) to investments in Kenyan maize research for the 1955–1988 period.[1] Equations representing maize area and yield as functions of inputs, research and extension expenditures, prices and 'weather and economic circumstances' variables were estimated using a linear two-stage regression analysis.

$$YD = 9.05 - 0.57AR + 0.25RE + 0.07EX + 0.08FT + 0.12HY + 0.24DM \quad (1)$$
$$(R^2 = 0.64)$$

$$AR = 8.47 - 0.26YD + 0.49RE + 0.07EX + 0.10MP + 0.02HY + 0.13DM \quad (2)$$
$$(R^2 = 0.86)$$

where YD is the average national maize yield in t/ha; AR is national maize area in ha; RE is maize research expenditure, lagged 10 years, in K£;[2] EX is maize extension expenditure in K£; FT is fertilizer imports in t; MP is average maize producer price in K£/t; HY is hybrid seed sales by the Kenya Seed Company (KSC) in t; and DM is dummy variable for weather and macroeconomic shocks. The maize research expenditure (1955–1988 period) is presented in Fig. 10.2.

The results indicate that improvement in maize yield and expansion of maize area are explained by, *inter alia*, increase in research and extension expenditures, and the spread of hybrid seed; yield is also positively influenced by use of fertilizer while maize area is stimulated by higher maize producer prices. The results also indicate that maize research, extension and seed programmes contribute positively to attainment of higher maize yields and expansion of maize area, and consequently to growth in output.

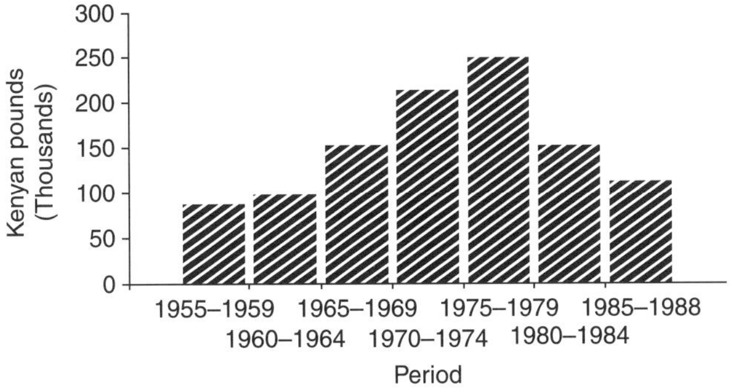

Fig. 10.2. Maize research expenditure in Kenya, 1955–1988.

The regression coefficients show that an increase in maize research expenditure of 1% increases maize yield by 0.25% and area expansion by 0.49% after 10 years.[3] However, the effect of area on yield and vice versa reduces the research impact on maize output by 0.35%, using a first order approximation. Therefore, when adjusted, an increase in research expenditure by 1% raises maize output by 0.40% 10 years later. This figure is converted to a change in maize output and its value in K£ was estimated. This resulted in an ROR of 68%, indicating that the public investment made in maize research in Kenya during the period 1955–1988 was productive.

Karanja (1996) used this information and data to test the hypothesis that the maize research programme was partly successful as a result of complementary effects of the extension programme of the Ministry of Agriculture, and the seed multiplication and distribution network of the KSC. To do this, the values of the exogenous variables, other than research, were set equal to the sample means for those variables. The only exception was the dummy variable representing weather and external macroeconomic shocks (DM), which was set equal to zero, so that the results corresponded to a 'normal' year. The interaction between yield (YD) and area (AR) is calculated using an exact solution to the simultaneous YD and AR equations (3) and (4) below:

$$\text{Ln(YD)} = 11.763 - 0.5705^*\text{Ln(AR)} + 0.2529^*\text{Ln(RE)} \quad (3)$$

$$\text{Ln(AR)} = 9.665 - 0.2623^*\text{Ln(YD)} + 0.4922^*\text{Ln(RE)} \quad (4)$$

Solving these equations simultaneously to express Ln(YD) and Ln(AR) as linear functions of Ln(RE) shows that:

$$\frac{dLn(AR)}{dLn(RE)} = 0.5008;\text{ and } \frac{dLn(YD)}{dLn(RE)} = -0.0328,\text{respectively.}$$

As, the annual maize output, Q, can be expressed as Q = YD × AR, therefore:

$$\text{Ln(Q)} = \text{Ln(Y)} + \text{Ln(A)} \quad (5)$$

Then, it followed that:

$$\frac{MLn(Q)}{MLn(R)} = \frac{MLn(Y)}{MLn(R)} + \frac{MLn(A)}{MLn(R)} \quad (6)$$

The value of equation (6) equals 0.5008 – 0.0328, which equals 0.468. Therefore, a 1% increase in research expenditure raised the maize output by 0.468% 10 years later. This number is referred to as the elasticity of maize output with respect to maize research. To quantify the effect of research on output in terms of weight, the sample average maize production, 1,692,417 t, was used. A 1% increase in research expenditure is equivalent to K£1543, which would increase maize output by 0.468% or about 7920 t. Using P_m = 22.58 K£/t, gives a ROR equal to 60.8%[4] (see Table 10.1).

Table 10.1. Summary of ROR calculations (%).

	Agricultural extension programme	
Seed programme	Effective	Half-effective
Effective	60.8	46.2
Half-effective	45.9	38.9

To estimate the effect of the agricultural extension service (AES) of the Ministry of Agriculture on the maize research impact, a scenario was constructed in which the AES was assumed to lose one-half of its effectiveness. This was done by halving the estimated coefficient of EX in the yield and area regressions (3) and (4), while those of other variables were not adjusted. The resulting equations, analogous to (3) and (4) are:

$$Ln(Y) = 11.397 - 0.5705^{*}Ln(A) + 0.2529^{*}Ln(R) \tag{6}$$

and

$$Ln(A) = 9.328 - 0.2623^{*}Ln(Y) + 0.4922^{*}Ln(R) \tag{7}$$

In this scenario, the projected increase in maize output from a 1% increase in maize research expenditure was 2738 t, a decline of about two-thirds from the base scenario. When applied to equation (3), the estimated ROR declined to 46.2%, indicating strong evidence of the role and impact of the extension service on maize research. Further, a second scenario assumed a half-effective hybrid seed multiplication and distribution programme (SMDP) of the KSC, and a fully effective AES. When the impact of seed on yield and area was halved, the resulting yield and area equations were:

$$Ln(Y) = 11.222 - 0.5705^{*}Ln(A) + 0.2529^{*}Ln(R)$$

and

$$Ln(A) = 9.586 - 0.2623^{*}Ln(Y) + 0.4922^{*}Ln(R)$$

The increase in maize production, ΔQ, due to a 1% increase in research expenditure is equal to 2678 t, resulting in an estimated ROR of 45.8%. Again, this demonstrates that the smallholder maize development programme SMDP had positive and complementary effects on he maize research programme. Lastly, a scenario in which each of the AES and SMDP were one-half as effective was used. The resulting equations were:

$$Ln(Y) = 10.855 - 0.5705^{*}Ln(A) + 0.2529^{*}Ln(R)$$

and

$$Ln(A) = 9.249 - 0.2623^{*}Ln(Y) + 0.4922^{*}Ln(R)$$

In this case, the increase in output from a 1% increase in maize research expenditure was reduced to 1644 t, yielding a ROR of 39%. The results presented here indicate that both the AES and the SMDP contributed to the success of the maize research programme. Both programmes had positive interaction with the maize research programme.

Institutional analysis

To explain and understand the success of Kenya's maize research programme, it was imperative to review the evolution of institutional and policy environments surrounding the maize industry since 1900. Karanja (1992) reviewed the past institutional history of maize research and categorized it into: (i) the Pre-hybrid Era: 1900–1963; (ii) the Hybrid Revolution: 1964–1980; and (iii) the Hybrid Plateau: 1980–1990. This study adds a fourth era, which can be referred to as the 'the Institutional Collapse: 1990–2000'. Table 10.2 summarizes the key developments in the maize research, support institutions and policy environment between 1900 and 1990.

The Pre-hybrid Era: 1900–1963

Britain became involved in East Africa in the 1870s, assumed territorial responsibilities in 1893 and colonized Kenya in 1920. A railway line was constructed from the Kenyan coast to Lake Victoria, to open up trade between Uganda and Britain. Because the railway operation could not cover its costs, the British government encouraged European settlement in Kenya, and encouraged agricultural production and trade mainly to finance the railway line and establish roads. The settlers alienated about 20% of Kenya's arable land, next to the railway line, imposed taxes on Africans to get cheap labour to work their farms, and controlled all monetary, transport, research and advisory services. Further, the administration banned Africans from growing the major cash crops, tea and coffee. In this way, African agriculture was relegated to a subsistence level, while settler farming was export-market oriented and protected from competition.

Maize, having been introduced by Arab traders at the coast, and moved interior by missionaries, appeared in exhibitions at the first agricultural show in Nairobi in 1902, and covered about 20% of the total food crop area by 1903 (Meinertzhagen, 1957). Many of the maize varieties whose origin was North America, came with settlers emigrating from South Africa. The crop became increasing popular among African farmers, but as there were no markets, there was no stimulus to increase production beyond subsistence.

However, the opportunity to increase maize production and its participation in the market came in 1942, when the colonial government enacted an Increased Production of Crops Ordinance. This ordinance mobilized both European and African farmers to increase production of all food crops, including maize and wheat, to feed troops during World War II. To facilitate this, the government established a Maize Control Unit, under the Agricultural Production Marketing Act (of 1936), to oversee the purchase of all maize in the colony. A government-appointed farmer association, the Kenya Farmers Association (KFA), was formed. Maize prices were guaranteed and, for the first time, short-term credit was issued in the form of Guaranteed Minimum Returns (GMR). Farmers responded by increasing production substantially: the hectarage under wheat and maize almost doubled.

The Maize Marketing Board (MMB) was established in 1959 and replaced the Maize Control Unit. The objectives of the MMB were to regulate, control and improve the collection, storage, marketing, distribution and supply of maize and maize products. Before creating its own marketing network, the MMB utilized the services of the KFA, and various provincial marketing boards were set up in the 1950s to reach out to all maize-growing regions. A sharp increase in agricultural production during the war aggravated problems of soil depletion and erosion. The extent of these problems on the African farms raised the concern of the colonial administration, which pledged to improve African agriculture, starting with emphasis on soil conservation. Between 1946 and 1954, the government invested substantial resources in soil conservation, livestock improvement, water supply experimentation on cultivation, fertilizers, crops and pasture.

A State of Emergency imposed in 1952 against the Mau-Mau rebellion, slowed the recruitment of new settlers and provided the stimulus to further develop African agriculture. In 1953, the colonial government agreed to provide employment to displaced Africans, raise their agricultural productivity and human and stock-carrying capacity of the land.

Table 10.2. An institutional analysis of maize research in Kenya, 1900–1990.

Stage	Period	Maize technology development	Maize support systems	Policy environment
Pre-hybrid	1900–1963	• Introduction of US maize varieties via South Africa • Scattered field trials and pre-selection of maize types Kenya White Complex • Screening maize rusts at Coastal Station, 1952 • Development of synthetic varieties • Hybrid research started 1955 • Rockefeller support to research, 1955 • Import of germplasm, 1959 • Expansion of maize research to Katumani, 1957 • Agronomic trials (ODA), 1963–1970	• Railway Network, 1890s • BEA Farmers' Association, 1919; renamed KFA, 1923 • Road networks, 1920s • Agricultural extension, first efforts 1910s; revamped 1954 • Agricultural Credit, 1942 through Land Bank • Kenya Seed Company, 1956; contract in 1963 • Maize Marketing Board, 1959 • Agricultural Finance Corporation, 1963	• European settlement • Agricultural Production Marketing Act, 1936 • Increased Production of Crops Ordinance, 1942 • Guaranteed government prices for maize, 1942 • Maize Control Committee, 1942 • Swynnerton Plan (smallholder agricultural development strategy), 1954
Hybrid Revolution	1964–1980	• First hybrid, H611, released, 1964 • Mid-altitude maize research, 1965 • 5000 demonstration trials, 1965 • FAO-sponsored fertilizer trials, 1970s • 11 hybrids and 3 OPVs, 1964–1980	• Agricultural Credit–Land Bank, AFC • Expansion of agricultural extension services • KSC: increased seed output and distribution • KFA: expanded fertilizer and farm inputs supply • Expanded maize marketing services, 1970s	• Independence in 1963 • Land transfer to Africans, 1960s • Expanded agriculture credit, extension, research • First oil crisis, 1973/74 • Second oil crisis, 1978/79 • Maize marketing control enhanced 1979 • Declining per capita food production, mid-1970s

			• Rural Access Roads Program • Maize price reviews, controlled marketing	• Declining economic performance, mid-1970s
Hybrid Plateau	1980–1990	• On-farm research • Focus on yield and marginal areas • Declining donor support and real research budget • High-protein maize (abandoned) • KNARP, 1986 • Reorganization of KARI, 1987/88 • Maize research review, November 1990 • 7 hybrids, 1 OPV, 1980–1990	• NEP pilot project, 1981; NEP I • KSC: output up, expanded network • Loan facilities, AFC, banks • Expanded NCPB's marketing and storage capacity • KSC research department, early 1980s • Guaranteed maize price and marketing • KSC: maize seed exports from 1984/85	• Food policy papers, 1981, 1986 • Major drought, 1984 • Focus on smallholder and marginal agriculture • Population pressure • Declining economic performance • Restructuring of NCPB, liberalized input prices • Declining per capita food production
The Challenge Ahead	1990–2000	• Emphasis: high/marginal maize • Breeding/agronomy; small/ large-scale • Increase real research budget • Enhanced networking: regional and international research systems	• NCPB: expand storage but liberalized market • Agricultural extension, collapsed! • Agricultural Credit, collapsed! • Liberalized seed markets, more efficient input supply, e.g. fertilizer • KGGCU, AFC and NCPB: suffered economic loss	• Political pluralism and ethnic clashes • Macroeconomic stress • Declining donor support

AFC, Agricultural Finance Corporation; NCPB, National Cereals and Produce Board; KGGCU, Kenya Grain Growers Cooperative Union; KARI, Kenya Agricultural Research Institute; BEA, British East Africa.

The colonial government then requested the Department of Agriculture to formulate a strategy for developing African agriculture. The response was the release of the Swynnerton Plan of 1954, compiled by the then Assistant Director of Agriculture, R.J.M. Swynnerton. The Plan outlined a comprehensive scheme for smallholder agricultural development. It proposed a review of the land tenure system, to enable consolidation and registration of land, and issuance of title deeds, which farmers could use to borrow loans, and improved agricultural extension, research, credit and marketing services. It argued that African farmers required credit and recommended creation of an agricultural credit bank, the Loan and Land Bank, for African farmers. The Plan also differentiated between the high potential and marginal regions, lifted restrictions on production of high-valued export crops by African farmers, recommended expansion of agricultural extension and research services to African farmers, and decentralization of marketing boards to improve marketing coverage. In response, the government expanded the agricultural extension programme and formed district survey teams to plan and advise farmers on land intensification through farm planning, and crop and livestock management. By 1962, the agricultural field staff had expanded to 179 agricultural officers and their assistants, 1107 agricultural instructors and 2347 assistant agricultural instructors. More than three-quarters of the staff were located in smallholder farming areas.

Maize research work started in 1930, at the Njoro Breeding Station, but was abandoned during World War II. Later, large commercial farmers in Trans Nzoia district pressured the government to start research in hybrid maize, which was already becoming popular in North America. Granting their request, the government hired a full-time maize breeder at Kitale Research Station. Using conventional breeding methods, synthetic varieties were formed from inbred lines developed from well-adapted Kenya Flat White complex (Gerhart, 1975). One of these synthetic varieties became parent to the first Kenyan hybrid seed, H611, which was commercially released in 1964. In 1963, with prospects of the first Kenyan maize hybrid at hand, the government entered into an agreement with KSC to multiply and distribute the hybrid seeds to farmers.[5] The KSC used the already established marketing network of the KFA to provide sufficient quantities of improved maize seeds.

The Hybrid Revolution: 1964–1980

Before Kenya's independence, the British government drew up plans to maintain agricultural output and transfer land to Africans, and promised aid for agricultural credit, technological and management services. In 1963, when Kenya attained its political independence, there was an exodus of settlers from large farms, and a smooth transition to African farming. The new government was committed to a comprehensive land resettlement programme that saw 35,000 families settled on 470,000 ha of land, from densely populated and resource-constrained African reserves, at a total cost of K£30 million. An additional 16,400 families were settled on 168,000 ha in subsequent settlement programmes.

The release of the first Kenyan hybrid maize in 1964 was timely amidst a looming threat of a major food deficit as a result of major changes in the agricultural sector. A survey of African-owned, large, mixed-farms in Trans Nzoia in 1967–1969 revealed that overall agricultural production fell by 80%. In response, the government further expanded the agricultural extension programme, provided credit, through the Agricultural Finance Corporation, and set up training centres for large-scale African farmers.

In 1965, the government expanded the maize research programme to cover medium-altitude maize-growing regions on the central highlands. The success of this expansion culminated in the release of medium-maturing maize hybrids and rapid farmer adoption. Between 1963 and 1970, the maize breeding programme was augmented by a comprehensive agronomy programme funded by the British government and the Rockefeller Foundation. Researchers generated recommendations on time of planting, spacing, weed management and fertilizer application rates. By 1972, the extension service had about 2600 agricultural and animal-health assistants and 5500 junior assistants; the latter group worked in direct contact with the farmers. In spite of insufficiently trained frontline staff, low farmer coverage and message-orientation towards large-scale farmers, the extension service played a key role in training all farmers in maize husbandry. Moreover, in 1965, extension agents had planted over 5000 demonstration plots across the country in major maize-growing districts and used them to train farmers. Further, in the late 1960s and early 1970s, the extension service conducted hundreds of fertilizer demonstrations in major maize-growing districts. In a survey on hybrid maize adoption in western Kenya in 1973, about 35% of the farmers interviewed indicated that they first heard of hybrids from extension agents and 64% said they would go to an extension agent for advice on maize (Gerhart, 1975).

The Agricultural Information Centre, established in 1966, also facilitated the hybrid campaign by supplying farmers with printed leaflets on hybrid recommendations in English and Kiswahili enclosed in every package of seed sold by the KSC. The KSC responded to demand for improved maize varieties by expanding its marketing network in the mid-1960s. While relying on the KFA to service the large-scale farms, the KSC contracted provincial marketing boards, farmer cooperatives and Dalgety Ltd, a private hides and skins merchant, to reach out to smallholder farmers in all maize growing regions. The company also appointed seed stockists among small-scale African shopkeepers, in almost every trading centre. The number of hybrid maize seed stockists increased from 103 in 1966 to 1171 and 2541 in 1971 and 1975, respectively. Hybrid seed sales increased from an average of 2100 t in 1966 to 6700 t in 1971 and 11,460 t in 1975. The company also contracted large-scale farmers and the government's Agricultural Development Corporation to multiply seed; the latter accounted for 50% of all the certified maize seed.

Use of fertilizer on maize never matched the adoption of hybrid seed. The reason commonly cited was that fertilizers were more expensive, bulkier and harder to store compared with the seed. Stockists lacked capital, storage space and transport to provide sufficient quantities of fertilizers to farmers in good time. To stimulate adoption, the government subsidized the cost of seed and fertilizer inputs. Farmers responded by increasing the use of both inputs, particularly the use of fertilizer among smallholder farmers.

Agricultural credit was cited by the Sywnnerton Plan as a critical input for agricultural transformation of African farming. However, the first government loans were issued to small farmers prior to the plan in 1948 by the African Land Utilization and Settlement Board, and later the African Land Development Board. The two Boards were combined, in 1963, to create the Agricultural Finance Corporation. Under the Agricultural Finance Corporation, short-term loans, such as the GMR, were advanced to small and large farmers. The provision of the GMR was coordinated through the KFA, which provided farm inputs on loan and deducted the loan from produce delivered at the end of the crop season. Although the GMR were provided to both large and small farmers, a survey conducted in 1972 revealed that only 1.5% of the official credit was used for maize production by the small farmers. Credit from private lending was limited, particularly, as the maize trading was restricted to the official marketing channel.

Guaranteed official prices and marketing facilitated Kenya's maize production boom in the 1964–1980 period. The MMB merged with several provincial agricultural marketing boards in 1966 to form the Maize and General Produce Board. Later in July 1979, this board merged with the Wheat Board of Kenya (established in 1952) to form the National Cereals and Produce Board. Between 1975 and 1980 maize purchases by the board were estimated at about 35% of the total marketed maize. Similarly, between 1964 and 1980, maize production increased from about 0.9 million t in 1964–1965 to 1.5 million t in 1974–1975 and 1.65 million t in 1979–1980. By 1980, there were 11 hybrids and five open-pollinated maize varieties available for farmers.

The Hybrid Plateau: 1980–1990

Kenya's macroeconomic performance in the 1980s was dismal. Food demand escalated due to rapid population growth and poor weather. Global market shares and prices for Kenyan export products declined sharply, and so did the increase in foreign debt obligations. This prompted the government to map out strategies to increase food production and economic growth. These strategies included: (i) investments in agriculture and agricultural research; (ii) expansion of agriculture into the vast marginal areas; and (iii) reduction in the population growth rate. However, the implementation of the Structural Adjustment Program, as urged by the International Monetary Fund and the World Bank, weakened the government's resolve to carry out these strategies. Instead, emphasis was shifted towards implementing major political and economic reforms. These changes were slow, and many of them never took place until the mid-1990s.

Meanwhile, despite the increases in maize output of about 30%, from 1.9 to 2.7 million t, and the rate of increase in maize area, yields declined between 1980 and 1981 and 1987 and 1988.[6] Seven more improved maize varieties were released to farmers between 1980 and 1990, and major structural transformations were made in the agricultural research and extension programmes.[7] The Science and Technology Act of 1979 converted the agricultural research department of the Ministry of Agriculture into a semi-autonomous institution, the Kenya Agricultural Research Institute (KARI), which became operational in 1987. Meanwhile, the extension programme embraced a new mode of extension, the training and visit system, after a pilot programme proved successful in several districts in 1981.

However, fewer maize varieties were released to farmers in the 1980s, especially in medium and marginal potential regions, and even the new maize varieties for the high potential areas had a lower yield advantage than their predecessors. In short, research yields were already exhibiting a 'plateau effect'. For instance, whereas H611 had a yield advantage of 40% over KS II, it was out-yielded by H622 by only 16%. In turn, H622 was out-yielded by H611C by 12%. In fact, H626 had only a 1% yield advantage over H625, which was released 8 years earlier and shared a close parentage with H626. Similar observations were made for the mid-altitude and coastal maize varieties. It took 15 years after the release of the Coast Composite in 1974 to release a new hybrid for the coastal region, but the new variety had only 0.78 t/ha yield advantage over the former. Similarly, for the mid-altitude region, H512 out-yielded H511 by only 0.45 t/ha and there has never been another release from that programme since 1970. In general, newer varieties lacked remarkable yield advantages over previous ones, thus prompting the question: are research yields approaching a peak? And if so, what are the implications for maize production in Kenya?

At that point, several additional trends were explored. Real (inflation-adjusted) expenditure on maize research increased gradually from 1955–1959 to 1975–1979, then declined in the 1980s. From an average expenditure of K£126,400 in the 1960s, real maize research expenditure increased to K£232,000 in the 1970s, but declined to about K£133,000 in the 1980s. This decline was likely to have had a profound effect on the productivity of the research programme. In spite of a large build-up of human and physical capital at KARI, insufficiency of operating capital has been cited as a major constraint on maize research (Rutto, 1992). A second equally interesting trend was the improved seed sales by the KSC. Whereas seed sales increased substantially in the 1960s and 1970s, the growth in sales during the 1980s levelled off. Seed sales increased from about 12,800 t in 1980/81, to about 21,000 t in 1988/89. The rate of increase in seed sales was approximately 2% per year between 1985 and 1990 compared with 40–70% in the 1960s and 20–30% in the 1970s. This indicated that adoption of improved seed was peaking. The KSC started its own research department to supplement the work done by KARI. The two partners jointly released H626 in 1989, while the KSC released Pwani Hybrid I in the same year. In summary, a maize production crises emerged in the 1980s as the rate of increase in maize area and yield, and therefore output, declined, sales of improved maize seed stagnated and real research funding declined.

Institutional Collapse: 1990–2000

During 1990–2000 Kenya changed from a single party to a multi-party democracy, allowing for a multiplicity of political parties and national leaders. Meanwhile, major reforms affected the economy and maize production. First, the foreign exchange markets were liberalized and the exchange rate freed from government control. This made imports more expensive and exports more attractive. Local manufacturers, who used foreign inputs, and were previously protected, suffered from high production costs, and lost markets to competitors in neighbouring countries.

Second, all agricultural input markets were liberalized in late 1992 causing the price of fertilizer to increase threefold, and that of maize seed to double. While conducting a national survey on maize production in 1993, it was evident that fertilizer use had been adversely affected by the price increase. Previously, the government subsidized both seed and fertilizer prices, the former through influence in the KSC, and the latter through foreign exchange protection and direct foreign subsidy. With liberalized input markets and controlled output markets, the government was under severe pressure to increase maize producer price beyond what the economy could support. Third, in 1994, the government liberalized the grain markets, thus ending a 52-year government control of grain markets. The results were devastating to maize farmers: whereas, on an average, input prices more than doubled, maize producer prices declined by more than two-thirds in the first 2–3 years. Moreover, almost all National Cereals and Produce Board depots were shut down, and there was no longer a guaranteed market for maize. Instead, farmers were left to the mercy of unscrupulous traders and middlemen, who took advantage of the situation and paid extremely low prices.

Fourth, under intense pressure from donors and complaints from farmers of poor quality maize seed, the government liberalized the seed markets in 1997. This officially allowed other seed companies to market their maize seeds, including some that had already submitted their varieties for local review. Previously, the seed industry was closed to outside competition with the dominant KSC forging a very close partnership with KARI, a relationship that was cemented by the fact that, at one time, the Chairman of the Board of Governors of the KARI was also the managing director of KSC.

Lastly, the training and visit extension programme of the World Bank was terminated in 1997, spelling doom to the national extension programme. The programme had become unpopular, because of its high capital nature and reliance on a continuous stream of research messages, which were becoming rare as the research institutions lacked the capacity to generate them. In fact, the impact of this change was so severe that in a survey of maize producers in 1999, access to the extension service in most maize growing regions had declined by more than 70%. Despite a decline in population growth rate from 3.8% in 1980–1984 to about 2.4% in 1996–2000, Kenya currently relies on maize imports from neighbouring countries to meet its maize deficit.

Implications for research and policy

Karanja (1990, 1996) attempted to estimate the economic payoff to maize research in Kenya and institutional arrangements and policy environment that contributed to the payoff. The findings showed that investments in maize research paid off over the 1955–1988 period, and that the seed multiplication and distribution role of the KSC and the AES of the Ministry of Agriculture played a crucial part in boosting maize production. There is no doubt that the aggressiveness of these complementary programmes enhanced the productivity of maize research, as evident from the countrywide spread and adoption of new maize varieties.

This chapter also highlighted other critical institutional arrangements and policies that led to maize technology take-off in Kenya. These include: (i) an effective railway and road network for facilitating farm input supply and produce marketing; (ii) an efficient marketing network for inputs, through the KFA, and produce, through the MMB; (iii) the guaranteed marketing and seasonal credit by the government; (iv) an aggressive seed processing and distribution company; and (v) a government commitment to increase food production by supporting the maize sector.

Several other important factors also led to the success of maize in Kenya. These include: (i) 50 years of local selection and adaptation preceded hybrid maize research and provided appropriate parental stock; (ii) the initial yield-advantage of the initial hybrids was overwhelming leading to rapid adoption of new seed; (iii) farmer demand for hybrids; (iv) importation of new genes from Latin America; and (v) the quality and continuity of the maize research programme was impressive from its inception in 1955 to 1973.

Lessons for the Future

The basic lesson that can be learned is that a profitable technology package is essential to meet Kenya's future food needs. Kenya was fortunate to have people and organizations that could be mobilized, adapted and co-ordinated to provide the services needed to make the new technology available to farmers from 1955 to the late 1970s. The issue of research funding is critical. Real (inflation-adjusted) funding for maize increased until the late 1970s, and declined thereafter. Agricultural research is an expensive venture with long gestation periods. Research priorities have been set and maize ranked at the top of Kenya's crop research agenda (KARI, 1992). If research is to meet the Herculean challenge ahead, then funding must be commensurate with the importance of maize in the Kenyan economy.

Notes

[1] The ROR is a single number that summarizes the pattern of research programme benefits and costs over time.

[2] One Kenyan pound (K£) is equivalent to about US$0.25.

[3] In Karanja (1990), '10 years' was used as the average research lag period.

[4] The difference between this result and that of Karanja (1990) is that this one uses a more accurate representation of yield, and area equations, whereas Karanja (1990) used a second-order approximation to the simultaneous equations.

[5] The KSC had been established in 1956 as a private company by a group of farmers in Trans-Nzoia district to develop improved pasture seed.

[6] Between 1970–1974 and 1975–1979, the average maize yield increased by 63% while maize area increased by 36%.

[7] This compared with 11 varieties released between 1964 and 1970, and five varieties released in the 1970s.

References

Gerhart, J. (1975) *The Diffusion of Hybrid Maize in Western Kenya.* Abridged version. CIMMYT, Mexico, D.F.

Hassan, R.M. and Karanja, D.D. (1997) Increasing maize production in Kenya: technology institutions and policy. In: Byerlee, D. and Eicher, C.K. (eds) *Africa's Emerging Maize Revolution.* Lynne Rienner Publishers, Inc. pp. 81–93.

Karanja, D.D. (1990) The rate of return to maize research in Kenya: 1955–88. MSc thesis, Department of Agricultural Economics, Michigan State University.

Karanja, D.D. (1996) *An Economic and Institutional Analysis of Maize Research in Kenya.* MSU International Development Working Paper No. 57. Department of Agricultural Economics, Michigan State University, East Lansing, Michigan.

Kenya Agricultural Research Institute (KARI) (1992) *Proceedings of a Workshop on the Review of the National Maize Research Program* held at the Golf Hotel, Kakamega on 19–23 November, 1990.

Meinertzhagen, R. (1957) *Kenya Diary: 1902–1906.* Oliver and Boyd Publishers, London.

Rutto, J. (1992) The coordination and management of the National Maize Research Program. In: *Proceedings of a Workshop on the Review of the National Maize Research Program*, Golf Hotel, Kakamega, 19–23 November, 1990. Kenya Agricultural Research Institute.

11 Impact of Maize Technology Development and Transfer in Ethiopia

T. ZEGEYE, G. TESFAHUN AND P. ANANDAJAYASEKERAM

Abstract: Over the last 40 years several open-pollinated and hybrid maize varieties were developed for different agro-ecologies in Ethiopia. This study employed the economic surplus method for an ex-post evaluation of maize technology generation and transfer investment made in Ethiopia from 1986 to 2000. The estimated rate of return for the maize technology development and transfer investment in Ethiopia was 29% and the net present value at the 10% discount rate was 193.5 million birr (1 birr is currently about US$0.11). In addition to these economic indicators, these investments have also contributed significantly to institutional development, human capacity building, food security and establishing linkages. The study has generally shown that maize technology development and transfer was profitable to the society. The study strongly implied that systematic ways of generating and documenting data are critical to assess the development impacts of agricultural research and extension investments. Studies should also be undertaken in the areas of developing methodologies for impact assessment under data scarcity and to estimate the returns of investments already made in the well established Ethiopian agricultural research system.

Introduction

Smallholder agriculture is the mainstay of the Ethiopian agrarian economy. According to Alemayehu and Tadele (2004) the bulk of agricultural output and value-added is generated by smallholder peasant production. In specific terms, agriculture accounts for 50% of the GDP, employs 85% of the labour force, generates 90% of the foreign exchange earnings, and supplies 70% of the raw material for food processing beverage and textile industries.

There is sufficient evidence to show that the potential of the agricultural sector in Ethiopia can be expanded considerably. The sector is dominated by subsistence production with low productivity. The average yield for grain crops has remained at about 1.1 t/ha, which is very low. Evidences reveal an average annual growth rate of 0.6% for major food crops compared with the corresponding population growth rate of about 3%. This shows that there had been an annual decline of 2.4% in per capita food grain availability from domestic production. These dismal characteristics are, in part, explained by Ethiopia's high dependence on rainfall and its rather limited use of modern agricultural technologies (Alemayehu and Tadele, 2004).

Despite considerable land degradation because of high erosion, Ethiopia is endowed with vast land potential for agricultural development. Agricultural production is predominantly rainfed, which depends on long and short rainy seasons, and is characterized by fragmented small plots of land due to soil degradation and population pressure.

Research can help in improving agricultural productivity through developing appropriate technologies that solve the major problems of the sector. Investments in research are expensive, risky and it takes a considerable amount of time before benefits are accrued. To get a reasonable result from the research investment, resources must be allocated to high priority research activities or programmes. Impact assessment is one of the techniques used in the planning process to assist efficient allocation of resources. Ex-post economic impact studies are also used to justify the allocation and use of additional resources.

This research focused on maize (one of the important cereal crops in Ethiopia) technology development and transfer (TDT) in order to assess the socio-economic impact of the whole process and outputs at different levels. This chapter is organized as follows. The background and rationale of the study will be presented first. Then follow the objective and the methodology of the study. The sections that come next will have results and discussions. Finally, we will present conclusions and recommendations.

Background and Rationale

The maize sub-sector

Maize was introduced to Ethiopia in the 16th to 17th centuries (Kebede *et al.*, 1993). Currently, maize is one of the most adaptable crops across the diverse agro-ecologies of the country. The bulk of production of maize comes from the Oromiya, Amhara and Southern regions. Maize is grown under a variety of cultural practices and in both short and long rainy seasons. Maize is primarily produced for human and livestock consumption. The grains are used for human consumption, whereas the leaves and stalks are used for cattle feed, fuel and construction.

The trend of total maize production is increasing. Yet, there is only a slight (0.5%) and statistically insignificant increase in productivity per unit area. The average annual maize output of about 1.5 million t in the early 1980s (1980/81) increased to 1.9 million t in 1996/97 denoting the largest share (28.4%) among cereals over the period 1992/93–1996/97. Maize output increased by nearly 2% per annum over the three decades. In 1995/96–1997/98, the mean maize output reached a record level of 2.6 million t and is expected to increase. Area under maize rapidly expanded after the political change in 1991, increasing from 808,900 ha in 1993 to over 1.4 million ha in 1997/98 (CSA, 1992, 1994, 1995, 1997, 1998, 1999).

According to Kebede *et al.*, (1993) the major constraints limiting maize production in Ethiopia are erratic rainfall, pests such as stalk-borer, diseases such as rust, blight, streak virus and downy mildew, weeds such as striga (*Striga hermonthica*, *S. aspera*, *S. asiatica*) and orobanche, and continuous use of land without proper soil and water management.

History of maize technology development

Research on maize was initiated at Jima College of Agriculture in 1952 and at Alemaya College of Agriculture (now Haramaya University) in 1953. Maize has also been an important research commodity in the later establishments of Bako and Awassa research centres in the western and central south of the country, respectively. Since then various maize germplasms have been introduced to the country. The research on maize focused on screening varieties and cultural practices such as seed rate, time of planting, spacing, fertilizer type and rate. Diagnostic and participatory research appraisal techniques have also been used to generate information on major maize production constraints and to make the research demand driven and client oriented. Technologies have been tested and evaluated at the farm level.

Over the last 40 years several open-pollinated and hybrid maize varieties were developed for different agro-ecologies of the country as indicated in Table 11.1. A number of agronomic and crop protection recommendations have also been made over the years as presented in Tables 11.2 and 11.3.

Policy makers and donors require information on TDT payoffs in order to assess alternative use of public funds. Research managers are frequently asked to economize and justify their budgets. Thus, the concept of impact assessment has become increasingly important in the agricultural TDT. This demand was the fundamental reason for this research. This chapter discusses in detail the socio-economic impact of the past maize TDT efforts.

Objectives

The overall objective of the study is to assess the socio-economic impact of maize TDT investment over the period 1986–2000. The specific objectives of the study are:

- discussing the evolution of maize research and the enabling environment in Ethiopia
- assessing the socio-economic impacts of maize TDT investments over the period 1986–2000
- identifying policy implications to improve the benefits of TDT investments in maize

Table 11.1. Maize varieties/hybrids developed by research and their areas of adaptation.

Varieties	Altitude (m)	Rainfall (mm)	Days to maturity	Year released	Experimental yield (q)	On-farm (q)
Open-pollinated						
A-511*	500–1800	800–1200	150	1970	50–60	30–40
UCB*	1700–2000	1000–2000	163	1975	50–70	40–45
Alemaya Composite*	1600–2200	1000–1200	163	1975	50–70	40–45
Katumani	1550	600–1000	105		60–70	40–45
ACV-3	1550	600–1000	110	1996	35–50	25–30
ACV-6	1550	600–1000	110	1996	35–50	25–30
Abo-Bako*	500–1000	1000–1200	150	1986	50–70	35–45
Kuleni*	1700–2200	1000–1200	150	1995	60–70	40–45
Gutto*	1000–1700	800–1200	130	1988	30–50	25–30
Hybrids						
BH-140*	1000–1800	1000–1200	140	1988	80–90	50–60
BH-660*	1600–2200	1000–1500	165	1993	90–120	60–80
BH-540*	1600–2000	1000–1200	145	1995	80–100	50–65
BH-530*	1000–1300	1000–1500	137	1997	80–90	50–60
Beletech*	1500–2000	800–1200	160	1990	50–70	40–45

Source: Mosisa *et al.* (2001).
*Under production.

Table 11.2. Agronomic recommendations for maize.

Practices	Recommendation
Land preparation	2–3 times ploughing with maresha (local plough)
Planting depth	5–7 cm, planted in rows
Spacing	80 cm × 50 cm, two plants/hill for full season varieties
Fertilizer rate	69 kg/ha N and 46 kg/ha P_2O_5 for open-pollinated varieties
Weed management	Hand weeding: twice hand weeding at 25–30 days and 55–60 days supplemented with slashing Herbicides: Premagram, 2 kg/ha Note: herbicide use should not obviate the need for supplementary hand weeding
Precursor crop	Noug followed by haricot bean
Crops suitable for intercropping and relay cropping	Haricot bean, sweet potato as well as forage legumes

Table 11.3. Protection recommendations for maize.

Pests and diseases	Recommendations
Stalk borer	Early planting after onset of rain Horizontal placement of the maize stalk in the sun for 4–6 weeks in the field
Storage pests	Drying of the grain to the optimum moisture level Use of insecticides such as primiphos-methyl dust
Diseases	Use of resistant/tolerant varieties Management practices • Proper tillage • Crop rotation • Timely weeding • Plant at the optimum sowing date • Optimum planting density • Balanced fertilization Crop sanitation • Removal of crop residues • Removal of subtle • Removal of alternate hosts • Seed dressing with chemicals Management of vectors that would transmit viral diseases such as streak virus

Methodology

Three analytical approaches can be employed to assess the impact of any agricultural research and development (R&D) investment: the econometric approach, which aims at estimating the marginal productivity of research over a long period time; programming method that aims at identifying one or more optimal technologies or research activities from a set of options; and economic surplus method, which measures the aggregate social benefit of a particular research project or programme (Masters *et al.*, 1996).

The economic surplus method is used more often to estimate rate of returns (RORs) to technology development as it is more flexible to specify the value of research in comparing the with and without situations. It can also handle side effects of technological changes such as income distribution and environmental consequences (Anandajayasekeram *et al.*, 1996). The economic surplus method uses the concept of supply and demand equilibrium. Supply represents producers' production costs, and demand represents consumers' consumption values. In most cases, equilibrium quantity and price levels result from the interaction of supply and demand. This study employed the economic surplus method for an ex-post evaluation of maize R&D investment made in Ethiopia from 1986 to 2000.

Data sources

Data for this study were obtained through extensive formal and informal interaction with researchers, extension personnel, the Ethiopian Seed Enterprise, Pioneer Hi Bred Seed Private Company, Fertilizer Industry Agency, International Maize and Wheat Improvement Center (CIMMYT) and Sasakawa Global (SG)-2000. Informal reconnaissance surveys, secondary data and in-depth interview with key informants at different levels were also used to collect the primary data.

Areas, production and yield of maize

The total area, production and yield figures for the without maize research and technology transfer scenario are based on national estimates of the Central Statistical Authority. Improved maize yield levels used in the impact analysis were estimated based on available literature and consulting members of the National Maize Research Program of the Ethiopian Institute of Agricultural Research (NMRP–EIAR).

Maize price

Surplus maize is produced when favourable climatic conditions (with good rainfall) prevail. After 1991, prices are set on the open market in response to supply and demand. There are no fixed prices for food crops and no government price subsidies to producers and consumers. Setting an appropriate price to value the increments of production because of improved technology depends on whether the amount produced is to be sold in the market. The structure of the market and the relative impact of increased supply are expected to have an influence on prices as well.

Although a small amount of cross-border trade is taking place, maize is not currently exported from Ethiopia. In this study, it was realistically assumed that Ethiopia is neither an importer nor exporter of maize, and any increase in the domestic price will not affect the world market price of maize.

In recent years, the increase in production of maize due to TDT has been relatively high and has effected on the supply and demand of maize. Despite the increase in maize production, the price stabilizing effect is not noticeable. Average prices of the Ethiopian Grain Trade Agency were used in this study to value the production of improved maize and to calculate the net benefits. In calculating the benefit from maize TDT, attempts were made to transform the nominal value of the net benefits streams into a real value, which is expected to control the effect of inflation. The net benefit stream was standardized by converting all nominal values to constant 1995 prices using the consumer price index.

Adoption of maize technologies

The national maize research programme has released nine composites and five hybrid maize varieties for the various agro-ecological zones of the country over the period 1970–1997 (Table 11.1). In addition, several site-specific crop management and protection recommendations were also released (Tables 11.2 and 11.3). In this study, it was assumed that any variety or crop management practice used prior to 1970 is 'old technology' and any variety and/or management practice recommended since 1970 is considered as 'new technology'. Technology adoption is one of the most uncertain fields/areas of the benefit–cost analysis approach. Seed companies in Ethiopia have multiplied maize hybrids and composites and seeds of other crops during the study period. The time-series data of seeds distributed by the Ethiopian Seed Enterprise and Pioneer Hi Breed Seed Company were used to estimate adoption rate of maize. If farmers are using retained seeds then this assumption will lead to the underestimation of the adoption rate.

Costs of technology development

There are three different sources of funds for agricultural research in Ethiopia; national treasury, local and international donations, and self-financing by research centres. The contribution of the national treasury (government financing) is for recurrent and investment (development) budgets. In this study, the government budget contribution and the cost of donor-funded maize projects research are considered.

Some donors are supporting a maize research programme through CIMMYT and SG-2000. The CIMMYT economics programme also occasionally provided technical assistance and limited recurrent budgetary support for specific activities. However, it was difficult to establish the exact expenditure associated with maize research. The costs associated with technology development by other organizations are not significant, and therefore are not included. The national research cost includes salaries, recurrent expenditures and investment costs. Because of an inadequate recording system, it was difficult to obtain annual salaries associated with maize research staff. However, salaries estimated by the maize research programme were incorporated in the analysis. The allocation of funds for 'replacement' is very low. It was also recognized that much of the replacement of vehicles and equipment was financed through development funds. Given this low expenditure and insufficiency of data, we do not include them in the analysis.

Costs of technology transfer

The Ministry of Agriculture and Regional Bureaus of Agriculture are the main institutions involved in transferring maize technologies developed by the national agricultural research system. The transfer of the technologies starts immediately after the National Variety Release Committee officially approves the release of the variety and the multiplication by the Ethiopian Seed Enterprise and Pioneer HI Breed Private Seed Company.

The cost allocated to extend a single commodity, such as improved maize and associated technologies, is very difficult to find especially when record keeping is highly aggregated. The yearly budget allocation and expenditure reports of MEDAC for 20 years were examined. However, it was not possible to disaggregate the cost of maize extension alone, because out of the fund allocated to Ministry of Agriculture, a considerable amount has been spent in extending different recommended technologies including maize, and the rest was spent in infrastructure development. The total budget allocated to Ministry of Agriculture by MEDAC was not considered in this analysis because of problems of disaggregating the cost associated with maize technology transfer.

The costs of maize technology transfer were estimated through the consultation of heads of Bureaus of Agriculture and professionals of Amhara, Oromiya and Southern regions. A number of assumptions were made in estimating the cost data for maize technology transfer.

Results and Discussion

Agricultural TDT has several impacts on the society. The impact of such an investment can be assessed in terms of direct product of the TDT as well as the intermediates, including institutional development and change in the enabling environment. The economic impacts of maize technology are measured in terms of rates of returns. A partial equilibrium analysis using the Akino and Hayami (1975) method was employed to estimate the economic ROR. The study focuses on the benefit accrued to the domestic producers and consumers. It is assumed that Ethiopian maize production would not have any effect on the international maize price. The calculation of research benefits includes the yield advantage of improved maize over the old ones, technology adoption, and the cost of TDT. Below are presented the results generated out of the analysis and the discussions thereof.

Costs and benefits of maize technology development and transfer

The annual benefits due to maize research and extension activities are estimated by subtracting the sum of annual maize technology development, transfer and adoption costs from the estimated annual gross benefits. One would expect during the early periods of the TDT process, the net benefit would be less than zero and increase gradually. Although the net gains are increasing over time, they also demonstrate a substantial fluctuation over years. This could be because most of the maize production in the country depends on rainfed agriculture (Fig. 11.1).

The value of investment over time was estimated by calculating the internal rate of return (IRR) and net present value (NPV) of the net flow. The real interest rate (social opportunity cost of capital) of 10% was assumed in this study. The estimated IRR was 29%. This means that, after recovering costs, every dollar invested in maize research and extension during the period 1986–2000 generated 29 cents in return. Given the current interest rate of 10% (the opportunity cost of capital), this result appears to be very attractive. The NPV of maize TDT investment at the 10% discount rate was 193.5 million birr and it was 97 million birr at the 15% discount rate. The NPV for 20% discount rate is 43.6 million birr (Table 11.4).

Owing to the lack of reliable data, several assumptions were made to estimate parameters for net benefit calculations. The assumptions considered in some cases might have underestimated the costs or overestimated the benefits, resulting in higher ROR. As some of the research and extension costs such as depreciation, administration and overhead costs were not included in the original estimates, it was decided to make a sensitivity test and re-estimate the NPV and IRR. Therefore, the analysis was repeated with 20%, 30% and 40% increases in research and extension costs.

When research and extension costs were increased by 20%, the estimated IRR with the Akino–Hayami method became 28%. The IRR fell to 27 and 26% when the TDT costs were increased by 30% and 40%, respectively. The NPVs for 5%, 10%, and 20% discount rates were 352,796,046, 181,117,157 and 37,698,888, respectively (Table 11.4).

Table 11.4. Estimated NPV (in million birr) and IRR (%) for the base scenario and for changes in maize research and development cost at different discount rates.

Base scenario			Sensitivity analysis						
				r = 5%		r = 10%		r = 20%	
r*	NPV	IRR (%)	% change in R&D cost	NPV	IRR	NPV	IRR	NPV	IRR
10	193.5	29	20	377.7	28	359.3	28	333.4	28
15	97	29	30	352.8	27	346.9	27	317.5	27
20	43.6	29	40	181.1	26	177.4	26		26

*r = discount rate.

Impacts of maize technology generation and transfer

In addition to the economic impact estimated with ROR, there are other impacts that could be explained qualitatively. This group of impacts includes increased capacity of research and extension to produce and deliver new or improved technologies to eventual customers. It also includes resource development such as physical infrastructure, financial resources, human resources and collaboration of the commodity (maize) programme with other national, regional and international research systems.

Availability of financial resources

As pointed out earlier, the major source of finance for the national maize research programme has been the government of Ethiopia. It was observed, however, that given the overall economic and financial situation of the country, the trend in the annual budget allocation was generally below the bare minimum requirements of maize research. Donor-assisted projects executed through international research systems such as CIMMYT have, therefore, strengthened maize research in Ethiopia. Quite a few economic and policy research activities were also undertaken with the assistance of CIMMYT in the 1980s and 1990s. From 1986 to 2000, the average capital budget allocated to the Ethiopian Insitute of Agricultural Research by the government was 32.3 million birr.

Human resources development

Several maize research staff obtained postgraduate degrees over the period 1986–2000. The disciplines in which the studies were undertaken included: breeding, agronomy, entomology and pathology. Several researchers and field officers have participated in short-term training. Ethiopian maize researchers also received short-term training at CIMMYT in Mexico and in Eastern African countries. They also attended periodic CIMMYT sponsored workshops on maize improvement. A total of 28 in service training were provided and 30 visiting scientists of the NMRP were trained at CIMMYT in Mexico from 1969 to 2000, with an estimated cost of US$438,200.

A number of training workshops were also conducted within the East African region. The number of maize researchers who participated in training, conferences and in travelling workshops between 1991 and 2001 was 214, 107 and 44, respectively. Within the same period, CIMMYT economists trained maize researchers and technicians of the national agricultural research system, which involved 214 in seed policy, 107 in natural resources management policy and 44 in gender analysis.

Organization of research and extension systems

The research staff in Bako, Awasa, Jima and Melkassa agricultural research centres are responsible for designing and conducting agronomic and crop protection trials in their respective agro-ecologies. The breeding work was co-ordinated by the Bako research centre and the breeders travelled to collaborative sub-programmes and centres to work on materials for the different agro-ecologies. The mid-altitude maize research subprogramme has been implemented at Bako, Jima and Awasa research centres. The Ambo research centre is responsible for breeding maize for high altitude areas of the country.

Currently, the maize research programme is carried out by an interdisciplinary team of researchers including breeders, agronomists, pathologists, entomologists, soil scientists and economists. Breeders and agronomists are attached to the maize programme, while other research staff provide professional assistance. The programme has an annual national planning and priority setting meeting. The systems approach has been introduced into the research planning and evaluation procedures. Diagnostic and participatory research appraisal techniques have been used to generate information on major maize production constraints and to make research demand-driven and client-oriented. Technologies are tested and evaluated at farm level. These are positive aspects that have emanated as a result of the systematic organization and functioning of the maize research programme.

Impact on linkages

Internally, the national maize research programme collaborates and shares experiences with the different programmes and projects of Ethiopian agricultural research organization and regional research centres. It has also created a strong relationship with institutions such as the Ethiopian seed enterprise, the Ministry of Agriculture, bureaus of agriculture in the regional states, Pioneer Hi Breed Private Seed Company and the National Variety Release Committee.

Externally, the maize research programme has a long history of linkage with CIMMYT in training and germplasm exchange. The maize research programme has actively participated in the activities of CIMMYT/CIDA-funded cereals improvement project in Eastern Africa. In addition, the programme has a very strong link with the SG 2000 project.

Environmental impact

In the absence of biophysical data and valuation of existing constraints, it is difficult to quantitatively measure the benefits and costs of environmental, natural and biological effects associated with maize (or any other commodity) technology generation and transfer.

The primary objective of the maize research programme in Ethiopia, as in most other developing countries, is to augment the productivity of natural resources, alleviate poverty and improve food security. No study has ever been conducted to address the impact of maize technologies on the environment and the natural resources. However, there are a number of positive impacts such as reduced soil erosion due to row planting. When maize is planted in rows, farmers pile up the soil on to the seedlings during weeding and small ridges are constructed across the field to control the flow of water. In addition, crop rotation, intercropping and under-ploughing of crop residues improve the soil texture, structure and soil fertility. As a heavy nutrient user, maize needs soil that is fertile. Farmers in Ethiopia, who don't use fertilizers, depend on trash and burn methods to increase the productivity of maize per unit area of land.

Food security

Food security refers to availability, accessibility and affordability of food at national, regional and grass-root levels. At the household level, food should be available at all times. At household levels, the surplus production of maize is marketed to generate income. The increased marketable surplus can be attributed to increases in productivity per hectare and expansion in area under maize production, which has shown an increasing trend over the last 10 years. The national average yield of maize has also increased from 1 t/ha in 1986 to 1.8 t/ha in 2000. The sample survey revealed that the adoption rate of improved maize for Oromiya was 39%, for Amhara 43% and for Southern nations 47%. Thus, yield increase and increase in marketing of surplus maize was as a result of improved maize technologies and strong extension and input delivery systems.

Conclusions and Recommendations

Since the inception of the maize research programme, 14 open-pollinated varieties and hybrids have been released. Several site-specific crop management recommendations were also made for various agro-ecological zones. The estimated ROR for the maize TDT investment in Ethiopia was 29%. In addition to the 29% ROR, the investment had also contributed significantly to institutional development, human capacity building, food security and establishing linkages. From the calculated IRR we have seen that maize TDT was profitable to the society. Therefore, continuity of the funding of research and the institutions involved in transferring the technology is important and rewarding so far as Ethiopia is concerned.

This analysis has pointed out a number of important issues, which need to be addressed. One key factor is the need for better data collection in order to document the performance of new technologies under farm conditions and to map the extent and spread of adoption. Currently, information on adoption of a specific variety/hybrid of maize is hardly available. Efficient data collection and management are key components in monitoring the impact of TDT and making more realistic estimates for ex-ante impact assessments.

Despite the high potential ROR to investment in maize research, Ethiopia is still challenged by various constraints that inhibit the adoption of new technologies and the achievement of significant research results. Foremost among these constraints are macro-level factors that are crucial in setting the scene for the overarching issue of enhancing the productivity and sustainability of maize production to address the bigger issue of food and nutrition insecurity and poverty. External economic shocks such as a drastic fall in international coffee prices undermine the capacity of the government of Ethiopia to invest in agricultural research, no matter how potentially profitable such investment may be.

References

Akino, M. and Hayami, Y. (1975) Efficiency and equity in public research: rice breeding in Japan's economic development. *American Journal of Agricultural Economics* 57, 10.

Alemayehu, S.T. and Tadele, F. (2004) *The structure of the Ethiopian Economy–a SAM-based Characterisation.* Addis Ababa University, Addis Ababa, Ethiopia.
Anandajayasekeram, P., Martella, D.R. and Rukuni, M. (1996) *A Training Manual on Research Evaluation and Impact Assessment of Investments in Agricultural and Natural Resources Research.* SADC-SACCAR, Gaborone, Botswana.
CSA (Central Statistical Authority) (1990) *Statistical Abstract.* CSA, Addis Ababa.
CSA (1992) *Statistical Abstract.* CSA, Addis Ababa.
CSA (1994) *Population and Housing Census of Ethiopia, Result at National Level. Volume 1. Statistical Report.* CSA, Addis Ababa.
CSA (1995) *Statistical Abstract.* CSA, Addis Ababa.
CSA (1997) *Statistical Abstract.* CSA, Addis Ababa.
CSA (1998) *Statistical Abstract.* CSA, Addis Ababa.
CSA (1999) *Report on Area and Production of Major Crops. Statistical Bulletin.* CSA, Addis Ababa.
Ethiopian Seed Enterprise (1999) *Seed System Development Project Status Report As of June 1999.*
Kebede, M., Gezahegn, B., Benti, T., Mosisa, W., Yigzaw, D. and Asefa, A. (1993) Maize production trends and research in Ethiopia. In: Benti, T. and Ranson, J.K. (eds) *Proceeding of the First National Maize Workshop of Ethiopia*, 5–7 May 1992, IAR/CIMMYT, Addis Ababa, Ethiopia.
Masters, W.A., Williams, A. and Sidibe, M. (1996) *The impact of Agricultural Research: a Practical Guide.* Department of Agricultural Research, Purdue University, Indiana.
Mosisa, W., Jemal Abdushukur, Leta Tulu, Haji Tuna, Legesse Wolde, Kassa Yilma, Wonde Abera, Aschalew Gutu, Sewagegne Tariku, Teshale Asefa, Tamirat Birhanu Yoseph Beyene and Habtamu Zeleke (2001) *Improved Germplasm Development for the Mid and Low Altitude Sub-Humid Agro-ecologies.* Paper presented at the Second National Maize Workshop of Ethiopia. 12–15 November 2001.
SG 2000 (2001) *Sasakawa Global 2000 Agricultural Project.* Project Concept Document. Addis Ababa.

12 Evaluating Agricultural Research and Extension in Tanzania: the Production Function Approach

A.C. Isinika

Abstract: Agricultural research in Tanzania began in 1923 on sorghum and cotton. Since then, the number of research stations and institutions involved in agricultural research has increased. However, there are very few evaluations of the performance of agricultural research. This study uses the production function approach to assess the efficiency of investments in agricultural research and extension from 1971 to 1992. Results of the regression analysis showed that the total factor productivity for crop production in Tanzania is responsive to expenditures on research while it is not responsive to expenditure on agricultural extension. The model also shows that total factor productivity for crop production was not significantly responsive to the literacy rate of farmers, rainfall and lagged export earnings. The study concludes that since the marginal rate of return of investment in agricultural research is greater than the social rate of return, continued funding of agricultural research by both government and private sector is encouraged. The marginal rate of return was found to be low. This calls for efforts to improve extension and enhance the linkages between agricultural research and extension.

Introduction

Agricultural research is an investment that must compete for resources with alternative investments. The competition is particularly high in developing countries where resources are acutely scarce. Policy makers often tend to give low priority to finance research programmes because of their long gestation period or the non-tangible nature of some benefits accruing from research. Proper and regular evaluation of investments in agricultural research is therefore very important, in order to assist policy makers to make informed decisions.

The evaluation of agricultural research differs in terms of coverage, timing and methodology. The production function approach was employed in this study to evaluate the returns from investment in agricultural research for crop production in Tanzania, between 1971 and 1992.

Background to the study

In Tanzania agriculture accounts for up to 60% of the gross domestic product (GDP), 80% of the employment and between 60% and 75% of the export earnings. Only about 12% of the

 Impact of Science on African Agriculture and Food Security (eds P. Anandajayasekeram, M. Rukuni, S. Babu, F. Liebenberg and C.L. Keswani)

land area is used for agricultural production, including grazing. While the political will to transform the economy into a semi-industrialized economy by the year 2025 has been expressed by the government, it is acknowledged that agriculture will remain the engine of economic growth for the near future. Investment in agricultural research is therefore necessary to complement the transformation process.

Smallholder farmers who account for over 80% of all the farm holdings characterize agriculture in Tanzania, but the average farm size is less than 2 ha per household. Approximately 10 million ha (outside national reserves) are considered to be of good agricultural potential in mainland Tanzania. It is, however, estimated that only about 3% of the land is cultivated in any given season, and out of this less than 5% is irrigated. The hand hoe is being used on approximately 85% of all the cultivated land. Oxen and tractors account for only 10% and 5% of the cultivated land, respectively. There are many problems which limit the use of land for agricultural production in Tanzania. These include shortage of soil moisture, low fertility, poor drainage, steep slopes, tsetse infestation, poor accessibility and degradation of land under cultivation.

The use of yield-enhancing inputs is extremely low, both in terms of quantities applied and the number of people using such inputs. It is estimated the only 15% of the farmers use chemical fertilizer, 27% use improved seed and 18% use pesticides.

Agricultural research in Tanzania

Research in Tanzania began in 1923 when research on sorghum and cotton was launched at Ukiriguru in western Tanzania. Since then, the number of research stations and sub-stations has increased to about 50, organized under seven agricultural research zones. During the period under review, most of the agricultural research was done by the Ministry of Agriculture (MoA), for which the official title as well as the composition has changed from time to time. However, other institutions, including Sokoine University of Agriculture, the Ministry of Natural Resources and Tourism and the private sector, complement the research efforts by the MoA.

Despite its long history, there are very few studies of the performance of agricultural research in Tanzania. This study was done between 1994 and 1995. The study has used the production function approach to assess the efficiency of investments in agricultural (crop) research and extension in Tanzania, between 1971 and 1992 (Isinika, 1995).

The production function approach

The production function approach, which is used to measure returns to agricultural research and extension, involves estimating parameters of a single commodity or an aggregate production function in which agricultural research and extension are explanatory variables (Annex 1). The choice of the functional form is guided by a number of factors, including the availability of data and the purpose for which the study is intended. Some of the functions that have been used include Cobb–Douglas type (Griliches, 1963; Evenson, 1967, 1968; Akino and Hayami, 1975; Davis, 1981), constant elasticity of substitution models (Griliches, 1964) and the transcendental logarithm (translog) models (Ball, 1985; Romano, 1987). In this study, the translog model was chosen to allow for interaction between some of the explanatory variables.

There are a number of basic factors, which must be taken into account for the effective use of the production function approach. One factor is the temporal distribution of benefits accruing from research. Investments in research do not generate immediate improved inputs or techniques. It is generally agreed that there are two levels of lags. First, from the time of investment ($T = 0$) up to when usable technology becomes available ($T = t$), second, between the time of availability to obsolescence of the technology ($T = t + m$). The distribution of returns between $T = t$ and $T = (t + m)$ may be symmetric or asymmetric (1968).

Total factor productivity

Before the production function approach can be applied, the total factor productivity (P_t) of the commodity, industry or sector being evaluated must be computed for each period. The total factor productivity may be represented as in equation (1)

$$P_t = \frac{Y_t}{X_t} \tag{1}$$

Where Y_t = an index of output during period t, X_t = an index of input during period t.

For ease of aggregation and inter-temporal comparison, the value of both inputs and outputs should be presented in real value terms. In this study, such values were presented in shillings, the currency used in Tanzania, deflated to 1971 real values. The conceptual model used in the study and the reduced form of the equation used for empirical estimation are presented in Annex 1.

Data sources and data aggregation

Data for both inputs and output were obtained from different sources, mainly: (i) various publications and records of the MoA; (ii) various FAO publications; and (iii) various World Bank publications. More specifically, the following data were collected and used for the analysis as has been described.

The annual output for 15 crops was collected from MoA and FAO records and the corresponding prices were obtained from MoA and World Bank sources. The annual nominal value for each crop was computed and deflated using the consumer price index. These data were used to compute divisia indices. Data on the quantity and prices of fertilizer (F), improved seed (S) and agricultural chemicals (C) were obtained from MoA and FAO records. The nominal values were then deflated using the consumer price index. Data on area under crops (A) and the value of farm implements were obtained from various FAO publications. Data on farm labour (L) were obtained from World Bank and FAO reports while data on the minimum wage, which was assumed to be the opportunity cost of labour, were obtained from various government documents.

Annual expenditures on crop research and extension were obtained from various government budget reports. The data were then deflated using the consumer price index. The cost of human capital for research and extension services was not included due to lack of consistent annual data. It was, however, assumed that the real value of human capital did not vary significantly from year to year during the period of the study. This none the less is a weakness of the study.

Data on the weather (W) and real lagged exports (Exp_{t-1}) were obtained from various government reports and records. An index of the weather was weighted by zone, to reflect spatial variations in crop productivity. Annual rainfall data for at least two stations from each of four agro-ecological zones were used to construct the weather index. The aggregate weather index for a given year was computed as a weighted average of the zonal contribution to six main crops: maize, beans, paddy, coffee, cotton and sisal.

Result of regression analysis

Results from the regression analysis are presented in equation (15) and Table 12.1. The model was tested for sensitivity over different periods and it was found to be very stable (Isinika, 1995):

$$P_t = -2.2194 - 2.7201 LnR_{t-3} - 13.9488 LnR_{t-4} + 11.511 LnEx_{t-1} + 5.3047 LnEd_t + 1.422 LnW_t$$
$$- 5.9129 LnExp_{t-1} + 16.9094 \frac{(LnR_{t-3})^2}{2} + 18.3474 \frac{(LnR_{t-4})^2}{2} + 14.9728 \frac{(LnEx_t)^2}{2} \quad (15)$$
$$- 29.9443 \frac{(LnW_t)^2}{2} - 28.0957 \frac{(LnExp_{t-1})^2}{2} - 5.343 LnT$$

Owing to the presence of quadratic and the interaction terms in the model, the parameter estimates and their corresponding signs must be interpreted jointly. The first derivative with respect to each variable is a measure of its elasticity of production, which is a measure of the proportional change in total factor productivity that would occur given a proportional change in the variable. The parameter estimates for research and extension represent the respective marginal product for that period. Differentiating equation (15) with respect to each independent variable, elasticities of production were calculated. Based on these, the marginal internal rate

Table 12.1. Results of the regression analysis.

Parameter Estimates			Elasticity of Production and MIRR			
Variable name	Parameter estimate	*t*-value	Independent variable	Elasticity of production	Returns to	Marginal internal rate of return (MIRR)
Intercept	–2.2195**	3.097	R_{t-3}	1.976	Agricultural research	33.2%
LnR_{t-3}	–2.7201	1.073	R_{t-4}	1.558		
LnR_{t-4}	–13.9488***	5.562	Ex	0.069		
$LnEx_{t-1}$	11.5108***	5.286	Ex_{t-1}	0.859	Agricultural extension	Approximately zero
$LnEd_t$	5.3047*	1.827	Ed	0.145		
LnW_t	–29.9443	0.346	W	–0.19		
$LnExp_{t-1}$	–5.9129	1.565	Exp_{t-1}	–0.363		
LnR^2_{t-3}	16.9094***	3.412	R_{t-3}	1.976		
LnR^2_{t-4}	18.3474***	3.744				
$LnEx^2_t$	14.9728***	5.068				
LnW^2_t	–29.9443***	1.030				
$LnExp^2_{t-1}$	–28.0957***	3.825				
LnT	–5.3429***	4.643				

BASIC STATISTICS OF THE MODEL
F value = 6.844***.
Adjusted R^2 = 0.7781.
DW statistic = 2.153 [LDW = 0.212 < 2.153 < UDW = 2.217].
Condition index = 28.
Sensitivity analysis = model is stable.
*The value is significantly different from zero at $\alpha = 0.1$.
**The value is significantly different from zero at $\alpha = 0.05$.
***The value is significantly different from zero at $\alpha = 0.01$.

of return for investments in crop research and extension in Tanzania was computed according to equation (16).

$$\text{Discounted } MPR - 1 = 0$$

$$\therefore = \sum\left[\frac{MPR}{(1+MIRR)^n}\right] - 1 = 0 \tag{16}$$

$$MIRR = \sum(MPR)^{\frac{1}{n}} - 1$$

Where MIRR = Marginal Internal Rate of Return of Research (or extension), MPR = Marginal Product of Research (or extension).

Discussion of results

The analysis shows that the model explains 77.8% of the variation in total factor productivity for crop production in Tanzania. The model does not indicate the presence of autocorrelation, and only moderate level of multicollinearity. Seven of the parameter estimates (including the intercept) are negative while nine of 12 independent variables are significantly different from zero. Of the negative parameters, those for the linear term for research lagged 4 years, the quadratic term of lagged export earnings, the interaction term and the intercept are significant.

The elasticity of production for research expenditure lagged 3 and 4 years are 1.558 and 0.069, respectively, resulting in a marginal internal rate of return estimated at 33.2%. This indicates that the total factor productivity for crop production in Tanzania is responsive to expenditures on research. This outcome lies within the range of results from other similar studies (Echeverria, 1990; Karanja, 1990). However, these results indicate that total factor productivity is not responsive to expenditure on agricultural extension, having an elasticity of production of 0.0693 and 0.859 for the current and lagged extension expenditure respectively, and a marginal internal rate of return that was close to zero. This outcome was not expected.

The low return on extension investments may reflect the fact that during most of the period under review, extension resources and agents were used for many non-extension activities, therefore implying that the actual cost attributed to extension may be lower than is reflected in government records (Isinika, 1995). Second, the results may reflect a possible specification error in the model or data problems, particularly considering that some of the elasticities of production were negative, contrary to expectations. According to this model, total factor productivity for crop production was also not significantly responsive to the literacy rate of the farmers, rainfall and lagged export earnings.

Conclusions

From the results and discussions, a number of conclusions could be drawn in terms of guiding the future R&D investments as well as in enhancing the effectiveness and efficiency. The key conclusions are summarized below:

- Despite problems that have been encountered by the research system in Tanzania, the study yielded a marginal rate of return that is greater than the social rate of return. This finding provides justification for continued funding of agricultural (crop) research in Tanzania. The government as well as the private sector are encouraged to invest in agricultural research as they will get a return on their investment that is at least greater than the cut-off discount rate for most lending institutions. Institutional reform designed to minimize overhead costs and other sources of inefficiency would raise the returns further.

- But the marginal rate of return for extension was found to be very low. While this may be attributed to specification error and data problems, it may also confirm the long-standing concern about the ineffectiveness of agricultural extension in Tanzania. This finding has two main implications: (i) efforts to improve the extension services need to be closely monitored in order to gauge the short- and medium-run effects of such investments; and (ii) linkages between agricultural research and extension should be forged and sustained, as high returns from research investments probably indicate that farmers have been able to obtain technologies from research through channels other than the extension system.
- A low elasticity of production with respect to educational level of farmers may imply that primary and adult education programmes in Tanzania did not contribute significantly to improvement in crop production during the period under review. Nevertheless, education acts as a stimulus for encouraging innovation. Primary education has been shown to increase productivity by up to 7.4% (Falk, 1992).
- By inference and based on other information, the study concluded that the limited supply of inputs during the study period coupled with poor infrastructure and inefficient institutions had a negative effect on crop productivity in Tanzania.

Annex 1

The conceptual model

Conceptually, measuring the effect of investments in agricultural research on productivity using the production function method involves a two-step process. First, a research production function is defined, where the output is 'incremental knowledge', simply defined as agricultural technical information emanating from research stations. This output is then incorporated in an agricultural production function as an independent variable. Alston *et al.* (1995) presented such a model as follows:

$$I_t = (R_t, R_{t-1}, R_{t-2}, \ldots\ldots\ldots \mathrm{R}_{t-k}, N_t, Z_t), \tag{2}$$

where I = increment in the stock of knowledge, R = investment in research, N = stock of knowledge, Z = a vector of other variables, t = a subscript denoting time, $t-k$ = lagged variables for $k = 1, 2, \ldots, K$.

The stock of knowledge that is created by the research process is in continuous decay, either due to the obsolescence of products and techniques or due to replacement by more improved technologies (Knutson and Tweeten, 1979). Investment in agricultural research is supposed to maintain or increase the knowledge capital stock. Thus, incremental knowledge (I_t) adds to the stock of knowledge (N_t), which depreciates at the rate (D_t) such that at any time (t) the stock of knowledge may be represented as in equation (3).

$$N_t = N_{t-1} + I_t - D_t \tag{3}$$

The implicit knowledge function then enters the agricultural production function (Y_t) as a flow of services (F_t). This flow of services, which is a function of the stock of knowledge (N_t), the relative prices of inputs (Pr_t), human capital (H_t) as well as the quality and quantity of extension services. The flow of services from research is further augmented by conventional inputs and other non-conventional inputs such as the weather and other random factors. The agricultural production function may therefore be presented as:

$$Y_t = f(X_t, W_t, Pr_t, H_t, N_t, Z_t, R_t, R_{t-i}, \ldots R_{t-k}, Ex_t, Ex_{t-i}) \tag{4}$$

where Y_t = level of agricultural production during period t, W_t = a weather variable during period t, all other variables are as previously defined.

It should be noted here that the rate of agricultural technical change depends on the level of private and public expenditure on research (R_t) while the rate of diffusion of technology depends on expenditures on agricultural extension (Ex_t) and the educational level of farmers (H_t), which improves farmers' technical and managerial skills (Evenson, 1967; Lu *et al.*, 1978). Overall, new technology, improved human capital and institutions complement each other in stimulating economic and agricultural development.

The theoretical model that is used here to estimate returns from research and extension is adapted from Davis (1991) and Wise (1986), as it is presented under equation (5).

$$Y = A\prod_{j=1}^{m} X_{jt}^{\beta_j} \prod_{i=0}^{n} R_{t-i}^{\alpha(t-i)} e^{u_t} \tag{5}$$

where Y = value of agricultural output, A = a shift factor, X_{jt} = conventional inputs j during period t, for j = 1, 2, ... m, R_{t-i} = research expenditures for the year $(t-i)$, β_j = a production coefficient on conventional inputs, $\alpha'(t-i)$ = a production coefficient on research, u_t = a random error, t = denotes time for i = 0, 1, 2, ..., n.

In this study, the theoretical model was operationalized in two stages. First, indices of output and inputs were calculated from which the total factor productivity (TFP) for each period was computed. Then, in the second stage, the rate of change of the total factor productivity with respect to time was regressed against expenditures on research and extension and other explanatory variables. This approach has also been used by a number of other researchers (Knutson and Tweeten, 1979; Romano, 1987; Scobie *et al.*, 1991).

Choice of functional form

In this study, the transcendental logarithm (translog) model was used because it allows for interaction between explanatory variables, in this case, interaction between research and extension expenditures. The divisia index, which is exact for a homogeneous translog function, was computed, based on the general functional form as it is presented under equation (6).

$$Lnf(Y) + \alpha_0 + \alpha_n \sum_{n=1}^{n} X_n + \alpha_n \sum_{j=1}^{n}\sum_{k=1}^{n} Y_{jk} LnX_j LnX_k \tag{6}$$

where $\sum_{n=1}^{n} \alpha = 1$, $Y_{jk} = Y_{kj}$, $\sum_{k=1}^{n} Y = 0$, for $j = 1,2,\ldots,n$.

The divisia index is based on price and quantity data. The rate of change of the index is equal to a weighted average of its components' share in the total value (Hulten, 1973). Unlike others, divisia indices are chain linked such that each year's prices are used as a base for estimating the rate of growth for the following year, thus overcoming the problem of changing prices. While the divisia index is continuous, in practice discrete alternatives such as the laspeyres or the Tornquvist approximation to the divisia index are used for empirical estimation.

The divisia index is derived from equation (6). After several steps of mathematical manipulation and substitution the input and output divisia indices can be written as presented in equations (7) and (8).

$$\frac{f(X^1)}{f(X^0)} = \prod_{n-1}^{n} \left(\frac{X_n^1}{X_n^0}\right)^{\frac{1}{2}(S_n^1 + S_n^0)} \tag{7}$$

$$\frac{Y^1}{Y^0} = \prod_{n-1}^{n} \left(\frac{Y_n^1}{Y_n^0}\right)^{\frac{1}{2}(m^1 + m^0)} \tag{8}$$

where s_n^i = the cost share for the *n*th factor during period (*i*), and m^i = the revenue share for the product (*Y*) during period (*i*).

The total factor productivity can then be computed as presented in equation (9).

$$P_t = \frac{1}{2}\sum\left(m_n^1 + m_n^0\right)Ln\left(\frac{Y^1}{Y^0}\right) - \frac{1}{2}\left(s_n^1 + s_n^0\right)Ln\left(\frac{X^1}{X^0}\right) \tag{9}$$

Specification of empirical model

Once the total factor productivity index for each period has been computed, the next step is to develop a model that accounts for changes in agricultural productivity from one time period to another. As stated earlier, total aggregate agricultural production is a function of two main set of inputs, namely conventional inputs and non-conventional inputs, which also include random variables.

$$Y = \prod_{j=1}^{m} X_j^{\beta_j} \prod_{i=1}^{n} Z_i^{\alpha_i}, \tag{10}$$

where Y = total aggregate output, X_j = the *j*th conventional input, Z_i = the *i*th non-conventional input, β_j = a production coefficient on conventional inputs, α_i = a production coefficient on non-conventional inputs.

Based on equations (1), (9) and (10), the total factor productivity can be expressed as follows:

$$P_t = \frac{Y_t}{\prod_{j=1}^{m} X_j^{\beta_j}} = \prod_{z=i}^{m} Z^{\alpha_i} \tag{11}$$

Equation (11) implies that, as any changes in the conventional inputs are accounted for when computing the divisia index (as it has described earlier), any variations in the total factor productivity are therefore accounted for by non-conventional inputs and random factors.

Explanatory variables that constitute the vector (Z) are based on the researcher's understanding of the industry being studied and other previous studies. In this study the factors were narrowed down to include annual expenditures on crop research and extension, the educational level of farmers (*Ed*), which was used as a proxy for human capital, and the weather (*W*) for which annual rainfall was used as a proxy. Real lagged export earnings (*Exp*) were also included in the model to represent the capacity of the economy to import a wide range of agricultural inputs. All these variables are specified in equation (12).

$$P_t = \prod_{i=1}^{n} Z^{\alpha_i} = \prod_{i=1}^{n}\left(R_{t-1}^{\alpha_{1i}}, Ex_{t-j}^{\alpha_{2j}}, Ed_t^{\alpha_{3t}}, W_t^{\alpha_{4t}}, Exp_{t-1}^{\alpha_{5t}} e^{\mu_t}\right) \tag{12}$$

For the purpose of estimation, equation (12) is transformed into logarithmic form as presented in equation (13).

$$LnP_t = \alpha_0 + \sum_{i=1}^{n}\alpha_{1i} LnR_{t-i} + \sum_{j=1}^{m}\alpha_{2j} LnEx_{t-j} + \alpha_3 LnEd_t + \alpha_4 LnW_t + \alpha_5 LnExp_{t-1} + U_t \tag{13}$$

where P_t = a divisia productivity indices for year *t*, R_{t-i} = real expenditures during year $(t-i)$ for $j = 1, 2 \ldots n$, Ex_{t-j} = real expenditures on extension during year $(t-j)$, for $j = 1, 2 \ldots\ldots m$, Ed_t = farmers' educational level during year *t*, W_t = the weather (using annual rainfall as a proxy) during year *t*, Exp_{t-1} = lagged export earnings, α_{1i} = a partial production coefficient on research expenditures, α_{2j} = a partial production coefficient on extension expenditures, $\alpha_{3\&4}$ = are parameter estimates for educational level and weather, U_t = the error term. All other variables are as previously defined.

Using the Almon lag as defined by Havlicek (1975), the lag structure of expenditures on crop research and extension in Tanzania was estimated at 4. Meanwhile, lagged extension expenditures were included in the model because it has been shown that most farmers adopt new technologies after learning from their neighbours and friends who are innovators or early adopters (Lupanga, 1986). However, research expenditure lagged for one and two periods and the linear term for extension expenditures were dropped due to autocorrelation. To further minimize problems arising due to autocorrelation, all interaction parameter estimates except that for research and extension, and some of the quadratic terms are assured to be zero. The interaction term between research and extension was assumed to be multiplicative. The complete model that was used for estimation is specified under equation (14). Estimation was done by ordinary least squares using SAS software.

$$LnP_t = \alpha_0 + \alpha_1 LnRt_{t-3} + \alpha_2 LnR_{t-4} + \alpha_3 LnEx_{t-1} + \alpha_4 LnEd_t + \alpha_5 LnW_t + \alpha_6 LnExp_{t-1} + \alpha_7 Ln\frac{R_{t-3}^2}{2} + \alpha_8 Ln\frac{R_{t-4}^2}{2} + \alpha_9 Ln\frac{Ex_t^2}{2} + \alpha_{10}\frac{W_t^2}{2} + \alpha_{11} Ln\frac{Exp_{t-1}^2}{2} + \alpha_{12} T_t + U_t \qquad (14)$$

Where the interaction term and all other variables are as previously defined.

References

Akino, M. and Hayami, Y. (1975) Efficiency and equity in public research: rice breeding in Japan's economic development. *American Journal of Agricultural Economics* 1975, 1–10.

Alston, J.M., Norton, G.W. and Pardley, P.G. (1995) *Science under Scarcity and Practice: Agricultural Research Evaluation and Priority Setting.* Cornell University Press, Ithaca, New York.

Ball, E.V. (1985) Output, input and productivity measurement in U.S. agriculture, 1948–79. *American Journal of Agricultural Economics* 67(3), 476–485.

Davis, J. (1981) The relationship between the economic surplus and production function approaches for estimating ex-post returns to agricultural research. *Review of Marketing and Agricultural Economics* 49(2), 95–105.

Echeverria, R.G. (1990) Assessing the impact of agricultural research. In: Echeverria, R. (ed.) *Methodology for Diagnosing Research Systems Constraints and Assessing the Impact of Agricultural Research*, Vol. II, *Assessing the Impact of Agricultural Research.* ISNAR, The Hague, pp. 1–31.

Evenson, R.E. (1967) The contribution of agricultural research to production. *Journal of Farm Economics* 49(1;1), 1415–1425.

Evenson, R.E. (1968) The contribution of agricultural research and extension to agricultural productivity. PhD thesis, University of Chicago, Illinois.

Falk, P. (1992) The contribution of training to the improved agricultural production of smallholder farmers in Sub-Saharan Africa. In: *Proceedings of an international conference on Agricultural Research, Training and Technology Transfer in the Southern Highlands of Tanzania: Past Achievements and Future Prospects.* 5th–9th October 1992, Mbeya, Tanzania.

Griliches, Z. (1963) The sources of measured productivity growth: United States agriculture 1940–1960. *Journal of Political Economy* 71(4), 331–346.

Griliches, Z. (1964) Research expenditures, education and aggregate production function. *The American Economic Review* 54(6), 962–974.

Hulten, C.R. (1973) Divisia Index Numbers. *Econometica* 41(6), 1017–1025.

Isinika, A.C. (1995) Assessing the effect of agricultural research expenditures on agricultural productivity in Tanzania. PhD thesis, University of Kentucky, Lexington, Kentucky.

Karanja, D.D. (1990) The rate of return to maize research in Kenya, 1955–1988. MSc thesis. Michigan State University, East Lansing, Michigan.

Knutson, M. and Tweeten, L. (1979) Towards an optimal rate of growth in agricultural production research and extension. *American Journal of Agricultural Economics* 70–76.

Lu, Y., Quance, L. and Liu, C. (1978) Projecting agricultural productivity and its economic impact. *American Journal of Agricultural Economics* 73(4), 1117–1125.

Lupanga, I.J. (1986) Linking research and extension in Tanzania: some communication behavior, dissemination, evaluation, linkage, institutional variables. PhD thesis, Cornell University, Ithaca, New York.

Romano, L. (1987) Economic evaluation of the Colombian agricultural research system. PhD thesis, Oklahoma State University, Stillwater, Oklahoma.

Scobie, G.M., Mullen, D. and Alston, J.M. (1991) The returns to investment in Australian wool production. *Australian Journal of Agricultural Economics* 35(2), 179–195.

Wise, W.S. (1986) The calculation of rates of return on agricultural research from production functions. *Journal of Agricultural Economics*, 151–161.

13 Impact Assessment of Sorghum Research in Zambia

MEDSON CHISI

Abstract: The resources for agricultural and natural resources research have declined in Zambia. Therefore, there is a need to generate evidence on the rate of return (ROR) on past investments in sorghum research. Zambia's sorghum research programme has released seven widely adapted varieties and management practices. The objective of this study is to assess the socio-economic impact of sorghum research in Zambia from 1983 to 2010 and to assist in identifying priority areas for future research. A multistage sampling procedure was used in the study and a total of 278 farmers comprising 241 small-scale, 25 medium-scale and 12 commercial farmers was interviewed. A surplus approach was used to estimate the ROR to research and development investment. The Akino–Hayami and simple benefit–cost analysis techniques were used to estimate the ROR to investment. The estimated adoption rate in the sorghum producing sector in Zambia was 33.5%. Farmers identified early maturity, drought tolerance and high yield as the most important attributes of improved varieties. About half of the respondents cited a lack of improved seed and information as reasons for non-adoption. The estimated ROR for investment in sorghum research and complementary services ranged from 12% to 19%, depending on the future adoption path.

Introduction

Although maize is the preferred food in Zambia, sorghum is of key importance to many Zambian households. The majority of the poorest farmers in Zambia are dependent on sorghum production for their subsistence. The bulk of the sorghum crop is retained by the household that produces it, and little is marketed through official channels. There is a preference, particularly among rural people, for sorghum and small grains to be used for brewing (Mckenzie, 1993). In parts of the Southern, Eastern, Copperbelt, Northern and Northwestern provinces, sorghum contributes substantially to the total calorie intake.

Social and political developments of the past four decades have marginalized sorghum. Therefore, the Government of the Republic of Zambia (GRZ) initiated a sorghum/millet improvement programme in 1983, with the support of the Swedish International Development Authority. The programme began with three professional staff to improve sorghum, pearl millet and finger millet by using improved germplasm from all over the world. The

 Impact of Science on African Agriculture and Food Security (eds P. Anandajayasekeram, M. Rukuni, S. Babu, F. Liebenberg and C.L. Keswani)

programme has developed a number of widely adapted varieties and hybrids, for different purposes in a short period of time. At present, sorghum is considered as one of the most viable alternatives for the crop diversification programme both by farmers and the GRZ (Chisi *et al.*, 1997). There is also considerable export potential.

Since 1987 the Zambian sorghum research programme has released seven widely adapted varieties and pre-released five new varieties (Table 13.1). These improved materials are being distributed to the farmers in various agro-ecological regions by the research services, Zambia Seed Company, as well as the various non-governmental organizations (NGOs) working throughout the country. No systematic information has been collected to assess the performance and current status of these varieties.

Table 13.1. Varieties and hybrids released and pre-released by the Zambian sorghum improvement programme, 1987–1995.

Name	Type	Status	Main attributes
White grain varieties			
Kuyuma (WSV-387)	V	Pre-released in 1987 and released in 1989	Early, short, white grain variety with tan plants and excellent milling properties; widely adapted to low rainfall areas with good resistance to all important diseases in Zambia, except for anthracnose
Sima (WSV-187)	V	Pre-released in 1987 and released in 1989	A medium-tall, medium-late, white grain variety with juicy and sweet stocks for dual purposes – grain and forage; well adapted to all three agro-climatic regions of Zambia and moderate resistance to all diseases
WSH-287	H	Released in 1987 but out of production	A widely adapted white grain, early hybrid with resistance to drought. Requires stagger planting of male and female parents in seed production during the main season
ZSV-12	V	Released in 1995	A semi-photoperiod sensitive, pigmented white grain variety with good resistance to soil acidity and anthracnose; well adapted to high rainfall areas
MMSH-928	H	Pre-released since 1993	A tan, short, early white grain hybrid with moderate yield potential for drought-prone areas
MMSH-1324	H	Pre-released in 1995	A tan, early, medium-height, white grain hybrid with good grain quality on tan plant and high yield; widely adapted to dry areas; resistance to all diseases except for anthracnose
MMSH-1256	H	Pre-released in 1995	A widely adapted high potential tan, medium-tall, medium-maturity, white grain hybrid with wide adaptation and good levels of resistance to most diseases, recommended throughout the country
ZSV-15	V	Pre-released in 1995	A widely adapted white grain variety of excellent grain quality and good levels of resistance to all diseases – an improvement over Kuyuma variety for the drier areas
WP-13	V	Pre-released in 1995	A photoperiod sensitive, high potential pigmented white large-grain landrace variety with good resistance to soil acidity and adaptation to high rainfall areas

(*Continued*)

Table 13.1. *Continued.*

Name	Type	Status	Main attributes
Brown grain varieties			
MMSH-375	H	Released in 1992	Very widely adapted high potential, brown grain hybrid with moderate levels of tannin and good resistance to all diseases of Zambia except for downy mildew; recommended for entire country
MMSH-413	H	Released in 1992	A widely adapted high potential brown grain hybrid with higher tannin content and excellent malting properties; resistant to most diseases
Forage varieties			
FSH-22	H	Released in 1995	A multi-cut high tillering forage sorghum for silage, hay and green chops; low HCN content and high leaf:stem ratio

V, variety; H, hybrid. Source: Sorghum and Millet Improvement Program, Ministry of Agriculture, Food and Fisheries, 1996.

Objectives

It has been a well-known fact that resources allocated to agricultural research and development (R&D) activities have declined. Both donors and policy makers are looking for evidence of success in order to mobilize additional resources for research. Thus, the research managers decided to assess the impact of the investment in sorghum research. The objective of this study was to assess the socio-economic impact of sorghum research in Zambia, for the period 1983–2010. The specific objectives were to: describe and understand the current sorghum production practices in Zambia; collect baseline information for sorghum production; assess the socio-economic impact of sorghum research and complementary services; and assist the sorghum R&D programme in identifying priority areas for future research initiatives.

Methodology

In discussing the impact of a programme one can identify three broad measures of success. In the first category, one looks at the direct output of the activity. For example, in the case of a breeding programme, the direct product of research may be a new variety or breed. The second measure goes beyond the direct product, and studies the effect on the ultimate users in the form of increased production, increased income, etc. This is often called the people-level impact and it includes production impact, economic impact, socio-cultural impact as well as environmental impact. The third measure of impact is an intermediate type that consists of changes in organizational structures, methods of conducting scientific research, and the availability and allocation of research resources. These changes constitute what is termed as institutional impact. Most of the ongoing research and impact studies address the people-level impact.

The current impact study, because of its timing, is an ex-post assessment. This study looks at the impact of investments in sorghum research and technology transfer since 1983. The Zambian sorghum improvement programme started in 1983 with the introduction and testing of over 6000 germplasm lines (Verma and Chisi, 1996). The study covers a number of technologies that have been developed and disseminated during this period.

Effectiveness analysis

In order to assess the effectiveness of the sorghum programme, it was vital to have quantified targets to compare with the actual achievements. Unfortunately, for the sorghum research programme, broader objectives were set, but no quantified time-specific targets were defined. Therefore, in this study, attempts are made to summarize the direct products in terms of recommendations that emanated from the programme.

Efficiency analysis

Efficiency analysis compares the costs and benefits of the research programme. Conceptually benefit–cost analysis is a procedure for comparing the 'net social' benefit of a project using a common measure. There were two important aspects to this type of analysis. One was to identify and quantify the likely benefits and costs, and the other was to express the likely benefits and the anticipated costs in common units. If the benefit is greater than the cost, then one could conclude that the project was efficient. However, the relative rate of return (ROR) is the criterion, which provides information regarding the 'best' use of the resource.

As major programme output from sorghum research was technologies, i.e. varieties and management practices, it was possible to estimate benefits derived from these technologies and products. Therefore a benefit–cost approach was considered to be the most desirable approach for this study. This enabled one to calculate the ROR for the combined effects of research, extension and complementary services.

The donors and public sector in Zambia have invested in sorghum research, extension, marketing and other support institutions (R&D), in order to improve the profitability of the sorghum industry. The scarce financial and human resources invested in the sorghum industry should yield returns that are higher than those in alternative investment opportunities. In the case of sorghum in Zambia, the public sector took the lead in R&D efforts.

Rate of Return Analysis

Several methods of evaluating returns to investments in agricultural R&D are available. These methods have been categorized into two groups, ex-ante and ex-post evaluations (Schuh and Tollini, 1979). The partial equilibrium approach, based on the concept of economic surplus, is followed to analyse the welfare effects of technological change. The Akino–Hayami model (Akino and Hayami, 1975) was used to estimate the total social gains from development and adoption of the new sorghum varieties in Zambia.

Estimates of price elasticities of supply and demand for sorghum are essential for the estimation of economic surplus. Unfortunately, this type of information was not available for Zambia. Instead, this analysis assumed 0.4 for both elasticities for sorghum as in Zimbabwe, as the demand and supply for sorghum are likely to be inelastic in Zambia as well (Anandajayasekeram *et al.*, 1995). Fortunately, social benefits are not sensitive to the choice of these parameters (Akino and Hayami, 1975). As part of the sensitivity analysis in this study, the simple benefit–cost model was used to estimate the gains from research. Any food gap in Zambia was likely to be closed by maize imports, rather than sorghum imports, because of the ready availability of white maize for human consumption in the world market. Hence, the social value attached to the increased sorghum production was the cost foregone from importing an equivalent quantity of maize. In this study, sorghum was evaluated at the import price parity of maize, as calculated by Howard *et al.* (1993).

The Market for Sorghum

The total demand for sorghum in Zambia has two components: the subsistence demand by small-scale producers, and the commercial demand for food and brewing. The commercial market includes both the domestic demand as well as the export demand. The national statistics on sorghum production and sales in Zambia suggest that the equilibrium quantity and price of sorghum in the domestic market were determined along the elastic downward sloping portion of the demand curve. The sale of sorghum has been increasing since the 1990/91 season. With the introduction of the improved varieties, the ability of the small-scale farmers to produce enough food and to market the surplus will be enhanced even further. Under such conditions, this demand structure will not affect the calculation of the social benefit from research and may only affect the distribution of benefits between consumers and producers. Therefore, the Akino–Hayami model and benefit–cost model were used to assess the return to sorghum R&D in Zambia.

Data Collection

The evaluation of Zambia's sorghum R&D programme was a data-intensive process, even when a simple benefit–cost analysis approach was used. The data types varied from trial data to define the direct impact of the new technology to macro-economic data on interest rates, inflation rates and subsidies. To define the technology adoption profile in Zambia for sorghum production, a sample survey was conducted. The main aim of this survey was to describe the production practices in smallholder, medium- and large-scale commercial farming areas, with the intention of getting farmers to identify the periods during which major shifts in production practices took place over the years.

Sampling

A multistage sampling procedure was used in selecting the farmers for the adoption study. The survey covered the sorghum producing districts in all nine provinces and all three groups of farmers: small (<5 ha), medium (5–20 ha) and large-scale (>20 ha). Within the group of smallholder producers, districts were chosen to represent all three agro-ecological regions, as well as the production systems. The three production systems identified were: shifting axe and hoe, semi-permanent hoe and semi-permanent hoe and ox-plough. The 1992 agricultural and population census was used to estimate the number of farmers to be interviewed within each category. The survey covered 241 smallholder farmers, 25 medium-scale farmers and 12 large-scale commercial farmers.

The large-scale commercial farmers were chosen from a list from the farmers' union. Originally, questionnaires were sent to 25 farmers, but only three returned the completed questionnaires. Subsequently, nine farmers (from the list of 25) were interviewed. Two questionnaires, one for small- and medium-scale farmers and the other for large-scale farmers, were designed. The survey was implemented during the period of April and May 1996, when the sorghum crop was just maturing. A separate questionnaire and a guideline were prepared for obtaining information from NGOs and the Zambia Seed Company.

Calculating the Benefits

The main feature of a benefit–cost analysis framework is the specification of the 'with' and 'without' scenario. The with-R&D case corresponds to what has been observed over the years. The without-R&D case was most difficult to construct. In cases of partial adoption of new technology, the without-technology case was constructed from non-adopter production trends. The solution is usually to seek expert opinion on what the trends in yields/production would have been, if there were no R&D initiatives. Using the before-research situation would grossly overestimate the benefits.

In order to assess the economic impact of research, the technology that was made available to farmers should be defined clearly. Sorghum technology is comprised of varieties and management practices. Adoption of the crop varieties was easy to establish from seed sales, area planted and seeding rates, provided the farmers were purchasing their entire seed requirements. Under normal circumstances, incremental yield gains were larger at the early stages, but over time the marginal gains became smaller and smaller. As the frequency of varietal release becomes shorter, the steps become smaller and eventually smooth out into a continuous straight line. The beginning of each step corresponds to the time when a new variety was recommended and made available to farmers.

The data on adoption of agronomic practices was more difficult to obtain, because of the step-wise adoption pattern of small-scale producers. The survey data includes the management practices of farmers in cultivating sorghum prior to the release of new technologies. From this, it was possible to construct the without scenario. The survey on sorghum production practices and adoption trends provided the essential information required to construct the adoption profile of production practices. The without-R&D value of production (quantity by price) was deducted from the value of production with-R&D to give an estimate of the flow of R&D benefits (gross).

Calculating research and transfer costs

The third important issue was the determination of technology development and transfer costs. These costs were broken down into three categories: research costs, adoption costs and transfer costs. The costs of conducting research include staff salaries and benefits, recurrent, administration and overhead expenditures, and provisions for the depreciation of capital assets. The sorghum research in Zambia is supported by the Ministry of Agriculture, Food and Fisheries as well as donors, including the Southern African Development Community/Southern African Center for Cooperation in Agricultural and Natural Resources Research regional sorghum and millet improvement programme. The research costs include all these inputs. Adoption costs were the costs that were associated with uptake of the new technology by farmers over time. In the case of sorghum, such costs arose from the differences in the seed costs between the new and old sorghum varieties. Thus, all the extra costs of seed, fertilizer, labour, implements and chemicals were considered in calculating adoption costs. What constitutes an extra cost is dependent on the way the new technology is defined. The important consideration in this process was to ensure that such costs were not underestimated.

The technology transfer costs were mainly extension and on-farm trial costs. The on-farm trial costs were part of the sorghum research costs, as these costs were not separated in the budget. In this study, the extension costs were those incurred by public sector extension services, as the contributions of the other organizations were assumed to be relatively insignificant. An estimate of the total cost flow was obtained by adding the research, adoption and transfer costs.

Calculating the internal rate of return

The incremental net benefit flow was calculated by subtracting the total cost flow from the net benefits. The incremental net benefit flow over the years was converted into values, which could be compared by discounting or compounding. This takes into account the time value of money. The research expenditures were lagged by a period long enough for the sorghum R&D programme to produce an impact at the farm level. In this study, the lag period was 5 years. The discount rate used in this study was an approximation of the social time preference rate. This is normally taken as the rate of interest on long-term government bonds. During the period 1983–1996, the lending rate, which was used to approximate the social time preference rate, averaged about 10% (World Bank, 1994). Rates of 5 and 15% were used as part of the sensitivity analysis. The estimated internal rate of return with the current level of demand and supply as well as for a series of sensitivity tests are presented in Table 13.2.

Survey results

The adoption survey covered a total of 278 farmers in the three agro-ecological regions of Zambia. The results of the survey were:

- The national average sorghum yield was about 7 bags per ha.
- In terms of total area under sorghum production, 88% was cultivated by small-scale, 8% by medium-scale and only 4% was cultivated by large-scale commercial farmers.
- The area under improved sorghum varieties by small-scale producers was 3% in 1987 and it increased steadily to 27% by 1995. The adoption rates of improved varieties by medium- and large-scale commercial farmers were 71%, and 100%, respectively. Overall the estimated adoption rate for the sorghum-producing sector in Zambia was 33%.
- Seventy-seven per cent of the farmers interviewed indicated that they planned to increase sorghum production, and 13% indicated that they will maintain their current level of production.

Table 13.2. Estimated IRR for R&D investments in sorghum technology development and dissemination in Zambia under three diffusion scenarios: 1983–2010.

	Diffusion levels		
Different scenarios	Current level	50% diffusion ceiling	75% diffusion ceiling
Base run with current demand and supply	11.82%	15.36%	18.67%
Potential expansion in production and utilization	18.49%	21.91%	25.18%
Sensitivity test			
Shorter time period to reach adoption ceiling (2004)			
Current demand and supply	NA	16.26%	21.05%
Potential demand	NA	22.73%	27.43%
Yield gains of 50%			
Current demand and supply	17.85%	21.10%	24.23%
Potential demand	23.83%	27.12%	30.33%

- Small-scale farmers in the agro-ecological region 2 planted sorghum as a sole crop. Intercropping of sorghum was practised by small-scale farmers in agro-ecological regions 1 and 3, and especially those in region 3. Large-scale commercial farmers grow sorghum as a sole crop.
- The three most important attributes of improved varieties preferred by farmers were early maturity, drought tolerance and high yield. The most popular varieties with small-scale farmers were Kuyuma (68%) and Sima (51%). Sima is a dual purpose variety, and 96% of the sampled commercial farmers were using this variety as a silage crop.
- Forty-five per cent of the sorghum farmers reported a lack of seed as a major reason for non-adoption, while 36% reported lack of information as a constraint. According to survey results, 40% of the adopters in 1995 depended on NGOs for seed, as compared with only 15% who depended on purchased seed from different sources, including ZAMSEED. Recycling and seed retention were reported by 15% of the farmers as their source of seed in 1995.
- Some small-scale farmers with livestock in region 1 used manure in their sorghum production. The majority of small- and medium-scale farmers used little or no chemical fertilizer. The commercial farmers reported using chemical fertilizer on their sorghum fields. They used an average of 137 kg/ha as basal dressing (mainly compound D) and 110 kg/ha as top dressing. The main reason for the low use of fertilizer was the belief that fertilizer was not needed for sorghum.
- Seventy-five per cent of the farmers sampled prepared their land in time, 95% used flat seedbeds and 86% of the farmers followed the recommendations with respect to planting time and planting method. Sixty-one per cent of the farmers planted within the recommended spacing and 70% of the farmers weeded their sorghum fields at least once.

Impact Assessment Results

- The estimated ROR for investment in sorghum research and complementary services ranged between 12% and 19%, depending on the future adoption path. Assuming the current level of adoption (33%) for the entire period of analysis (1983–2010), the estimated ROR was 12%. A 50% and 75% adoption ceiling will yield ROR of 15 and 19%, respectively.
- Based on the anticipated potential level of production of 150,000 t by the year 2010, the ROR from R&D increased substantially to 18.5%, even with the most conservative estimates of diffusion. When higher levels of diffusion ceilings (50 and 75%) were considered, the ROR increased to 22% and 25%, respectively. The net present value of the annual net returns was positive at all diffusion paths, even with relatively high discount rate of 15%.
- If one assumes that the adoption ceiling will be achieved in year 2005 instead of year 2010, then the ROR for the current demand and supply situation for 50% and 75% ceilings are 16% and 21%, respectively. The RORs for the potential demand, i.e. increased level of production to 150,000 t, is in the range of 23% and 27% for the two assumed adoption ceilings. The analysis indicated that if the adoption ceiling was reached by the year 2005, the ROR would be increased slightly.
- If the yield gain were increased from 36% to 50%, then the estimated ROR for the current demand and supply situation would be 18%, 21% and 24%, respectively, for the three adoption levels. With such yield gain, the ROR increases substantially from 24% to 30% if the potential demand is realized.
- Using the simple benefit–cost method resulted in slightly lower RORs than those estimated using the Akino–Hayami model. The benefit–cost approach provides the most conservative estimates of social gains.

- The number of staff working in the sorghum research programme has increased substantially since the inception, and they were also better qualified. At the time of this study the programme had six professional staff and 12 support staff. However, the national budget for sorghum research remained the same throughout this study.
- The sorghum research programme contributed to interdisciplinary approaches to sorghum research and strengthened the planning, priority setting, and resource allocation process. A system of evaluating research projects and budgets has been initiated, and the use of the logical framework has become the central part of this system.
- A large proportion of the small-scale producers were not using any chemical fertilizer or pest control measures. In addition, fertilizer recommendations for sorghum were much lower than those for maize.
- ZAMSEED technical personnel were trained in various aspects of producing good quality seed. The infrastructure developed by the programme at Golden Valley was being used by other commodities also. In addition, some of the sorghum varieties produced in Zambia were also being used in Namibia, Botswana, Mozambique and Zimbabwe. The professional staff in the programme is also working on other related commodities. Thus, there was considerable spill-over across commodities and national boundaries. These spill-over effects are not included in the analysis.

Conclusions and Implications

Despite the success of the sorghum research programme in producing a range of new varieties for farmers, the benefits have been modest. This is particularly true of the smallholder sorghum sector in Zambia where the diffusion process is still far from optimum. There is still room for achieving higher yields, as the yields in the small-scale sector are extremely low. The reported research indicates that yield potential for the new varieties is up to 5 t/ha. If farmers can obtain only half of that potential yield, the research benefits will be tripled.

At present, the market for sorghum is limited. Additional research is needed to develop production and processing technologies in order to expand the utilization of sorghum in bakeries, breweries and livestock feed industries. Marketing, policy and utilization studies are important to provide information to private entrepreneurs and policy makers. The sorghum research team should work more closely with economists and post-harvest technologists to address these issues.

All new sorghum varieties have higher yield potentials under experimental conditions. However, these yield potentials have not yet been realized under farmers' conditions. In addition, blanket agronomic recommendations are used. Farmers in regions 1 and 3 intercrop sorghum, but there are no recommendations for intercropping. There is a need for site-specific recommendations, to reflect the resource constraints and cropping systems of the farmers. Therefore, in order to increase the productivity of sorghum, it may be more logical to focus on site-specific agronomic research, especially fertility management technologies. Given the current scenario, future breeding work on sorghum should be targeted to address the critical constraints facing the small-scale farmers. Varietal development should focus on drought tolerance, striga resistance and nitrogen use efficiency.

The performance of the seed market in providing sufficient seed, and extension efforts are critical in increasing the adoption level of the available technologies. Availability of improved seed and the lack of information were identified as the two major limiting factors for adoption of available varieties. Therefore, these two areas need immediate attention.

At present, the recurrent expenditure on the sorghum research programme is largely from the Swedish International Development Authority. The government of Zambia covers the salaries and benefits of staff, and a small proportion of the operational costs. Therefore, to sustain the programme, there is a need to increase the national contribution to sorghum research.

Limitations of the Study

The major limitations of this study are listed below:

- The benefits generated by the introduction of improved sorghum varieties in Zambia proved to be difficult to estimate because of the shortage of some critical data and the uncertainties of future demand for the crop.
- The analysis revealed that there was considerable cross-commodity and cross-boundary spill-over effects arising from the sorghum R&D activities. In addition, the programme invested a substantial proportion of its budget in long-term human capital development. The benefit of such investment flows beyond the period of analysis and was difficult to estimate. Thus, the overall benefit of the R&D investment is an underestimate.
- In the recent past, substantial macro-economic policy changes have taken place in Zambia, largely through the structural adjustment programme and the agricultural sector investment programme. The effects of these changes were not explicitly included in the current analysis. Despite these limitations, the estimated ROR for sorghum research ranged from 12% to 19%.

References

Akino, M. and Hayami, T. (1975) Efficiency and equity in public research: rice breeding in Japan's economic development. *American Journal of Agricultural Economics* 57, 1–10.

Anandajayasekeram, P., Martella, D.R., Sanders, J. and Kupfuma, B. (1995) *Report on the Impact Assessment of the SADC/ICRISAT Sorghum and Millet Improvement Program*, Vol. I. SACCAR, Gaborone, Botswana.

Chisi, M., Anandajayasekeram, P., Martella, D., Ahmed, M. and Mwape, M. (1997) *Impact Assessment of Sorghum Research in Zambia.* SACCAR, Gaborone, Botswana.

Howard, J.A., Chitalu, G.M. and Kalonge, S.M. (1993) *The impact of investments in maize research and dissemination in Zambia.* MSU International Development Working Paper No. 39/1, East Lansing, Michigan. Department of Agricultural Economics, Michigan State University.

Mckenzie, J. (1993) *The Potential for Sorghum in Zambia's Crop Portfolio.* Paper presented at the workshop on 'Rural Business Opportunities, Sorghum Processing and Utilisation – A New Look at Traditional Crops', June 2, Lusaka.

Schuh, G.E. and Tollini, H. (1979) *Costs and Benefits of Agricultural Research: State of the Arts.* World Bank Staff Paper No. 360. World Bank, Washington, DC.

Verma, B.N. and Chisi, M. (1996) Progress in Sorghum Breeding in Zambia, its Impact and Future Needs. A poster presentation at the International Conference on Genetic Improvement of Sorghum and Pearl Millet, Lubbock, Texas, September 23–27.

World Bank (1994) *Infrastructure for Development. World Development Indicators.* World Bank, Washington, DC.

14 The Economic Impact of Investments in Cassava Research in Uganda

A. Bua, G. Acola, R.L. Adupa, G.W. Otim-Nape, Y.K. Baguma, D. Sserunkuma, V. Manyong and O. Coulibaly

Abstract: Cassava plays a major role in both household and national food security in Uganda. However, cassava production in Uganda has been threatened by cassava mosaic disease. The government has made concerted efforts to evaluate, multiply and distribute new cassava planting materials in most districts. The new varieties outperformed the local varieties and quickly replaced them. An economic assessment of this programme using the benefit–cost approach revealed that the internal rate of return for this investment was 167%. This high rate of return can be partially explained by lack of data on some costs incurred by other development partners. The success of this investment was due to the joint contribution of complementary organizations. The biggest lesson learned is that it is important to strengthen the linkages between the research programme and other government institutions especially training, marketing, infrastructure development and political institutions.

Background

Cassava is now Africa's second most imported food crop (Nweke *et al.*, 2002). Since cassava was introduced in Uganda in 1962, it has spread to most parts of the country. Cassava now contributes up to 60% of the basic food requirements of Uganda (Bua *et al.*, 1991). Cassava plays a major role in the household and national food security. The key factor encouraging the spread of cassava is its relatively high productivity per unit of land and labour compared with other food crops. Over the last decade, cassava has been recognized as one of the most important food crops in Uganda, second to bananas in terms of area cultivated, total production and per capita consumption (Vanegas and Basasha, 1991, in Otim-Nape, 1993). Analysis of cassava production and area trends indicates that except for 1985, the production and the area of the crop in Uganda has been rising from 1981 to 1989. The increase in cassava production is attributed particularly to the introduction of new high yielding varieties through the efforts of the national research institute and farmer initiatives in eastern and northern Uganda.

However, the production of cassava in Uganda declined in 1989 followed by a dramatic fall from 1992 to 1994. The causes of this decline are complex and may be attributed to some or all of the following: poor extension services, acute shortages of agricultural

 Impact of Science on African Agriculture and Food Security (eds P. Anandajayasekeram, M. Rukuni, S. Babu, F. Liebenberg and C.L. Keswani)

inputs, political and socio-economic strifes that befell the country and the pandemic of cassava mosaic disease (CMD) (Ocitti p'Obwoya and Otim-Nape, 1986; Bua, 1989; Otim-Nape and Zziwa, 1990). The most pressing issue, however, is the prevalence of CMD that threatens food security. Comprehensive surveys conducted in 1990–1992 and again in 1994 in all cassava growing districts revealed that CMD had spread throughout the country (Otim-Nape *et al.*, 1997a). By 1992, the average incidence of CMD in Uganda was 56% and in 1994 this had reached a peak of 97% in the districts of Lira, Apac and Rakai as compared with a national average of 65%. This coincided with average disease severity of 3.4 on a scale of 1–5. Some of the most susceptible cassava cultivars have been virtually eliminated since the outset of a severe epidemic of CMD in Uganda in the 1980s (Jameson, 1964).

The overall incidence and severity of CMD have been changing rapidly in many parts of the country. The severity of CMD ranges between 2.5 in Soroti and 4.2 in Kiboga and Mubende districts. This emphasizes that the disease is widespread and that none of the widely grown local varieties at the time was highly resistant. Moreover, over 500 local cassava genotypes are threatened with extinction and special conservation measures have been required to protect them.

The prevalence of CMD in many districts confirms the serious problem posed by the disease in Uganda, as it can cause up to 80–100% yield loss in individual plants depending on the sensitivity of the variety, the stage when infection occurs and symptom severity (Otim-Nape *et al.*, 1997a). By the beginning of 1998, the epidemic had spread far southwards along a broad front at a rate of 15–20 km/year (Otim-Nape *et al.*, 1997b). Consequently, the annual losses due to CMD were estimated as 60,000 ha of cassava, which was equivalent to 600,000 t worth US$60 million (Otim-Nape *et al.*, 1997b). This signifies the economic importance of CMD in Uganda and the benefits that can be derived from increased research and extension efforts to achieve more effective disease control.

Objectives

The broad objective of this study is to evaluate the importance of cassava research and development in generating cassava technologies that increase cassava production. Specifically, the study was designed to document approaches used to control CMD and compare the costs and benefits of implementing cassava research and development programmes.

Adoption of cassava varieties

The term 'adoption' means the extent to which a new technology is used by the farmer at one point in time, while diffusion is the spread of new technology across a population over time. This was the premise assumed for the adoption and diffusion of new cassava varieties in Uganda. Increasingly, adoption of technological innovations in agriculture has attracted considerable attention among development economists because the majority of the population of developing countries derives its livelihood from agricultural production. Many argue that new technology offers an opportunity to substantially increase production and income of the farmers. Indeed, it was expected that the adoption of cassava varieties could increase cassava production and subsequently increase average farm incomes. One of the objectives of this study is to evaluate the diffusion and adoption patterns of introduced cassava varieties and to assess the impact of these technologies on the production and consumption of cassava.

Cassava yields

Cassava yields were estimated by the crop cutting method in plots of 5 m × 5 m. The new varieties outperformed the local ones. However, there are clear differences in the yield gaps between locations probably due to the prevalence of CMD and other factors. However, the yield gaps between new and local cassava varieties reveal that the new varieties could improve food security and the welfare of farmers in various districts.

Adoption of Local and Improved Cassava Varieties

There was a slight expansion of the use of clean local varieties from 1997 to 1999. This is probably because local varieties are good only where CMD was not an acute problem. The uptake of these varieties demonstrates that the distribution of clean planting material of local varieties provides a short-term solution to cassava production in Uganda (Bua *et al.*, 1997a,b, Bua, 1998).

Investments in cassava R&D

The adoption and spread of new cassava varieties on such a scale contributed to increased household and national food security and improved incomes for cassava farmers. With such success, there was a need to evaluate the economic returns on the investments. Decision makers require information on research payoffs in order to assess alternative uses of public funds. Anandajayasekeram *et al.* (1993) note that to maximize the social value of project outputs, resources should be allocated where their benefit is greatest at the margin of expenditure. In this section we shall analyse the returns to investment in research, development and transfer of cassava technologies designed to increase cassava production in Uganda in the face of the CMD epidemic.

Data Source and Methodology

Data and information used in this study were assembled from primary and secondary sources. Data on yield, production costs and farm gate prices of cassava from 1993 to 1999 were obtained using standard questionnaires that were administered in the 1995, 1996, 1997 and 1999 surveys. The data for yield, production cost and farm gate prices for 1990–1992 were obtained from annual crop returns from district extension staff that were part of cassava research and development teams in Uganda. In this analysis, estimates of annual labour inputs were obtained by interviewing group and individual farmers in different districts on the costs of producing 1 ha of cassava in one cropping season for the period of 1993–1999. The imputed labour costs (in terms of Uganda shillings/ha, using local wage rates and standard estimates of labour on per ha basis) covering the period from 1990 to 1992 were obtained from the annual crop returns from district extension staff. The average rural wage rates were used in the calculation, recognizing that the opportunity cost of labour in Uganda is low due to limited or lack of labour market in the region.

Owing to communal land tenure systems in most of the districts surveyed and abundant land availability in some districts, no opportunity cost was attributed to land. In the same way the costs of planting material were considered to have negligible opportunity cost among farmers. However, the cost of planting material, multiplication and distribution incurred by various organizations that implemented development and transfer of cassava technologies was included as input cost in the benefit–cost analysis, but considered as a non-traded good.

Information on the area under new cassava varieties was obtained from annual crop returns from district extension services and NGOs who closely monitored, recorded and reported the spread of the new varieties to the National Cassava Programme. The costs of research and dissemination were assembled from reports of the various institutions and organizations that participated in the research, development and transfer of cassava technology in Uganda. Meanwhile, the consumer price indices used to deflate the total project cost and the gross benefits of the projects were obtained from the Ministry of Finance, Planning and Economic Development. The 1989 consumer price index was used as a base year because the Uganda government considered it to be a normal year when the economy was stable.

The benefit–cost approach (Gittinger, 1982) was used to estimate the rate of return for the R&D investments. The benefits of the research and dissemination programmes are approximated by the actual net value of production (net of research, adoption and production costs) from introduced cassava varieties. In most districts studied, cassava was completely destroyed by CMD before research and development interventions were instituted. Cassava was considered as a non-traded good, thus the average annual farm gate prices that vary across years were used for the analysis. Moreover, the majority of the farmers interviewed market their cassava in the form of fresh roots directly from the field mostly to other farmers (consumers) for domestic consumption. Research, adoption and production expenditures are treated as investments from which the internal rate of return on investment was estimated. The flow of cassava production is treated as the benefits stream.

Results

The internal rate of return to research in the development and transfer of cassava technologies in Uganda as a whole has been estimated at 167%. The investments resulted in a benefit–cost ratio of 18.2. These results suggest that investment in cassava research and development has been successful and profitable to the cassava producers and the society. The high internal rate of return to investment in cassava research and development may be partly explained by a lack of data on some costs of research and development incurred by other organizations and donors such as the International Development Research Centre (IDRC) contribution in the mid-1980s. However, the fundamental reason for the high internal rate of return is the large yield gap (between the 'with' and 'without' scenario), which narrowed when new cassava varieties are grown.

Implications

The results of the study imply that cassava research and development efforts were profitable to Uganda. It is clear that the success of cassava participatory research approaches in development, transfer and adoption of cassava technologies was due to the joint contribution of these complementary organizations. This implies that when research develops better varieties, there is a need to have efficient testing and transfer systems to enhance adoption and diffusion. Despite some limitations the general results hold that investment in cassava research and development in Uganda has yielded a large, positive internal rate of return.

Lessons learned and recommendations

Several lessons about agricultural research can be drawn from this study. The first lesson is that a profitable technology resulting from high yields and profits can improve agricultural productivity.

Second, there is a need to have complementary organizations to facilitate adoption. Third, research and development need to focus on critical issues in order to ensure that scarce resources are spent on priority problem areas. Fourth, the breeding programme should focus on diversification of cassava varieties with diverse attributes to reduce production risks and increase profitability on a recurring basis.

Acknowledgement

The authors wish to acknowledge the financial contribution of Inernational Institute of Tropical Agriculture (IITA) and the assistance of Victor Manyong and Ousman Coulibaly. We wish to extend our unreserved thanks to all the Uganda Cassava Program staff, particularly Titus Alicai and William Sserubombwe for their contribution to data collection, collation and preliminary analysis.

References

Anandajayasekeram, P., Rose, R.N. and Holland, P. (1993) *Applications of Ex Ante Evaluation Procedures to Research.* Australian Bureau of Agricultural and Resource Economics, Canberra.

Bua, A. (1989) Implementation of the farming systems research (FSR) methodology: the case of root crops-based farming systems in Lira and Apac districts, Northern Uganda. MSc thesis, University of East Anglia, Norwich, UK.

Bua, A. (1998) Evaluation of participatory research approaches in the development, transfer and adoption of cassava technologies in Uganda. PhD thesis, Wye College, University of London, London.

Bua, A., Otim-Nape, G.W., Ebyau, J. and Masembe, W.K. (1991) *Cassava in Uganda. A Preliminary Report of the Collaborative Study of Cassava in Africa* (COSCA). Uganda Root Crops Programme, Namulonge Research Station, Kampala.

Bua, A., Otim-Nape, G.W., Acola, G. and Baguma, Y.K. (1997a) The approaches adopted for an impact of cassava multiplication in Uganda. In: *Progress in Cassava Technology Transfer in Uganda. Proceedings of the National Workshop on Cassava Multiplication.* Masindi, 9–12 January, 1996. Natural Resources Institute, NARO and Chatham, UK.

Bua, A., Otim-Nape, G.W., Byabakama, B., Acola, G. and Baguma, Y.K. (1997b) The uptake of improved varieties in the six Gatsby Project Districts. In: *Progress in Cassava Technology Transfer in Uganda. Proceedings of the National Workshop on Cassava Multiplication.* Masindi, 9–12 January, 1996. Natural Resources Institute, NARO and Chatham, UK.

Gittinger, J.P. (1982) *Economic Analysis of Agricultural Projects.* John Hopkins Press, Baltimore, Maryland.

Jameson, J.D. (1964) Cassava mosaic disease in Uganda. *East African Agricultural and Forestry Journal* 29, 208–213.

Nweke, F.I., Spencer, D.S.C. and Lynam, J.K. (eds.) (2002) *The Cassava Transformation: Africa's Best Kept Secret.* Michigan State University Press, East Lansing, Michigan.

Ocitti p'Obwoya, C.N. and Otim-Nape, G.W. (1986) *Report on On-Farm Survey of Root Crops-Based Farming System in Soroti, Kumi and Tororo Districts, Eastern Uganda.* Uganda National Root Crops Improvement Programme, Soroti. Serere Research Station.

Otim-Nape, G.W. (1993) Epidemiology of the African Cassava Mosaic Geminivirus Disease (CMD) in Uganda. PhD thesis, University of Reading, Reading, UK.

Otim-Nape, G.W. and Zziwa, S. (1990) *Cassava as a Major Staple Food Crop in Uganda.* Report of Phase I of Collaborative Study of Cassava in Africa (COSCA). COSCA Working Paper No. 3. Collaborative Study of Cassava in Africa. IITA, Ibadan, Nigeria.

Otim-Nape, G.W., Thresh, J.M. and Shaw, M.W. (1997a) The incidence of severity of cassava virus disease in Uganda: 1990–92. *Tropical Science* 37, 1–7.

Otim-Nape, G.W., Bua, A., Thresh, J.M., Baguma, Y.K., Ogwal, S., Semakula, G.N., Acola, G., Byabakama, B. and Martin, A. (1997b) *Cassava Mosaic Virus Disease in Uganda.* Natural Resources Institute, Chatham, UK.

15 Impact Assessment of Groundnut Research in Zimbabwe

E.P. MAZHANGARA, P. ANANDAJAYASEKERAM, M. MUDHARA, D. MARTELLA AND M. MURATA

Abstract: Groundnut research in Zimbabwe dates back to the 1950s. Makulu Red is a long season variety and it was released to farmers in 1966. Following its release, a considerable amount of research was undertaken that led to the release of four improved long season varieties, five short season varieties and one medium season variety. The objective of this study was to assess the impact of groundnut research and extension in Zimbabwe over the period 1966–2000. The specific objectives were to describe and understand the groundnut production system, assess the socio-economic impact of groundnut research and complementary services. The study utilized primary and secondary data. A simple benefit–cost analysis approach (a derivative of the surplus approach) was used to calculate the rate of return to groundnut R&D investments. In addition to the rate of return analysis, a qualitative assessment of institutional, environmental and socio-cultural impacts of groundnut R&D investment was carried out.

Background

The decline in public and donor funding for agricultural research since the mid-1980s has culminated in increased pressure for greater accountability and improved management of Agricultural and Natural Resources (ANR) research resources. Beginning in 1993, the Southern African Center for Cooperation in Agricultural and Natural Resources Research initiated a set of activities related to impact assessment. This was based on the realization that there was a need for developing appropriate methodologies, the capacity and means of implementing impact evaluation and agricultural policy assessment both at the national and regional level.

Objectives

The general objective of this study was to assess the impact of groundnut research, extension and the enabling environment in Zimbabwe from 1975 to 1996. The specific objectives were: to describe and understand the existing groundnut production systems in Zimbabwe; to trace the evolution of research and enabling environment with respect to groundnut technology development and transfer; to assess the impacts of groundnut research and enabling environment; to conduct preliminary analysis of the effects of the recent liberalization of the groundnut industry; and to assess policy implications for future investments in research and enabling environments.

Groundnuts in Zimbabwe

Groundnuts were a major commercial crop in Zimbabwe prior to the UDI in 1965 as a result of the formation of the groundnut processing industry, which acted as a sink for all groundnuts produced in the country. Subsequently, the crop found its way into the smallholder production sector and became a popular crop. The *Cercospora* scourge of 1966–1972 and the emergence of substitute crops in the oilseed industry led to the decline in the production of groundnuts. Since then, it has become more of a smallholder crop than a commercial crop.

The major groundnut growing area in Zimbabwe is Natural Region (NR) II due to its high rainfall, followed by NR III. A number of players are involved in the groundnut industry of Zimbabwe (see Fig. 15.1). Both the large-scale commercial and the smallholder farmers are involved in the production of groundnuts. Smallholder farmers include communal area farmers, small-scale commercial farmers, and resettlement farmers. Previously, two farmers unions represented crop producers, i.e. the Commercial Farmers Union through the Commercial Oilseed Producers Association representing the large-scale commercial farmers, and the Zimbabwe Farmers Union representing the smallholder farmers. In recent years, a third farmers union, Indigenous Commercial Farmers Union, was formed and has been approved by government

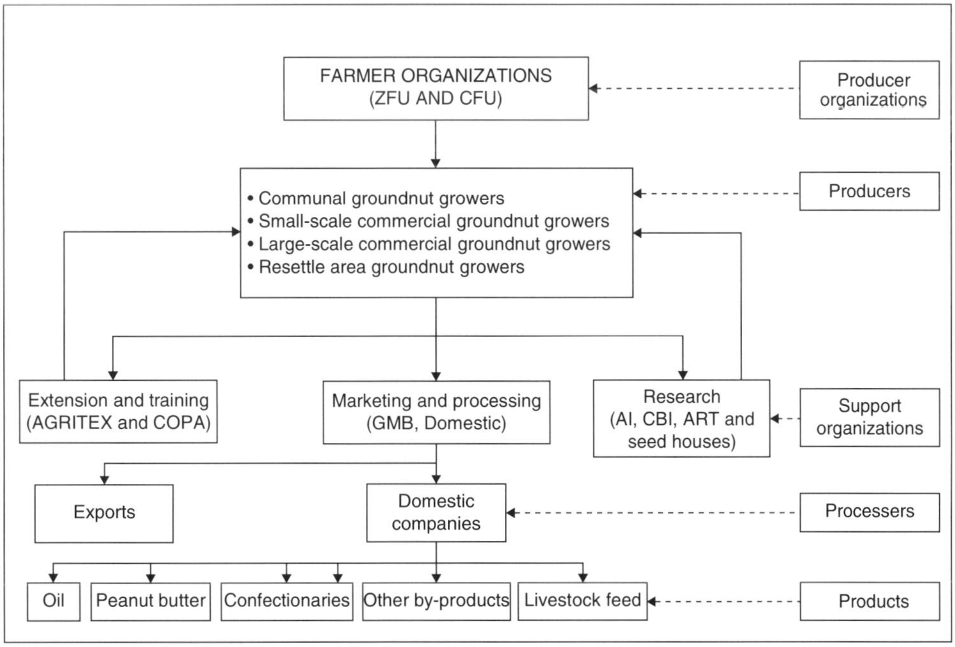

Fig. 15.1. Groundnut industry in Zimbabwe.

to represent the interests of the emerging black large-scale commercial farmers. The majority of these farmers grow tobacco and maize and very few, if any, grow groundnuts commercially.

Groundnuts are produced by nearly all farmers in NRs I, II and III, and by a significant number of farmers in NRs IV and V. Johnson *et al.* (1994) observed that the yield levels have been declining for the following reasons:

- Lack of suitable seed for farmers in this sector. Smallholder farmers use retained groundnut seed, which has been recycled for many years.
- The price of seed for the new varieties is too high.
- Leaf spot cercospora disease.
- Low soil calcium and sulphur levels, and low pH levels in NR II.
- High incidence of the aflatoxin disease.

History of groundnut research

Groundnut research is conducted by both public and private research organizations. Among the private research organizations are: the Agricultural Research Trust (ART), which works mainly on agronomic aspects; the various seed firms that do breeding work; and the universities that mainly conduct agronomy research. Public research on groundnut in Zimbabwe is conducted by three institutions: Agronomy Institute for crop management, Crop Breeding Institute for breeding, and Plant Protection Research Institute for plant health aspects. Public research on groundnuts dates back to the 1940s. Groundnut research, which led to the release of Makulu Red in the late-1950s, contributed significantly to the growth of groundnut production in the commercial sector. Between 1960 and 1972, no new varieties were released. Groundnut yields suffered considerably from cercospora leaf spot disease. Commercial farmers affected by cercospora shifted into producing cotton, wheat and tobacco, which were profitable and enjoyed new production technologies and crop protection techniques. It was not until the 1970s that research managed to address the problems in producing resistant varieties and better disease control measures suitable for the commercial farmers. The varieties being marketed by the Seed Company (formerly the Seed Cooperative) are Natal Common, Valencia R2, Plover (short season), Swallow (medium season), Makulu Red, Egret and Flamingo (long season). Natal Common is an old variety that has been grown for many years. Valencia R2 and Plover originated from South America, while Makulu Red originated from Zambia. Makulu Red was introduced in Zimbabwe around 1960 and released in 1966. Swallow and Flamingo were bred and released in Zimbabwe, while the variety Egret was released from local selections. Research has also been successfully done on important groundnut pests and diseases.

Methodology

Social benefit–cost analysis

Calculation of the rate of return (ROR) involves the estimation of benefits and the costs of research and transfer of technology to society. In this analysis, any cost or benefit accrued to individuals or societies outside Zimbabwe is not included. The prices and costs used in the analysis are opportunity costs and prices. An extreme case of the surplus approach, the benefit–cost framework is used in the analysis. The advantage of this method is that it does not require any elasticity measures.

Calculating the gross benefits

The main features of a benefit–cost analysis framework is the specification of the 'with' and 'without' scenarios. The 'with'-R&D case corresponds to what has been observed over the years. The 'without'-R&D case is difficult to define. In this study, the mean average yield over the period 1965–1969 is used to define the 'without' scenario. Using the before-research situation would grossly overestimate the benefits. In order to assess the economic impact of research, the technology that was made available to the farmers during the period of analysis should be clearly defined. In this case, several varieties and agronomic recommendations have been generated and disseminated to the farmers. What is being measured is the combined effect of these technologies.

Calculating the research and transfer costs

The technology development and transfer costs can be broken down into research costs, adoption costs and transfer costs. The cost of conducting research includes staff salaries and benefits, recurrent expenditures, administration and overhead expenditures, and provision for the depreciation of capital assets. Adoption costs are associated with the uptake of the new technology by farmers over time. Changes in the production practices involve changes in input use (including varietal change), such as fertilizers, herbicides, insecticides, equipment/implements, etc. Thus, all extra costs involved in using the technology should be considered in calculating adoption costs. The important consideration is that all costs are included, ensuring that such costs are not underestimated.

The technology transfer costs are mainly the extension costs, including the on-farm research costs. If chemical companies, seed companies and NGOs are involved in technology transfer, their costs should be included. The incremental net benefit flow is calculated by subtracting the total cost flow from the incremental gross benefits. The incremental net benefit flow over the years is converted into values that can be compared by discounting and/or compounding. This takes into account the time value of money. The research expenditure costs are lagged by a period of 5 years in this study.

The discount rate used in this study is an approximation of the social time preference rate. This is normally taken as the real interest rate on long-term government bonds. During the period 1983–1996, the lending rate averaged about 10% (World Bank, 1994). Rates of 5% and 15% are used as part of sensitivity analysis. The summation of the discounted incremental net benefits yields the net present value (NPV) of groundnut R&D, which indicates positive impact when it is greater than zero. Of interest is the discount rate that reduces the sum of discounted incremental net benefits to zero. This discount rate is the average internal rate of return (IRR) of groundnut R&D. When this rate is greater than the social time preference rate, then past investments in groundnut R&D have been socially profitable.

Data Collection

Primary data on the current management practices and the adoption of various technological components over time were collected using a questionnaire. Personal interviews were conducted to collect the information from communal farmers and a mail survey was used to collect data from the commercial farmers. Information on household characteristics, farm characteristics, details on groundnut cultivation practices over time and various aspects of marketing and support services were collected. In addition, various individuals involved in the groundnut R&D activities were also interviewed to obtain specific information. Cost information, evolution of the groundnut R&D programme and information on various technologies were collected from various published and unpublished reports. Information regarding the consumer price index, prices, production, acreage and yields were obtained from the Central Statistical Office (1989, 1996).

Sampling

Different procedures were used to select the sample farmers for the survey for the two sectors. In the communal areas, a multistage sample technique was used for drawing up the sample of groundnut growers. The district-level production statistics, with figures provided by the Early Warning Unit of Agricultural Technical and Extension Services (AGRITEX), were used in identifying districts producing groundnuts. Ten districts were selected from the list of districts that produced at least 5% of the total communal area production. An average production figure from two production seasons, i.e. 1994/95 and 1995/96, was used. In each selected district, two wards were randomly selected from the list of wards in that district. Ten groundnut growers were randomly selected and interviewed in each ward. Another 24 households were selected from the Hurungwe district. The final sample was composed of 62, 62 and 100 farmers in NR II, NR III and NR IV, respectively.

For the large-scale commercial sector, a sampling frame of all registered large-scale commercial groundnut growers (640) was provided by the Commercial Oilseed Producers Association. Using a systematic sampling procedure, a total of 150 farmers were selected for the mail survey. Questionnaires were sent to the 150 farmers selected. Forty-seven farmers returned the completed questionnaires. This is about 7.3% of the population and 31.3% of the sample selected.

Impact of Past Groundnut Research, Extension and Enabling Environment: 1966–2000

Past investments in research and development programmes of agricultural commodities have several impacts on society. These include changes in the enabling environment, socio-cultural impacts, environmental impact as well as production and income effect on the target groups. This section analyses the impacts associated with groundnut R&D activities in Zimbabwe.

People-level impact: economic impact

The main focus of this section is on the calculation of the ROR using a simple benefit–cost analysis that uses no elasticities. This involves estimation of benefits, and research and transfer costs.

Benefits: deriving the 'without' technology scenario for groundnuts

The first improved groundnut variety was released in 1965. Therefore, the first benefits from research would have been experienced in 1970. Some of the early varieties are still being grown. The mean of the average yields obtained during the period 1965–1969 were taken as the yields of the 'without' the technology scenario. It was also assumed that the yields would have continued at this level in the without technology phase.

The difference between the average base yield and the actual yield at time *t* gave the incremental yield. In other words:

$$Y_{\mathrm{It}} = Y_{\mathrm{At}} - Y_{\mathrm{Wt}}$$

Where:
Y_{It} = incremental yield at time *t*
Y_{At} = actual yield at time *t*
Y_{Wt} = base yield 'without' R&D yield at time *t*

To get the incremental production, the incremental yield was multiplied by the area in time *t*. This was multiplied by the price to give the incremental value of production.

Calculating research and transfer costs

Research and transfer costs can be broken down into research, development, adoption and transfer costs. Research development includes salaries and benefits, recurrent costs, overhead and administration, and depreciation. For groundnuts, research costs were incurred by the Department of Research and Specialist Services (DR&SS) through the Agronomy Institute, Crop Breeding Institute and the Plant Protection Research Institute. Each of these institutes conducted research on a number of commodities including groundnuts. On the basis of advice from the institutes, the groundnut component for each institute was determined. Salaries and allowances for each institute are included in the DR&SS allocation. To get each institute's component of salaries and allowances, the DR&SS allocation was multiplied by the proportion of the particular institute's number of professional staff to the total number of professional staff in DR&SS.

Extension costs

AGRITEX was the organization mainly involved in transferring technologies on groundnuts to farmers. As in the case of research costs, the extension costs were not broken down by commodity. The following steps were taken in estimating the annual budgets for groundnuts extension during the period 1971–2000.

- The total AGRITEX budget was divided into livestock and crops budget using the ratio of the value of crop production or livestock production to total value of agricultural production.
- The crop budget was multiplied by the ratio of the area under groundnuts to the total area under crop in order to estimate the groundnut extension budget.

Adoption costs

The adoption percentage was multiplied by the estimate of the additional expenditure that had to be incurred when using the new technology and the total area under groundnuts. In addition it was assumed that smallholder farmers would retain some of their crop for use as seed. Therefore, 10% of the area under improved technology was taken as having been planted to fresh seed.

Estimating groundnut area and yield beyond 1996

The area planted to groundnuts beyond 1996 was assumed to continue the trend observed over the 1970–1996 period. Yields beyond 1996 were computed using a weighted yield based on the probability of three season types occurring and the related yield realization from each season type.

Considering these season types, the yield in Y_t is given by:

$$\begin{aligned} Yield_t &= Yield_E \times ((0.5) + (0.3 \times 0.6) + (0.2 \times 0.5)) \\ &= Yield_E \times (0.78) \end{aligned}$$

Where $Yield_E$ was estimated from a single regression equation of yield realized over the period 1970–1976.

Payoff to investments in groundnut research and development

Subtracting the research, extension and adoption costs from the gross benefits gave the net benefits. The net benefits to groundnuts were very volatile after the initial negative flows experienced before the introduction of the technology in 1970. The best way to determine whether there was a payoff to the investments in groundnuts research and extension is by calculating the ROR, and the NPV of the net benefit stream. In the study, a social time preference rate of 10% was used for determining the NPV of the net benefit flow, and for comparison with the IRR. The ROR of 59% was well above the real interest rate. It was even comparable with the commercial bank interest rate. During the period, at a 10% discount rate, research and development into groundnuts generated Z$547.7 million. Raising the real interest rate to 15% reduced NPV to Z$254.2 million. The groundnut sector contributed immensely to the national economy.

Sensitivity analysis

As the IRR figure is calculated from estimated values, it is necessary to test the strength of the figure in the face of changes of the assumption used. In the smallholder farming sector the adoption percentage was raised from the current 45% to 75% in the forecast period 1997–2000. This resulted in the NPV falling to Z$199.3 million and Z$104.2 million at an interest rate of 10% and 15%, respectively.

Intermediate impacts

Intermediate impacts focus on the ability of the research system to efficiently produce the research output. These are changes that occur in the enabling environment of TDT, which can improve or worsen the efficiency and effectiveness of the research system. These include human resources development, availability of financial resources, changes in procedures and techniques, and the strength of linkages with the collaborating institutions.

Human resources

The groundnut research programme spans three institutes: Crop Breeding Institute, Agronomy Institute and Plant Protection Research Institute, which are headed by separate managers. The groundnut breeding and agronomy programmes have been well staffed by senior scientists. After independence in 1980 many young scientists replaced departing senior researchers. Within the Plant Protection Institute, subject matter specialists are available to work on a range of crops. In the 1990s, the institute was mainly providing advisory services to the Crop Breeding Institute and Agronomy Institute. In the 1980s, there was a scientist working 50% of their time on groundnut pathology while in the 1970s there was a full-time pathologist working on leaf spot disease.

Financial resources

The main budget for DR&SS and its institutes is from the national treasury. Over the years, the recurrent budget has been declining in real terms. This is a major cause of concern. In recent years, the contributions from non-treasury sources have been increasing. The national programme has benefited from regional cooperation with the Southern African Development Community/ International Crops Research Institute for the Semi-Arid Tropics Groundnut Program.

Limited funding and technical cooperation has been obtained through the collaboration with such regional bodies.

People-level impact: others

Socio-cultural impact

Groundnuts are regarded as a women's crop in more than 80% of smallholder households. In such households, the women work in the fields, market, and use the money to buy kitchenware and family clothing. The decline in groundnut production in the large-scale commercial farmer sector was met with the rise of production by smallholders. This has resulted in increased incomes in the smallholder sector. Peanut butter produced by the women has had a large impact on nutrition, particularly among children. Peanut butter sauce added to vegetables and consumed with main meals is a very important source of protein.

Summary Survey Results

- Groundnut production fluctuated between 1970 and 1994 because of the vagaries of the weather and the cercospora epidemic.
- The current adoption rates of new groundnut varieties by smallholders and large-scale commercial farmers are 45% and 100%, respectively.
- The most common traditional varieties grown by the communal farmers are Kasawaya, Tumbe, Kabowe, Chingungwana, Bandahwe and Chitununu. The most commonly grown improved varieties by communal farmers were Natal Common, Makulu Red and Valencia. The improved varieties grown by commercial farmers were Flamingo, Falcon, Plover and Egret.
- The average seed rate used by the communal farmers is almost half of the recommended rate.
- The commercial farmers apply fertilizer to groundnuts, whereas more than 80% of the communal farmers do not apply fertilizer either on improved or traditional varieties.
- In the communal areas, 'locals' constitute the greatest market for nuts and peanut butter. Fifty-five per cent of the commercial farmers sold groundnuts directly to the harvesters.
- Sixty-one per cent, 58% and 75% of the smallholder farmers interviewed are going to continue growing groundnuts in NR II, III, IV, respectively.
- Commercial farmers rely more on farmer-to-farmer and chemical companies for advice than on AGRITEX and/or DR&SS. AGRITEX is the major source of extension service for the communal farmers.
- Ninety-eight per cent of smallholder farmers do not use credit in producing groundnuts.

Summary Impact assessment results

- The estimated ROR for investment in groundnut research and complementary services of 59% shows that the investments were worthwhile. This rate is higher than returns from alternative investments of public funds.
- Results of environmental impact assessment of groundnut R&D indicate a positive impact. The use of groundnut as a nitrogen fixing legume has a bearing on chemical fertilizer use.

Underground water pollution is minimized as less chemicals are used. Soil nutrient mining is also reduced in groundnut-based crop rotations.

- The majority of the communal farmers are growing groundnuts, and a good portion of the production is retained for home consumption. Peanut butter is consumed in the household, and the surplus is sold.

Conclusions

The main objective of this study was to assess the impact of groundnut research and extension in Zimbabwe over the period 1960–2000. The specific objectives were to describe the current groundnut production system, and assess the socio-economic impact of groundnut research and complementary services.

Since Makulu Red variety was released to farmers in 1966, crop breeding research has released four improved long season varieties and one medium season variety. Agronomic and plant health research generated improved crop management and crop protection practices and extension services has promoted the uptake of the new groundnut technologies. Investments in groundnut research and extension in Zimbabwe yielded an impressive 59% rate of return.

An ROR of 59% compared with a social time preference rate of 10% implies investments in groundnut research and extension in Zimbabwe has generated a high pay off. There is, however, a need to develop improved seed to increase crop yield in rural areas. Research should also corroborate with seed companies and the Zimbabwe Farmers Union to facilitate this process.

Limitations of the study

This study has several limitations. The main limitation is related to the availability and quality of data used in this assessment because the groundnut R&D programmes were designed without due consideration for the need to collect information to be used for impact assessment.

- The yield levels without groundnut R&D, which are important in the calculation of benefits, had to be estimated. Given the absence of baseline data, the process was extremely difficult. Extrapolations and opinions of key informants were used to calculate the 'without' technology yield levels.
- The response from the mail survey was much lower than expected; this reduces the level of confidence in interpreting the results for the commercial farmers.
- The absence of appropriate data made it impossible to adequately assess the impact of groundnut R&D on the environment, gender, employment and food security.

References

Central Statistical Office (1989) *Statistical Year Book.* CSO, Harare.

Central Statistical Office (1996) *Quarterly Digest of Statistics.* CSO, Harare.

Johnson, P., Zitsonza, N., Sukume, C. and Rukuni, M. (1994) *The Oilseed Sub-Sector in Zimbabwe.* World Bank, Washington, DC.

16 The Rate of Return to Wine Grape Research and Technology Development in South Africa[1]

Rob Townsend and Johan van Zyl[2]

Abstract: This chapter estimates the rate of return (ROR) to investments in research and technology development in the wine grape industry in South Africa. The analysis illustrates the applied and adaptive research conducted in the industry and estimates RORs of about 40% for R&D and extension. These high RORs should provide significant incentives for further investments in R&D. Given the nature of the industry, more emphasis should be given to soliciting private funds for research activities, subsequently moving away from dependence on public funds.

Introduction

The importance of agriculture in the process of economic growth highlights the critical role of sustained advances in farm production practices by improving the quantity and quality of farm products (Alston *et al.*, 1995). In this context, investment in improved agricultural technology continues to be an important avenue, for not only government assistance to specific sectors of the economy, but also for resources from industries and farmers. However, the increased resource pressure facing the governments, donors and recipients, have emphasized the need for the assessment of potential benefits of research investments, at national and farm level, to assist in planning and management.

Based on the general global evidence that the rates of return (ROR) to agricultural R&D investments have been high, and indicative of unnecessary conservative investment, further support for agricultural research has been advocated by numerous authors. Studies on South African agriculture suggest an ROR of between 44% and 58% (Khatri *et al.*, 1995, 1996; Arnade *et al.*, 1996), which justifies increased investment in research. However, increasing monetary and fiscal pressure and demand from other sectors has resulted in strong competition for limited government funds to agriculture research. This increasingly constrained environment has placed greater emphasis on determining 'value for money in science', not only at the aggregate agricultural level, but also at the individual crop level.

Disaggregated studies concentrating on individual crops, for example wine grapes, can provide useful information to decision makers seeking to determine the optimal allocation of their scarce resources. This chapter attempts to provide empirical evidence on the value of research for wine grapes in South Africa. The wine grape crop accounts for about 14% of the

 Impact of Science on African Agriculture and Food Security (eds P. Anandajayasekeram, M. Rukuni, S. Babu, F. Liebenberg and C.L. Keswani)

value of the horticultural sector, which is currently enjoying the highest growth rate in South African agriculture. It also generates a significant amount of foreign exchange with the value of exports almost five times higher as compared with 1990. South Africa ranks eighth in the world in terms of wine grape production, producing 3.7% of world production (KWV, 1996).

Objectives

Within this context, the specific aims of this study were to evaluate and assess the impact of research and technology development in the wine grape industry in order to: determine the ROR to R&D; and make specific recommendations on funding arrangements.

The wine grape market is highly competitive, and relatively high local production costs and decreasing margins necessitate research for developing cutting-edge technology to improve productivity and maintain competitiveness in the industry. While there has been a general understanding of the importance of research, the ROR on research and technology development in the wine grape industry in South Africa is unknown, thus limiting the information for decisions on whether to increase, reallocate or maintain investment in research. This chapter concentrates on production-related R&D and does not include wine-making. The presentation is organized as follows: review of expenditures on research and extension, in the wine-making industry, methodological issues of evaluating and assessing the impact of agricultural research are then discussed; the data used in the analysis, and deviations in yield of wine grapes over time are presented; the appropriate lag structures used in the modelling are then determined, followed by calculations of the ROR to R&D. Some recommendations conclude the chapter.

Background to research and extension in the wine grape industry

Two institutions are responsible for the bulk of the R&D and extension within the South African wine grape industry. While there are no clear demarcations in their activities, both institutions being involved in R&D and extension, most of the research on wine grapes is conducted at the Nietvoorbij Institute for Viticulture and Oenology of the Agricultural Research Council (ARC), while extension is almost exclusively conducted by the 'Kooperatiewe Wynbouersvereniging' (KWV). For purposes of this chapter, these two categories, although not necessarily correct, are used as the only sources of R&D and extension relevant to the wine grape industry. This approach is not without problems. International research conducted elsewhere in the world may also impact on the South African wine grape industry. Therefore, an analysis of the returns to R&D should allow for both private spill-overs, from abroad and for the technology provided by multinational seed, chemical and machinery companies, which may or may not be performing their research in South Africa. This is usually done by using international patent counts, or other measures to construct a 'knowledge stock' variable (Khatri *et al.*, 1996). In this case, it is assumed that these effects are negligible compared with the impact of local R&D, and it is therefore ignored.

Indeed, as was noted by Griliches (1973), if agricultural inputs were supplied by a monopolist and the input statistics took proper account of quality adjustments, technical change emanating from the private sector input industries would be included in the input series. Such technological changes are in the farm inputs sector, not the farm sector itself (Kislev and Peterson, 1982), and would not present any difficulties. It is only to the extent that the input suppliers are monopolistic competitors and that the statistical sources fail to measure inputs in efficiency units, that allowance must be made for private R&D expenditures. Owing to these two factors, not all technical change in the input industries is correctly measured at source and there will be some spill-over that is caught instead in the measures of agricultural productivity. Thus, in estimating the returns to R&D, all the non-market public expenditures should be included on the cost side and some proportion of private expenditures should be added. However, in case of South Africa, and particularly wine grapes, statistics and market structure in the agricultural input industries are such that these effects can be considered negligible (see Van Zyl and Groenewald, 1988, for a discussion of market structure in input industries). Therefore, the assumption that Nietvoorbij and the KWV are the only sources of R&D and extension relevant to the wine grape industry has merit.

The increase in real expenditure on extension activities (by KWV) relative to R&D (by Nietvoorbij) can be seen clearly in Figs 16.1 and 16.2. KWV's expenditure (on extension) has the highest growth rate of the variables under consideration with an annual growth rate of 11.6%, even though this slowed towards the end to 6.9% per annum (between 1987 and 1996). In contrast, Nietvoorbij's real research expenditure has a fairly constant growth rate.

The ARC is the principal agricultural research entity in South Africa. It oversees the operations of 16 research institutes, with a network of experimental farms and a 1998 staff component of 4150, including 672 professionals. The infrastructure is spread throughout the entire country and undertakes R&D on all the major agricultural commodities in South Africa (with the exception of sugarcane). The Nietvoorbij Institute for Viticulture and Oenology is one of these research institutes, with responsibility for research on viticulture and oenology. Nietvoorbij has developed over the past 42 years into a world-renowned one-stop research facility, generating leading-edge technology for South African grape, wine and brandy industries. Research priorities are driven by producer needs and market demand. These priorities are determined in conjunction with the industries. It is situated within the major wine region, Stellenbosch, with experimental farms in Paarl, De Doorns, Robertson and Lutzville, trial

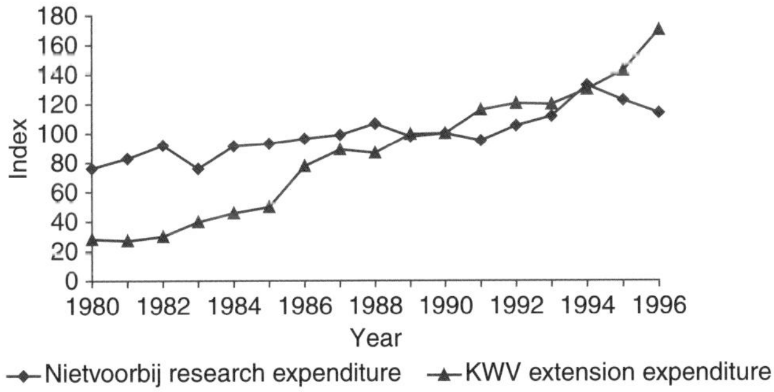

Fig. 16.1. Indices of Nietvoorbij's research expenditures and KWV's extension expenditures (real values).

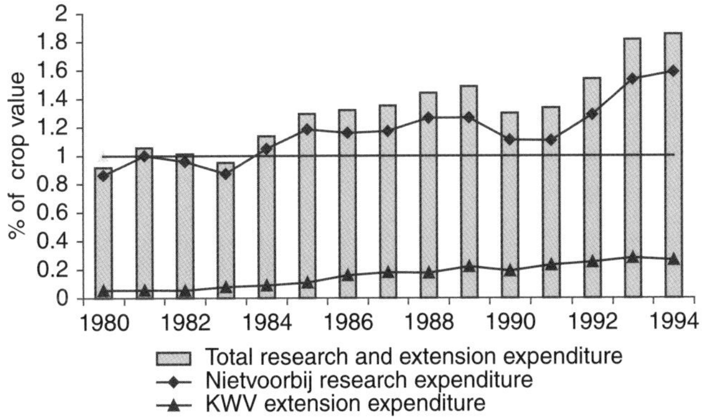

Fig. 16.2. Research and extension expenditures as a percentage of the value of the wine grape crop.

plots near Upington and co-operative trials on producer farms. ARC institutes have relatively little administrative autonomy.

Decisions regarding research priorities within the ARC are taken at the council level and funding for each institute comes via the central administration of ARC. Up to 1992, the government funded the total research effort in full. Since April 1992, commercial agriculture, associated industries and input suppliers had to pay in part for research conducted on their behalf. Presently, commercial agriculture pays for 22% of the research budget. However, the total external income budgeted for in 1997/98 amounted to 32% due to income from the sale of farm produce, services rendered and consultancies. The short-term aim is to receive 30% of total project cost from commercial agriculture. The Council receives an annual appropriation from the government, which still accounts for the bulk of its funds.

Nietvoorbij distinguishes itself for being probably the only facility in the world where all research pertaining to grapes and wine is executed under one roof. A capacity to serve small-scale farming, on a multidisciplinary basis, has also been established. Nietvoorbij has five research divisions, namely soil science, wine grapes, table and raisin grapes, plant protection and oenology. The duration of a typical project is 3–4 years. The research conducted serves the wine, table and raisin grape industries of South Africa. Nietvoorbij employs a staff of 215, of whom 30 are researchers and 34 are research technicians. The balance includes laboratory assistants (13), data personnel, public relations officers, security officers, administrative and farm personnel and general assistants (138).

The KWV was established in 1918 and represents more than 4600 wine grape producers. It produces and markets wine and related products domestically and internationally. It also provides a wide range of specialist products and services to wine producers, the wine community and consumers. As a service organization it also provides the wine industry with: extension on the growing of wine grapes; multiplication of the best plant material; and consultancy services to all producers, co-operative cellars, and the broader wine industry by specialists on wine grape production, soil science, agricultural economics and manpower management. These services are provided at cost. In addition, it publishes a specialized magazine aimed at the wine farmer.

Methodology

The majority of the empirical ROR models are based on either economic surplus calculations, following Griliches (1958), or on econometric estimation of the output elasticity of R&D derived from the production relationship. This study is based on the production approach.

In the production function approach, the R&D elasticity is derived directly from estimation of a production, cost or profit function, with R&D included as a variable, or by first constructing a total factor productivity index, changes in which are then explained by the technology-related variables. In either case, it is the effect of the technology input on output or on the productivity of the sector adopting the technology that is measured.

A simple production function model is used in this study to evaluate the return to research expenditure. The more complex and data intensive profit function could not be used due to data limitations. Both aggregate and farm management data are used in the analysis. The aggregate data were obtained from the Abstract of Agricultural Statistics, the South African Wine Industry Statistics, published by KWV, and unpublished data from KWV and the Nietvoorbij Institute for Viticulture and Oenology of the ARC. The farm surveys consisted of farm production data collected by the KWV in the Orange River, Olifants River, Klein Karoo, Malmesbury en Paarl, Robertson, Stellenbosch and Worcester regions over a 10-year period from 1987 to 1996. This provides a (7 by 10) panel of data with variations between regions and over time.

In this study, agricultural output is explained by three traditional inputs (land, labour and livestock), two modern inputs (fertilizer, pesticides and machinery), investments in infrastructure (irrigation and machinery), public R&D, extension and education, and the real protection coefficient, which represents the policy environment.

Similar to most conventional production function studies, this analysis used yield as the dependent variable and inputs and technology variables to explain yield deviations. The plot of the wine grape yield (t/ha) index showed no substantial increase over the time period under consideration. This, at first appearance, suggests that the technology effects on wine grapes have been minimal.

However, as is common to many crops, research is also focused on increasing the quality of the product, as opposed to concentrating on yield. An example of this quality-improving research conducted at Nietvoorbij is the testing of wine grape cultivation methods under different soil and climate conditions, along with identification of regions with a specific potential for the production of high-quality wine of different types. In order to capture this improvement, the yield series must be adjusted for quality changes. A quality-adjustment index was created by taking into account the increasing proportion of 'good wine' produced. This was done by dividing the value of 'good wine' by the total value of wine. An increase in this ratio is a reflection of higher quality. This index was then normalized to 1 in 1987 and multiplied by the yield to derive a quality adjusted yield index. Both yield series show substantial annual variations. This is to be expected, as weather has a major effect on yields. Data obtained from the KWV to construct a weather index support this notion. The weather index incorporates several relevant variables, including rainfall, humidity, temperature, wind, etc. These variables have been used by the KWV in the past to predict the total size of the wine grape harvest, with a high degree of reliability, and are used in a similar manner to construct the weather index. The fluctuations in the weather index are seen to be highly correlated with the yield indices, which is a reflection of the large impact weather has on wine grape yields.

Modelling the lag structure

There is usually a lag between R&D expenditures and their impact on productivity growth, and several methods have been developed to model these lag structures. Estimation of the lag coefficients of R&D, using an unrestricted functional form with many lag terms, will probably give positive and negative coefficients because of collinearity problems. However, providing that the ordinary least squares (OLS) assumptions are satisfied, the sum of the unrestricted lag coefficients should be an unbiased estimate of the total elasticity. To overcome the collinearity problem, a distributed lag structure is often assumed. This is normally an inverted V or a second degree Almon polynomial lag, which is an inverted U-shape. The polynomial lag is popular, due to its empirical simplicity, providing a smooth and feasible form. However, the specification may require restrictions, and the validity of these restrictions has been questioned, particularly end-point restrictions (Hallam, 1990). There are also suggestions that these models may lead to biased estimates of the effects of research spending. To avoid these biased results, less restrictive forms, such as the beta and gamma distributions (derived from the Pearson representation), as well as the unrestricted model, can be used.[3]

The equation used to determine the appropriate lag structure for this analysis is:

$$\ln YIELD_t = \ln \alpha_0 + \ln \alpha_1 WEATHER + \sum_{i=1}^{n} \beta_i \ln RD_{t-i} + u_t \qquad (1)$$

where *YIELD* is the tons per hectare of wine grapes, *RD* is the research expenditure and *WEATHER* is the weather index. β is the elasticity of R&D at various lag lengths, where n is the maximum lag of R&D, that affects *YIELD*. u_t is the residual, which accounts for all the deviations in yield not explained by the model. Conventional inputs, such as fertilizer, may also be expected to affect yields. However, due to a degrees of freedom problem, all the conventional inputs could not be included. Although ROR calculations in this chapter were made on the basis of estimating equations, which do not include conventional inputs (see Akgungor *et al.*, 1996 for a recent example), Alston *et al.* (1995) correctly argue that conventional inputs should normally be included. This is a potentially serious problem as it could bias the R&D elasticity upwards and inflate the ROR. To mitigate this problem panel data were used for the analysis.

Panel data estimation

Equation (2) was used to determine the lag structure, which was imposed on the data, and used in the panel estimation to calculate a ROR to R&D. This provides more degrees of freedom, and allows the conventional inputs to be taken into account, which is important as the technology developed may result in less inputs being used.

Panel data estimation has the advantage of giving ample degrees of freedom. In addition, both the cross-sectional and time-series variances help to determine the parameter estimates. The disadvantage of panel data estimation is in terms of imposing restrictions, as the three alternative models below show. The pooled OLS equation,

$$Y_{it} = \alpha + \beta X_{it} + u_{it}, \quad \text{i.e. } \alpha, \beta = \alpha_{it}, \beta_{it} \tag{2}$$

implies that both the intercepts, α, and slope coefficients, β, are the same for all wine regions (which is unrealistic given the variation between regions). Furthermore, the fixed effects model,

$$Y_{it} = \alpha_{it} + \beta X_{it} + u_{it}, \quad \text{i.e. } \alpha_{it}, \beta = \alpha_{it}, \beta_{it} \tag{3}$$

assumes that the intercepts, α_{it}, vary, but the slope coefficients, β, are the same. (This is equivalent to least squares with dummy variables.) The random effects model allows for a random region-specific effect to enter the equation, through the error term,

$$u_{it} = \mu_i + e_{it}, \quad \text{where } Var(\mu_i) = \sigma^2_\mu, Var(e_{it}) = \sigma^2_2 \tag{4}$$

It is clear that the first model (the pooled OLS equation) is not appropriate where regions differ considerably in their yield potential. However, all three of these models were estimated, using the sample survey data for 10 years. A production function was estimated with yield as the dependent variable and intermediate inputs (fertilizer, pesticides, etc.), labour, machinery, miscellaneous inputs (water, electricity, etc.), weather and R&D expenditures as independent variables. The latter two variables were assumed to be similar for all regions, while the first four variables were region specific.

Calculating the rate of return

In order to derive an ROR, the elasticities were converted to value of marginal products, using the procedure in Thirtle and Bottomley (1988) described below. In the case of the time-series analysis, each lag coefficient, β_i, is the output elasticity of R&D for that year:

$$\beta_i = \frac{\partial \ln YIELD_t}{\partial \ln RD_{t-i}} = \frac{\partial YIELD}{\partial RD_{t-i}} \cdot \frac{RD_{t-i}}{YIELD} \tag{5}$$

Thus, the marginal physical product of R&D was the elasticity multiplied by the average physical product:

$$MPP_{t-i} = \frac{\partial YIELD_t}{\partial RD_t} = \beta_i \frac{YIELD}{RD_{t-i}} \tag{6}$$

Replacing $YIELD/RD_{t-i}$ by its geometric mean and changing from continuous to discrete approximations gave:

$$\frac{\Delta OUTPUT_t}{\Delta RD_{t-i}} = \beta_i \frac{\overline{YIELD}}{\overline{RD}_{t-i}} \tag{7}$$

Then, multiplying by the increase in the value of output, divided by the change in quantity converts from output quantity to output value. Thus, the value marginal product of R&D in the period $t - i$ can then be written as:

$$VMP_{t-i} = \frac{\Delta VALUE_t}{\Delta RD_{t-i}} = \beta_i \frac{\overline{YIELD}}{\overline{RD}_{t-i}} \cdot \frac{\Delta VALUE_t}{\Delta YIELD_t} \tag{8}$$

where $YIELD/RD_{t-i}$ is an average and $\Delta VALUE_t/\Delta YIELD_t$ was calculated as the average of the last 5 years, minus the average for the first 5 years, for both variables. Thus, these were constants, but β_i varied over the lag period, giving a series of marginal returns, resulting from a unit change in R&D expenditure. The value of output, $\Delta VALUE_t/\Delta YIELD_t$ was the geometric mean calculated using the value of output at constant prices. Similarly, $YIELD/RD_{t-i}$ was a constant-price geometric average. The marginal internal ROR was calculated from:

$$\sum_{i=1}^{n} \frac{VMP_{t-i}}{(1+r)^i} - 1 = 0 \tag{9}$$

Where, n is the lag length, by solving for r (the marginal internal ROR). This methodology was applied to the panel data estimates.

Results

The polynomial lag model was estimated with no restrictions, and near end, far end and both end points restricted to equal zero. These restrictions were applied to second, third and fourth order polynomials for a range of lag lengths. The Schwarz and Akaike Criteria indicated a 4-year lead time, with a second degree polynomial with both end-point restrictions for the unadjusted yield equation. For the quality adjusted yield equation, these criteria suggested a second degree polynomial, with end-point restrictions and a lag of 7 years. Simple t-tests of the significance of alternative lags were consistent with these results.

Results of the polynomial lag distribution reveal a 4-year lead time for research expenditure to impact on unadjusted yields. When the yields were adjusted for quality, the lag structure changed and there was no lead time. With research already having an impact in the current year, this could reflect the value of the more direct impact of improving quality of the crop, as opposed to improving only yields. As mentioned earlier, Nietvoorbij had specific projects to improve quality: such as the development and testing of wine grape cultivation methods, under different soil and climatic conditions, together with identification of regions with a specific potential for the production of high-quality wine of different types.

The research impact reaches a peak after about 3–4 years, after which it declines. This decline relates only to the expenditure in year 0. The expenditure in year 1 will have the same structure, and so will the peak in year 4–5. This will maintain the research effect on yield at a peak, as the effect of the expenditure in year 0 is superseded by the equation in year 1.

As expected from the panel data estimation, the pooled OLS model was rejected on the basis of the poor statistical fit in favour of the fixed effects model/least squares with dummy variables and random effects models. The coefficients on the R&D variables were subsequently retrieved from the latter two models, and ROR were calculated. The estimated ROR ranged between 21 and 42. The quality adjusted yield gave a much higher ROR than the observed yield.

Conclusions

Several important conclusions were drawn from the analysis. The relative short lead time and lag structures clearly indicated that R&D in the wine grape industry was problem-oriented and adaptive in nature, i.e. there was a relatively short time lapse between investment in R&D and subsequent impact. The example of the integrated pest management project on pheromones is a case in point, where applied research yielded positive effects within a relative short time span. There were large differences in the ROR to research derived from unadjusted yield and quality adjusted yield. This suggests that much of the research undertaken has also resulted in improved quality of the product, and not only the yield alone. This, for example, is the case in the terrain identification project, where wine grape quality was linked to soil and other adoptive factors.

These results compare well with other studies, determining the time lag between R&D investment and positive yield effects, and measuring ROR to agricultural R&D, both in South Africa and elsewhere in the world. The lead time and lag structure for wine grapes are shorter than that for both South African and British agriculture as a whole and the returns to R&D and extension are high relative to other crops in South Africa (see Khatri *et al.*, 1996; Townsend, 1997; Van Zyl *et al.*, 1997).

The results have several policy and other implications. The relatively high RORs indicate that R&D expenditures could be increased substantially. In addition, a relatively low percentage of the total value of wine grape production was invested in R&D (see Fig. 16.2). This was lower than that for agriculture as a whole, and lags considerably behind R&D expenditures of countries such as the USA and Australia. The RORs were higher than that usually required from public investments, indicating that there is a case for increasing expenditure on research. However, given the nature of the industry and the benefits, more of these investment resources should be coming from the beneficiaries. Indeed, it seems to make good economic sense for wine farmers to contribute more to R&D, thus reducing the reliance on public funds. At the level of research focus, the major effects of weather on yields indicated that the research minimizing the negative effects of weather should be encouraged, as it will impact positively on the industry, particularly the farmers.

The findings also have implications for future work in the wine grape industry (and other industries). While there are several well-developed methods for measuring the impact of and return on agricultural R&D, data availability usually provides a problem. This also applied to the wine grape industry. The lack of time-series and project-specific data was a serious limitation to rigorous economic analysis. This should be addressed in two ways: (i) existing data should be systematically gathered and compiled into appropriate series, while there is still some institutional memory on project details in order to facilitate analysis of the impact of R&D; and (ii) all aspects of existing and future projects should be adequately recorded in a systematic way, as an important part of each project. This will also allow for analysis of specific projects, which is not always possible given the present paucity of data. Proper documentation of data will yield reliable analysis, which in turn could form the basis for the better allocation of scarce resources between different initiatives, both within the industry and between industries.

Notes

[1] This chapter is based on an article published in *Agrekon* (1998; 37(2), 189–210).

[2] When the article was written the authors were respectively Lecturer and Professor, Department of Agricultural Economics, Extension and Rural Development, University of Pretoria. The authors wish to thank Colin Thirtle and Leopoldt van Huyssteen for their support and comments on earlier drafts of this report.

[3] The polynomial function has been fitted to data for US agriculture (Evenson, 1967, 1968; Cline and Lu, 1976; Lu *et al.*, 1978, 1979; Evenson *et al.*, 1979; Knutson and Tweeten, 1979; White and Havlicek, 1982); to Australian data (Hastings, 1981); to UK agriculture (Doyle and Ridout, 1985; Thirtle and Bottomley, 1989); and to the commercial sector farm data in Zimbabwe (Thirtle *et al.*, 1993). Khatri *et al.* (1995, 1996) and Townsend (1997) used this approach for South African case studies.

References

Akgungor, S., Makanda, D., Oehmke, J., Myers, R. and Choe, Y. (1996) A dynamic analysis of Kenyan wheat research and rate of return. *Proceedings of the Global Agricultural Science Policy for the Twenty First Century Conference*, Melbourne, Australia.

Alston, J.M., Norton, G.W. and Pardey, P.G. (1995) *Science Under Scarcity: Principles and Practice for Agricultural Research Evaluation and Priority Setting*. Cornell University Press, Ithaca, New York.

Arnade, C., Schimmelpfennig, D., Thirtle, C. and Van Zyl, J. (1996) Short and long-run returns to agricultural R&D in South Africa, or will the real rate of return please stand up? *Proceedings of the Global Agricultural Science Policy for the Twenty First Century Conference*, Melbourne, Australia.

Cline, P.L. and Lu, Y.C. (1976) Efficiency aspects of the spatial allocation of public sector agricultural research and extension in the United States. *Regional Science Perspectives* 6, 1–16.

Doyle, C.J. and Ridout, M.S. (1985) The impact of scientific research on UK agricultural productivity. *Research Policy* 14, 109–116.

Evenson, R.E. (1967) The contribution of agricultural research to production. *Journal of Farm Economics* 49, 1415–1425.

Evenson, R.E. (1968) The contributions of agricultural research and extension to agricultural productivity. PhD thesis, University of Chicago, Illinois.

Evenson, R.E., Waggoner, P.E. and Ruttan, V.W. (1979) Economic benefits from research: an example from agriculture. *Science* 205, 1101–1107.

Griliches, Z. (1958) Research costs and social returns: hybrid corn and related innovations. *Journal of Political Economics* 66, 419–431.

Griliches, Z. (1973) Issues in assessing the contribution of R&D to productivity growth. *Bell Journal of Economics* 10, 92–116.

Hallam, D. (1990) Agricultural research expenditures and agricultural productivity change. *Journal of Agricultural Economics* 41, 434–439.

Hastings, J. (1981) The impact of scientific research on Australian rural productivity. *Australian Journal of Agricultural Economics* 25, 48–59.

Khatri, Y., Thirtle, C. and Van Zyl, J. (1995) South African agricultural competitiveness: a profit function approach to the effects of policies and technology. In: Peters, G.H. and Hedley, D.D. (eds.) *Agricultural Competitiveness: Market Forces and Policy Choice.* Dartmouth Publishing Company, Aldershot, UK, pp. 670–684.

Khatri, Y., Thirtle, C. and Van Zyl, J. (1996) Public research and development as a source of productivity change in South African agriculture. *South African Journal of Science* 92, 143–150.

Kislev, Y. and Peterson, W.L. (1982) Prices, technology and farm size. *Journal of Political Economy* 90, 578–595.

Knutson, M. and Tweeten, L.G. (1979) Toward an optimal rate of growth in agricultural production research and extension. *American Journal of Agricultural Economics* 61, 70–76.

KWV (1996) *Statistics in the Wine Industry, 1996.* KWV, Paarl, South Africa.

Lu, Y., Quance, L. and Liu, C.L. (1978) Projecting agricultural productivity and its economic impact. *American Journal of Agricultural Economics* 60, 976–980.

Lu, Y., Cline, P. and Quance, L. (1979) *Prospects for Productivity Growth in US Agriculture.* Agriculture Economics Report No. 435, USDA, Washington, DC.

Thirtle, C. and Bottomley, P. (1988) Is publicly funded agricultural research excessive? *Journal of Agricultural Economics* 39, 99–111.

Thirtle, C. and Bottomley, P. (1989) The rate of return to public sector agricultural R&D in the UK, 1965–80. *Applied Economics* 21, 1063–1086.

Thirtle, C., Atkins, J., Bottomley, P., Gonese, N. and Govereh, J. (1993) Productivity of commercial agriculture in Zimbabwe, 1970–89. *Journal of International Development* 5, 193–205.

Townsend, R. (1997) Policy distortions and agricultural performance in the South African economy. PhD thesis, University of Pretoria, South Africa.

Van Zyl, J. and Groenewald, J.A. (1988) The effects of protection on South African agriculture. *Journal of Agricultural Economics* 39(3), 387–401.

Van Zyl, J., Thirtle, C. and Botha, C.A.J. (1997) Productivity change in South African commercial agriculture: an evaluation of the returns to investment in public agricultural extension. *South African Journal of Agricultural Extension* 25, 1–11.

White, F.C. and Havlicek, J. (1982) Optimal expenditures for agricultural research and extension: implications of underfunding. *American Journal of Agricultural Economics* 64, 47–55.

17 Socio-economic Impact of Lachenalia[1] Research in South Africa

J.G. NIEDERWIESER

Abstract: This chapter presents the results of a comprehensive impact assessment of the lachenalia R&D programme at Agricultural Research Council-Roodeplaat in South Africa. The lachenalia programme has developed flower bulbs for the international flower market for more than 30 years. The production of bulbs in South Africa will contribute to employment generation. The financial analysis showed that the institute would be able to cover the annual cost of the programme from royalty income, but that it will not recover the total cost of the investment since the beginning of the R&D in 1965. Results of the economic analysis were negative unless increased productivity and market penetration can be realized and the gestation period is reduced.

Introduction

Floriculture has emerged as a high value industry in many sub-Saharan African countries (Matter and Reitenbach, 1996). South Africa (SA) exported flowers valued at $30.3 million in 1998. It was estimated that the value of exports can be increased by focusing on producing the right products (initial emphasis on the indigenous plants), ensuring industry-wide marketing, establishing a strong selling network, and achieving economies of scale in logistics and the creation of a sophisticated enabling environment. The SA floriculture industry has a number of competitive advantages, including the exchange rate, reverse seasons, availability of land, institutional support from the Department of Trade and Industry, a good product mix, infrastructure and expertise (Kaiser Associates, 2000).

SA is rich in unique indigenous flower bulbs (approximately 2700 species of 15 families). A number have been commercialized in foreign countries. *Gladiolus*, *Freesia*, *Zanthedeschia* and *Ornithogalum* are grown world-wide as cut flowers. Commercial bulb growing is concentrated in regions with a maritime climate and the cultivated area is estimated to be 33,000 ha (Kleijn and Heybroek, 1992). The hectarage in the Netherlands amounts to approximately 50% of the total area in the world. The production value of flower bulbs was estimated at $765 million in 1992 (Kleijn and Heybroek, 1992). The value of Dutch production was $500 million, Japan $66 million, the UK $63 million and the USA $41 million (Kleijn and Heybroek, 1992).

 Impact of Science on African Agriculture and Food Security (eds P. Anandajayasekeram, M. Rukuni, S. Babu, F. Liebenberg and C.L. Keswani)

Tulips, lilies and gladiolus are by far the most important flower bulbs. Noteworthy is the trend that the sale of 'other bulbs', which includes hyacinths, grape hyacinth, lily of the valley, etc., increased in the period 1980–1990. In the early spring, flower bulbs have little competition with other plants in the garden. However, greater competition from other products, short flowering period, short time period within which these bulbs are available and the failure of bulbs to bloom at times, are threats to the 'other bulb' group. Kleijn and Heybroek (1992) recommended that the range of products needs to expand to meet the changing needs of the growing flower bulb market.

Lachenalia is a genus of flowering bulbs that occur naturally in SA's winter rainfall areas. The genus consists of approximately 110 species. Lachenalia plants are perennial plants with small bulbs and show a definite, seasonal growth pattern. The fact that the flower colour and shape of the different species show wide variation makes the genus suitable for the development of a commercial product. Small, dormant bulbs are suitable for export, also by sea. Lachenalias, which were available before 1997 on the global market, included selected species and intraspecies crosses of *Lachenalia aloides*, which are grown by collectors and gardeners. Lachenalia varieties (hybrids) are more robust in growth and blooming than natural species. In addition they offer an improvement in range, colour and flower shape. The technology developed at Agricultural Research Council (ARC)-Roodeplaat to produce potted bulbs with an extended flowering time will enhance the market potential of the product. A disadvantage of this product is its susceptibility to virus infection. ARC-Roodeplaat's lachenalias are produced from stock plants propagated by means of tissue culture from disease indexed, nuclear plants. The current purchasing power of the SA Rand will deter propagators in the Northern Hemisphere to produce lachenalias. Although the genus *Lachenalia* is taxonomically related to well known, and well researched, genera such as *Hycynthus* and *Scilla*, they differ in respect to optimal storage and growing temperatures.

Lachenalia bulb production is labour intensive, particularly because the bulbs are propagated by means of leaf cuttings. It is anticipated that breeders in Europe, the USA and Japan will start breeding programmes if ARC-Roodeplaat's varieties prove to be successful in the international market. ARC-Roodeplaat will, therefore, need to develop new and improved varieties continuously and make the most of the available and unique genetic material. Agronomic research will have to concentrate on continuously improving production and quality of bulbs so that local production will maintain its market share in the international market. Diversification, such as the development of cut flower production under local conditions, will strengthen the local industry and increase ARC-Roodeplaat's return on investment. ARC-Roodeplaat and the local growers will not be able to maintain the market share without the continuous development of technology, effective promotion of the products in the main consumer countries, as well as reliable knowledge of market trends.

Background

The lachenalia research programme was started in 1965 and its policy was broadly in accordance with the policy of the Department of Agriculture. Research was funded by the state and the industry was involved in the selection of hybrids. The anticipated role of the institute was to develop the new varieties, and the local bulb growers were to commercialize the product.

When the ARC was formed in 1992, Roodeplaat scientists had to make a difficult decision regarding the future of the lachenalia research programme: 34 varieties, which attracted a lot of attention from local and foreign floriculture experts, were registered with plant breeders' rights, disease-free nuclear plants were being maintained, a considerable number of species accessions were being maintained in the gene bank and some basic information regarding cultivation practices existed. One of the strategic mandates of the ARC is to conserve genetic resources, in particularly indigenous species. The funding policy of the ARC, however, implicitly stated that 30% of the cost of research must be funded by industry, or geared towards developing new basic knowledge (basic research). Fides (the agent for commercialization at the time) performed a market study in 1993, which indicated that the potential market for lachenalia cultivars in Europe alone was 20 million bulbs/year, at a price of Hlf0.25 per bulb, and a royalty rate of 3.5-guilder cent per bulb. That means a potential royalty income of R1.6 million/year for the institute.

Based on these considerations, the institute decided to continue, and in fact, expand the research programme to generate the required information to support growers: propagation, bulb production and forcing,[2] evaluation of existing varieties, and to create a climate for commercialization. Income from the sale of propagation stock was thought to be sufficient to fund the programme until sufficient income was generated from royalties. Until the 1996/97 financial year, the institute was able to fund the recurrent cost of the programme through this type of funding. As the commercialization process was slower than anticipated, the resources invested in the lachenalia research and development were reduced until signals from the market required increased R&D.

The research effort in lachenalia has been going on for more than 30 years. Given the emerging challenges of the new government, there is a need to prioritize and reorient the R&D activities of ARC-Roodeplaat. The lachenalia R&D programme is unique in that a new product (a differentiated product) with potential in the international flower bulb industry has been developed from indigenous species. As the commercialization process was slower than required, external funding from royalty income was not forthcoming and the income from selling propagation stock was limited, particularly as the research results were disseminated to commercial growers.

Given this scenario, the research team and the management of ARC-Roodeplaat realized that there was a need to assess the socio-economic impact of this programme in order to determine the future direction and research investment potential of this programme, as well as product development in future. Thus the ARC Department of Impact Analysis and the research managers at ARC-Roodeplaat decided to study the impact of the lachenalia R&D programme.

A number of changes have taken place in the lachenalia R&D programme, as well as in the commercialization process since the results of the ex ante study have been published (Niederwieser *et al.*, 1997, 1998; Marasas *et al.*, 1998). These changes were the result of adoptions of recommendations made in the study, from market signals and the addition of the second propagator, Vosbol.

Objectives

The objectives of this study were: to trace the evolution of the lachenalia R&D programme at ARC-Roodeplaat, to establish the potential market for lachenalia products; to estimate the benefits and costs of lachenalia R&D activities at ARC-Roodeplaat, assess the cost recovery of the programme; and assess the socio-economic impact of the lachenalia R&D programme for the period 1965–2010.

Methodology

A conceptual framework was developed to look at the various impacts of the lachenalia R&D programme, as well as to conduct a financial analysis of the investments at ARC-Roodeplaat.

In addition to the available secondary data, primary data on the cost of propagation and market potential were collected from various sources using guidelines and questionnaires. Research costs and technology development costs were estimated using available information from various institutes involved. A benefit–cost framework was used to estimate the benefits to the society in terms of rate of return (ROR) on the investment. The other type of impacts, such as environmental impacts, socio-cultural impacts and intermediate institutional impacts were identified and assessed in the study.

Results

Evolution of the lachenalia R&D programme at ARC-Roodeplaat

Collection and conservation of *Lachenalia* species began in 1965. Since then, the programme has gone through five developmental phases. It should be noted that the breeding cycle of lachenalia is 3–4 years and that bulbs are produced within 2–3 years.

- Phase I: scoping study (1965–1972)
- Phase II: development of improved hybrids (1973–1983)
- Phase III: evaluation for commercial potential (1984–1992)
- Phase IV: revitalization of research (1993–1997)
- Phase V: commercialization (1997 onwards).

Phase I: Scoping study (1965–1972)

During this phase, the potential for a breeding programme was investigated. Starting with 17 gene bank accessions, basic procedures for pollination, pollen storage, maintenance and storage of bulbs, propagation and cultivation, germination and seedling growth were established. The total number of species–species crosses made was 177 and the gene bank increased to 59. The SA Bulb Growers Association assisted the institute to select promising hybrids for further evaluation. This phase was concluded by the recommendation of the Bulb Growers Association that the first hybrids had commercial potential. Roodeplaat subsequently decided to expand the programme by appointing a flower breeder to increase the number of accessions in the gene bank and to develop improved cultivars through breeding and selection.

Phase II: Development of improved hybrids (1973–1983)

The objectives of the programme were to develop pot plant cultivars for export in the international flower bulb market. The emphasis during this period was to make as many as possible species–species and hybrid–hybrid crosses and to select those that were suitable for the target market (Lubbinge, 1980). Basic criteria such as growing period, flowering time, uniformity and appearance of the inflorescence were used for hybrid evaluation. Propagation of the early selections by growers led to the realization that lachenalias are very susceptible to infection by virus (later shown to be ornithogalum mosaic virus). Subsequently, meristem culture for virus elimination and tissue culture propagation of virus-free material was initiated (Klesser and Nel, 1976). Preliminary studies on leaf cutting production, as well as techniques for flower manipulation were done during this period; however, these were not well documented. Hybrids of this phase were impressive. In fact all the cultivars currently in production originated from this period. Commercial growers were involved in the selection of the first five hybrids to be registered for plant breeder's rights in 1980.

Phase III: Evaluation for commercial potential (1984–1992)

The personnel turn-over in the civil service was high in the 1980s and it had a negative effect on the lachenalia R&D programme. The programme leaders as well as the research technician left the civil service in 1981 and inexperienced scientists were appointed to take charge of the programme. The average number of crosses per year dropped to less than 25 and apart from a cytogenetic study that was used for a postgraduate study (Hancke, 1990) little progress was made with the breeding programme. A study on flower initiation later developed into one on flower manipulation (Louw, 1993). Other positive developments during Phase III include a detection method for ornithogalum mosaic virus (Burger and Von Wechmar, 1988), the initiation of a plant improvement scheme, improvement of *in vitro* propagation methods (Van Rensburg and Vcelar, 1989; Niederwieser and Vcelar, 1990) and the application thereof. In 1983, 17 additional hybrids (resulting from work of the 1970s) were registered for plant breeder's rights and at the same time they were tested by local growers.

Although the evaluation trials by SA growers were positive, particularly with regard to the products' potential as potted bulb, commercial production of bulbs showed very little progress during Phase III. Initially, eight growers received bulbs for trials. However, by the end of the phase only three growers were still in possession of lachenalia bulbs. Several reasons could be attributed to the slow progress in commercialization. These are:

- Limited amounts of propagation material were supplied to growers and recommendations for virus control were not available at the time. As a result, plants became heavily infected with virus.
- ARC-Roodeplaat was not in a position to make recommendations regarding large-scale cultivation as no agronomic trials were carried out by the institution.
- Local growers did not have the expertise or funding to conduct in-house trials. Virus infection was the single most important production problem to overcome.

On recommendation of the Flower Growers Association of SA, Roodeplaat appointed a large chrysanthemum breeder (Fides BV) as agent for Roodeplaat's lachenalias. By 1992 it became clear that little success had been made apart from evaluation trials that were carried out in various locations. The turning point for the lachenalia R&D programme was when the breeder investigated new product development in the Netherlands (NL) and returned with a clear message that the breeder of a new flower product needs to supply supporting technology to prospective growers. The management of the institute was thus faced with a difficult decision regarding the future of the programme. It was clear that no funding was forthcoming from the floriculture industry for lachenalia research, yet the ARC funding situation required that all projects had to be supported financially by a client.

Phase IV: Revitalization of research (1993–1997)

Fides (appointed agent) performed a market survey to determine the market potential for Roodeplaat's lachenalia varieties in 1993. Bulb experts in the NL estimated that 20 million bulbs could be sold in Europe alone. The management of the institute thus decided to continue, and in fact, to expand the research on lachenalia. The research team realized that an enormous amount of information had to be generated in a short time. A committed multidisciplinary team cross-subsidized the programme through contract (commercial) research over a period of 5 years.

Priorities were focused on flower manipulation studies, cultivation methods for pot plant production, improved propagation methods and the production of disease free propagation stock, evaluation of hybrids, technology transfer and the establishment of a network of agents for commercialization. Other aspects included characterization of approximately 1000 gene bank accessions, genetic and cytogenetic studies, phenological studies, plant nutrition studies, plant reproductive studies, virus detection methods and 250 crosses per year. Working group meetings facilitated information exchange between the pioneer growers and the research team as well as the opportunity to identify priority research fields.

The period after 1992 was also characterized by the termination of old partnerships and the formation of new ones. As the ARC and Fides could not come to an agreement on the royalty split, the agreement was terminated. Royalty International (RAI) was appointed as the royalty administration agent in 1995. Hobaho Intermediaries was appointed as the international sales agent in 1996. Two licensees withdrew from the programme in 1993 and a third one was liquidated in 1994. Langberg Nursery remained as the only propagator/grower. Good progress was made with the preparation for commercialization: ten varieties were registered for plant breeder's rights in NL and evaluation trials were carried out by the research team. A commercial pot plant trial was carried out in 1996. Towards the end of 1996, a substantial amount of critical information was generated, a plant improvement scheme established and 200,000 disease-free bulblets were produced for further use by the propagator. The phase was concluded by a reduction of the size of the research team. The focus of the team switched to improving cultivation methods.

Phase V: Commercialization since 1997

The R&D input of the programme was reduced to a few critical areas, namely: gene bank maintenance, evaluation of remaining hybrids, improvement of production per hectare. A commercial potted bulb grower in NL, Van Der Vossen BV, started to introduce lachenalia as a new product in the potted bulb sector. Buyers responded positively, but the rate of the market development has been slower than predicted as the product is unknown to consumers. A second licensee (Vosbol) was appointed in 1999. Langberg Nursery was liquidated in 2000. Propagation is done once more by a single propagator. Vosbol, however, appointed three subcontractors to grow bulbs to marketable size. The institute is thus less dependent on the propagator for production. Vosbol (two partners were part of the research team until 1998) applied research done at Roodeplaat and increased the yield to 1 million bulbs/ha. The 1996 projections for the production period were found to be conservative. Vosbol has been very successful with lachenalia production and more than 1 million bulbs were exported in 2000.

Flow of lachenalia technology and products to the markets

The flow of lachenalia information, technology and products between different role players is summarized in Fig. 17.1. New varieties are developed by ARC-Roodeplaat using accessions in their gene bank. Production-related technology is developed primarily by the ARC. Technology development is supported by basic research, which is done at ARC and through postgraduate studies at universities. In order to sustain production of disease-free bulbs, ARC sells a limited amount of propagation stock to licensees annually. Licensees multiply bulbs and the cultivation of the bulbs to marketable size is done by licensees and their subcontractors. Two agencies, RAI and Hobaho, assist in the sale of bulbs in the international market. RAI is responsible for royalty administration, and Hobaho acts as mediator during the sale of bulbs in foreign countries. The market channel for export bulbs is primarily through wholesalers in NL, from where bulbs are distributed throughout the world.

ARC-Roodeplaat receives royalties on varieties for which they have plant breeder's rights or plant patents (in the USA). The royalty rate is a fixed rate per bulb and varies in different countries. The royalty rate in SA is R50/1000 bulbs, in Europe Hfl 35/1000 bulbs and in the USA, $20/1000 bulbs. Twenty-five per cent of the royalties earned is paid to RAI as an agency fee. The role of Hobaho is to mediate sales between the licensee and the bulb trade. Hobaho

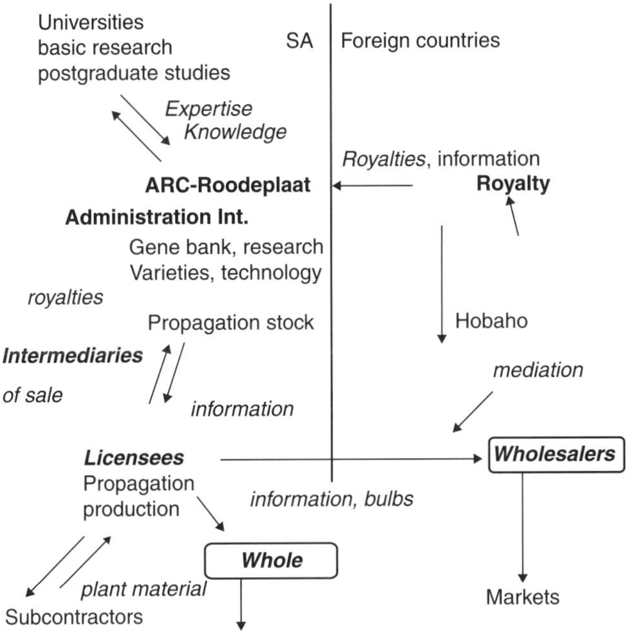

Fig. 17.1. Flow of lachenalia products, technology and information.

has extensive knowledge of the international bulb industry and their involvement is valuable to identify trustworthy buyers and to mediate in cases of disagreement between the buyers and the licensee. The flow of technology and products is expected to change as the commercialization process proceeds and the market becomes established.

Market demand for lachenalia cultivars

The original objective of ARC-Roodeplaat was to develop lachenalia varieties for the pot plant market. The market potential for lachenalia varieties was estimated at 20 million bulbs/year by Fides BV (NL) in 1993 after interviews with 14 different players in the international flower bulb market. The growth of the market, local and export to Europe, was estimated by the propagator, export agent and research manager (Table 17.1). The views of the three groups are substantially different and the most conservative projection (of the export agent) was used in the financial and economic analysis.

Although the supply of bulbs from Vosbol has been enough for the projected market (1 million in 2000), market development in Europe was slower than anticipated. Less than 200,000 bulbs were sold in 1999 and, in 2000, less than 500,000. However, commercial trials have been initiated in the USA in 1999, in Japan in 2000 and in NZ in 2001. Commercial pot plant growers are hesitant to buy large numbers of bulbs of this new crop before they have done a commercial trial. Bulbs for several commercial trials in Europe and the USA have been ordered. Dry bulbs and pot plants have been sold in SA since 2001. Vosbol also started to develop a market for cut flowers in SA.

Financial analysis

To evaluate the commercial viability and cost recovery potential of the programme from the viewpoint of the ARC, a financial analysis was performed. The ongoing market rate at the

Table 17.1. Market potential for lachenalia products.

	Domestic sales			Export market (millions of bulbs)			
				Europe			USA and Japan†
Year	Dry bulbs	Pot-plants	Total bulb equivalents	Research	Propagator	Export agent	
1997	4,000	1,200	5,200	0.05	0.04	0.04	
1998	35,000	3,000	38,000	0.10	0.10	0.15	
2000	170,000	6,000	176,000	1.00	0.75	0.55	
2002	560,000	12,000	572,000	4.00	5.00	2.30	
2004	685,000	15,000	700,000	5.00	6.84	2.85	
2006	840,000	20,000	1,040,000	6.75	11.76	4.81	3.75
2008	1,220,000	20,000	1,240,000	8.25	16.68	8.13	7.49
2010	1,500,000	20,000	1,570,000	10.00	20.30*	13.74	11.20

*The market demand for Europe is estimated at 20 million bulbs pa.
†Based on the estimate by the sales agent.

time of the study was 8.3% after adjusting for inflation. The ROR and net present value was both negative for all the scenarios considered. The 32-year gestation period was largely responsible for the negative results.

Cost recovery is an important priority setting criterion for the institute as the funding policy of the ARC required that a considerable portion of the cost of research must be recovered from external sources, in this case from royalties on the sale of bulbs. Conservative scenarios were considered in the analysis. Cost recovery through sale of propagation stock and royalties started in 1997 at a rate of 5%. This could increase to 84–200% after 2010. As the cost recovery rate was at its maximum when all export markets (Europe, USA and Japan) were realized, commercial trials in the USA and Japan were initiated in 1999. Under all scenarios, the institute would recover more than 30% of the cost from 2000. ARC-Roodeplaat thus decided to continue with the R&D programme and to focus on the development of improved varieties to maintain their market share. The cost recovery rate of the programme since the study was more than 30%. However, as a result of decreased funding, the minimum external funding rate was increased to 50%. The institute thus had to make a difficult decision once again. Given the fact that the second propagator was successful in meeting the targets for production, the institute will continue to maintain a research team and include universities in the research.

Economic analysis

A simple benefit–cost approach was used to estimate the economic impacts of the lachenalia R&D programme. Lachenalia is a new product and the benefit derived from the R&D investment is largely dependent on the market potential. The commercialization of the product is largely new with 40,000 bulbs exported in 1997 and 1,000,000 in 2000. Thus the analysis is largely an ex ante type. The calculation of research benefits has different components including the anticipated market potential, cost of technology development, cost of programme materials and cost of technology transfer and marketing. Economic activities related to commercialization of lachenalia taking place outside SA were not included in the analysis.

The major cost component of the analysis was the R&D cost. Staff salaries and benefits accounted for 60% of the cost in 1993. This was expected to decline to 45% in 2010. In this study, the real interest rate (6.7%) was used as discount rate. This was derived from the long-term government bond rate adjusted for inflation. Benefits pertain to the sale of propagation material, bulbs and potted plants, as well as royalties from domestic and foreign sales. Benefits resulting from the sale of potted bulbs, royalties, agency commission, etc. in foreign countries were excluded.

Sensitivity analysis included the following: discount rate, expansion of foreign markets, reduced gestation period and increased productivity. Because of the uncertainties involved, in choosing the discount rate, a sensitivity estimate of 5% was included. The analysis included scenarios which included the European market only, expansion of the market to the USA and Japan in 2010 and in 2005. The market agent estimated the market shares of the three export markets to be 55% for Europe, 20% for the USA and 25% for Japan. The gestation period for the lachenalia programme was 32 years. Researchers were of the opinion that given the breeding cycle of 3 years, a reasonable timeframe for the development of new varieties and commercialization is about 15 years. Several factors contributed to the delay of commercialization in the 1980s, which included management decisions. The analysis was thus done for gestation periods of 15 and 30 years. In 1996, the estimated yield at full capacity was 650,000 bulbs/ha. Based on research results, researchers felt that the productivity could be increased to 1 million bulbs/ha. The analysis was thus repeated with the more optimistic projection as well. Production results by Vosbol in 1999 and 2000 indicated that this estimation was indeed realistic.

The economic impact of the lachenalia R&D programme was negative (Table 17.2) unless a shortened gestation period was assumed and increased productivity and expanded markets. Under conditions of improved productivity and expanded market, a ROR of 7–8% was realized. This increased to 9–12% if a decreased gestation period was assumed. The net present value ranged between 2 and 13 million Rand.

Social impact

In addition to human development discussed, the greatest social impact of the R&D programme is employment creation. Given the labour-intensive nature of lachenalia bulb production, it was

Table 17.2. Economic analysis: estimated net present value and internal rate of return for lachenalia R&D, 1965–2015.

	Net present value 5% discount rate		Internal rate of return	
Scenario	Propagator*	Export agent†	Propagator	Export agent
European export market				
Domestic market – fully realized	(2,065,770)	(4,374,204)	0.03	–
Domestic market – 50%	(2,519,593)	(4,792,311)	0.02	–
European, Japanese and USA markets				
Domestic market – fully realized	(679,153)	(2,491,571)	0.04	0.02
Domestic market – 50%	(2,911,245)	(2,945,394)	0.02	0.02

*Propagator: based on propagator's estimate.
†Export agent: based on export agent's estimate.
Source: ARC-Roodeplaat analysis.

anticipated that the programme could contribute significantly to employment creation. This was one of the important considerations of the institute to implement the research. The new lachenalia industry can create employment by directly employing people as well as through auxiliary services such as transport, packaging and retail sales of bulbs. In the impact study, only the direct employment generated in the bulb industry was estimated.

During the period 1993–1995 the institute employed a full-time technician and six assistants to propagate lachenalia through tissue culture. Fifty per cent of these employees' time was allocated to lachenalia. From 1996 onwards the need for tissue-cultured plants decreased as a result of improved methods of propagation by means of leaf cuttings at Roodeplaat. The result was that employment in the bulb propagation unit increased to 0.5-person year of a technician, two full-time assistants and seven part-time labourers. For 2 years, Roodeplaat produced more than 1 million bulblets. The situation has been adapted once again after propagation by the current licensee has been established. Currently, Roodeplaat produces only 200,000 bulblets per year to assist the commercial growers to maintain the quality of propagation stock and to be able to produce sufficient numbers of bulbs of new varieties in future.

Estimation of employment generation at commercial production units was based on information provided by Langberg Nursery in 1997, ten full-time labourers and 20–30 seasonal labourers for a period of 3 months were used to produce bulbs on 1 ha. The estimated total demand for lachenalia bulbs based on the estimate of the exporters. Using this assumption, and given the anticipated expansion plan for commercial bulb production, the employment generated in terms of full-time equivalents was calculated. Assuming that the propagator is meeting the domestic demand and European market demands for the lachenalia bulbs, the estimated total labour requirement is eight in 1997 and this number would increase to 576 in the year 2008. It was estimated that if new markets could be exploited in Japan and the USA, then 440 full-time and 1054 seasonal labourers could be employed. This is the direct employment opportunity. Additional employment would be created in the retail and wholesalers markets also.

In terms of human capacity building, the lachenalia R&D programme made a special contribution. A project of the Reconstruction and Development Program of the South African Government was initiated in 1997 in Nieuwoudtville. With the support of the Provincial Department of Agriculture, the project exported 40,000 bulbs in 1999 and 200,000 bulbs in 2000. It is expected that the project will be able to export 1 million bulbs/year. These activities resulted in income generation and human capacity building of the project members in terms of production of bulbs, project and production planning and knowledge and expertise of bulb exports requirements, etc. Currently, the Nieuwoudtville project is a subcontractor of Vosbol.

Expertise and knowledge of bulb production at Roodeplaat was further disseminated to the private sector with the formation of the second propagator, Vosbol, in 1999 by two former researchers in the lachenalia R&D team. The success of Vosbol in bulb propagation and production can probably be ascribed to the combination of first-hand knowledge and expertise in lachenalia production and full-time commitment to the production of one product, namely lachenalia.

Institutional impact

Previous to this programme, R&D to develop new floral crops was mainly restricted to countries in the Northern Hemisphere. This programme enabled SA to develop its indigenous capacity. The experience of the laboratory managers is disseminated to the private sector through training and consultation. Expertise and knowledge of plant improvement schemes have been disseminated to growers. This knowledge can be disseminated to growers of other bulbs who are currently dependent on Dutch supply of propagation material. Since the completion of the impact study in 1997, research activities at Roodeplaat have been reduced to breeding and selection and are being done by postgraduate students in crop production (University of Pretoria and University of the Orange Free State (UOFS)) and genetics (UOFS). Projects are conceptualized by scientists at Roodeplaat who then act as co-leaders for these studies. In this way the knowledge and expertise has been disseminated to the wider scientific community in SA. Research at the UOFS now also includes other indigenous flower bulbs.

As a result of the lachenalia programme, better linkages have been established between the research team, propagators and the market. The experience also showed the importance of researchers participating in the commercialization of the product and research priorities are demand driven based on the signals of the market.

Programme activities have contributed to more than 30 scientific publications at the present moment and numerous contributions at scientific congresses. This number is expected to increase as a result of the involvement of postgraduate students and researchers at universities. Thus the benefit of the lachenalia programme has begun to flow from the institute through tertiary training to the wider scientific community in SA.

Environmental impact

One of the most important consequences of the lachenalia R&D programme is the preservation of biodiversity. Collection, characterization and maintenance of an extended gene bank contribute to the conservation of one of SA's many endemic genera. Breeding and commercialization enhances the economic utilization of the indigenous flora. The number of gene bank accessions increased from 17 when the programme was started to 1000 at the time of the study. In addition, 250 crosses were made per annum to commercialize the biological potential of this flower bulb.

As lachenalia production takes place under shade net or in the open field (without additional heating, cooling or lighting) and uses relatively little pesticides, lachenalia production can have a minimum effect on the environment.

Developments after Completion of the Study/Evaluation

After completion of the study in 1997, a number of recommendations were followed through:

- The number of propagators has been increased to two after the previous project manager and crop scientist resigned to become propagators (Vosbol cc). This proved to be an extremely valuable development as serious problems with virus infection developed at Langberg shortly after the completion of the study. Langberg went into liquidation in 2000 and propagation is carried out once again by only one propagator. More propagators can be appointed when market penetration has improved growing-on bulbs.
- Increased productivity proposed by researchers in 1997 was realized during the 1999 and 2000 production seasons at Vosbol.
- Research activities were scaled down to maintenance of the gene bank, breeding and selection (to be able to introduce new varieties in order to maintain Roodeplaat's lead in this new product), and agronomic research in co-operation with universities.
- The expertise and knowledge that existed at Roodeplaat regarding indigenous bulbs have been diffused to the private sector (Vosbol) and at least two universities (Pretoria and UOFS). Publications in scientific journals (Marasas *et al.*, 1998; Niederwieser *et al.*, 1998, 2001; Kleynhans *et al.*, 2001) and contributions at scientific meetings (Niederwieser *et al.*, 1998,

2000; Kleynhans *et al.*, 2000) made a contribution towards the diffusion of the experience gained and lessons learned, to the local and international scientific community.

- Production of propagation stock has been increased to more than 1 million bulbs per annum shortly after the completion of the study to satisfy the demand from propagators. However, it has recently been decreased to 200,000 because Vosbol has been very successful with the multiplication and production of bulbs since 1999.

Conclusions

The lachenalia R&D programme is a new crop that was developed from one of SA's many unexplored indigenous flowers. The objective of the programme from its inception was to develop a new export product and to generate employment in SA. The commercialization of the product in the international flower bulb market, however, proved more difficult than initially anticipated. ARC-Roodeplaat has learned that the breeder of a new product should be intimately involved with the development of production methods and the commercialization of the products.

As a public investment, development of new flower crops can be profitable, providing that the gestation period is as short as possible and production is as high as possible. This can be achieved by agronomic research at the inception of the breeding programme and the early introduction of varieties into the market. The socio-economic impact assessment of the lachenalia R&D programme proved to be an extremely valuable research tool for the ARC-Roodeplaat management.

Notes

[1] '*Lachenalia*' refers to the botanical genus to which the plants belong. The name 'lachenalia' refers to the common name for the crop.

[2] Forcing is the term used in the bulb industry to describe the out-of-season production of cut flowers and pot plants by growing bulbs in greenhouses and at specific temperatures to 'force' them to flower.

References

Burger, J.T. and Von Wechmar, B. (1988) Rapid diagnosis of *Ornithogalum* and *Lachenalia* viruses in propagation stock. *Acta Horticulturae* 238, 31–38.

Hancke, F.L. (1990) Cytogenetic study on seven *Lachenalia* species for use in a flower breeding program (In Afrikaans). MSc thesis, University of Pretoria, South Africa.

Kaiser Associates (2000) *South African Floriculture Cluster Study*. Kaiser Associates, Cape Town.

Kleijn, E.H.J.M. and Heybroek, A.M.A. (1992) *A View of the Competitiveness of the Flower Bulb Industry*. Rabo Bank, Eindhoven, The Netherlands.

Klesser, P. and Nel, D.D. (1976) Virus diseases and tissue culture of some South African flower bulbs. *Acta Horticulturae* 59, 71–76.

Kleynhans, R., Niederwieser, J.G. and Hancke, F.L. (2000) *Lachenalia: development and commercialisation of a new flower bulb crop*. 8th International Symposium on Flowerbulbs. Kirstenbosch, Cape Town, South Africa, 28–31 August 2000.

Kleynhans, R., Niederwieser, J.G. and Hancke, F.L. (2001) *Lachenalia*: development and commercialisation of a new flower bulb crop. *Acta Horticulturae* 570, 81–85.

Louw, E. (1993) Morphology of *Lachenalia* cv Romelia inflorescence development. *Journal of the South African Society of Horticulture Science* 3, 59–63.

Lubbinge, J. (1980) Lachenalia breeding. 1. Introduction. *Acta Horticulturae* 109, 289–295.

Marasas, C., Anandajayasekeram, P., Niederwieser, J.G., Coetzee, M., Martella, D. and Pieterse, B.J. (1998) The future of wild flower research and development in South Africa. The lachenalia case study. *Agrekon* 37, 597–609.

Matter, A.J. and Reitenbach, A. (1996) *Profits from petals: the development of cut flower exports in Southern Africa.* Technical paper prepared as part of the Southern Africa Regional Agribusiness Study. The World Bank, Washington, DC.

Niederwieser, J.G. and Vcelar, B.M. (1990) Regeneration of *Lachenalia* species from leaf explants. *HortScience* 25, 684–687.

Niederwieser, J.G., Anandajayasekeram, P., Coetzee, M., Martella, D., Pieterse, B.J. and Marasas, C. (1997) *Socio-economic impact assessment of the Lachenalia research programme.* SACCAR, ARC/SACCAR report, Gaborone, Botswana.

Niederwieser, J.G., Anandajayasekeram, P., Coetzee, M., Martella, D., Pieterse, B.J. and Marasas, C. (1998) Socio-economic impact assessment of the *Lachenalia* research programme at Roodeplaat as a research tool. *Journal of the South African Society of Horticulture Science* 8, 80–84.

Niederwieser, J.G., Kleynhans, R. and Hancke, F.L. (2000) Development of a new flower bulb crop in South Africa. Keynote address. 8th International Symposium on Flowerbulbs. Kirstenbosch, Cape Town, South Africa. 28–31 August 2000.

Niederwieser, J.G., Kleynhans, R. and Hancke, F.L. (2001) Development of a new flower bulb crop in South Africa. *Acta Horticulturae* 570, 67–73.

Van Rensburg, J.G.J. and Vcelar, B.M. (1989) The effect of the sucrose concentration on the initiation and growth of adventitious buds from *Lachenalia* leaf tissue. *South African Journal of Botany* 54, 196–202.

18 Impact Assessment of Sunflower in Zimbabwe

M. Murata, P. Anandajayasekeram, M. Mudhara, D. Martella and E. Mazhangara

Abstract: The objective of this study was to assess the impact of sunflower research, extension and the enabling environment in Zimbabwe from 1976 to 2000. The study concentrated on comparing the social costs and returns of the sunflower R&D investment. Both primary and secondary data were used in the analysis. Using a standard questionnaire and multistage sampling, information was collected from communal and commercial farmers in Natural Regions II, III and IV. Results of the survey indicate that sunflower is a universally adopted crop in the communal areas across all three natural regions, and the area under sunflower cultivation has increased substantially over the years. However, communal area farmers' adoption of improved varieties and better management practices is low. The study points out the need for research to explore the constraints on the adoption of improved varieties and improved agronomic practices. One of the weaknesses of the R&D programme on sunflower has been the limited release of improved varieties. The study estimated the rate of return (ROR) to be 12% for sunflower R&D in Zimbabwe over the period 1976–2000. A sensitivity analysis of the estimated ROR indicated that ROR is sensitive to changes in research costs. The net environmental effect of sunflower R&D was difficult to assess, in the absence of biophysical data to conduct a systematic analysis. The crop, however, was found to have a positive impact on food security, as it is grown successfully in marginal environments, and the income generated is used to meet the daily needs of the households. The results of this study clearly indicate that the ROR on past investments in sunflower R&D in Zimbabwe is attractive enough to warrant continued investment.

Introduction

Sunflower production in Zimbabwe is directed mainly towards oil extraction. Sunflower is largely produced in Natural Regions (NR) II, III and IV by smallholder farmers, who include communal, small-scale commercial farmers and resettlement farmers. Large-scale commercial farmers (LSCF) contribute less than 5% of total production (CSO, 1996). The popularity of the crop among smallholders is due to its drought tolerance. In the commercial sector, due to the poor yields of open pollinated varieties, sunflower was grown when the hybrids became available in the 1980s (COPA, 1986). The average yield of sunflower is about 0.5 t/ha for smallholder farms and 1.0 t/ha for large-scale commercial farms (CSO, 1996). The low average

 Impact of Science on African Agriculture and Food Security (eds P. Anandajayasekeram, M. Rukuni, S. Babu, F. Liebenberg and C.L. Keswani)

yields are partly explained by the following factors: (i) the crop is generally treated as a secondary crop; (ii) hybrid seed costs four times more than the open pollinated cultivar 'Peredovik'; (iii) old seed; and (iv) poor soil fertility management. Support services to sunflowers are provided by various organizations. Public research and extension is undertaken by the Department of Research and Extension (DR&E) while the Agricultural Research Trust provides crop management advice to commercial farmers. However, commercial farmers depend more on private organizations, their associations and consultants for advice. Both the smallholder and the LSCF sectors receive advice from seed houses.

Agronomic research on sunflower has been conducted at the Harare Research Station since 1912 (Weimann, 1972). The first results, mainly on cultivar and spacing evaluations, were reported in the then Salisbury Agricultural Experiment Station (1920–1922) and Gwebi Experiment Farm Annual Reports of 1915–1922. Cultivars grown were developed elsewhere and had a variable oil content of between 28% and 35%. Cultivars introduced at a later stage, in the 1960s, had a higher oil content (up to 40%). 'Peredovik' (an open pollinated cultivar) is by far the longest surviving cultivar grown by smallholder farmers. The sunflower improvement programme was officially established at Salisbury Research Station in 1975 as a part of the Oilseeds Breeding Team. Its aim was to develop cultivars that have the potential of producing high yields of oil per unit area of land, under varying environmental conditions. There are only two locally bred hybrid cultivars on the Zimbabwean market. The first product of a breeding effort was hybrid 'Msasa', which was released in 1985 (Mtindi, 1985) followed by the release of the hybrid 'Mopane' in 1990 (Mudzana, 1990).

While breeding of new sunflower varieties has traditionally been done by the Crop Breeding Institute in the DR&E, various seed houses (Seed Co., Pannar, Pioneer and Cargill) have also begun doing some breeding work. Despite this development, improved sunflower seed is seldom found in rural retail shops. The wholesalers and retailers claim that the demand for improved seed is very low. Some of the factors limiting the demand for improved sunflower seed are: (i) the hybrid seed is over four times more expensive than open pollinated seed; (ii) hybrids are more demanding in terms of water and management; and (iii) as there is no grading for oil content at Grain Marketing Board depots, farmers have no incentive to adopt the improved varieties with a higher oil content.

Contrary to the downward trend in the production of the other oilseeds, sunflower production increased from 9000 t in 1979 to 50,000 t in 1996. This increase was a combination of both an increase in area under production as well as productivity; both went up by 27% and 20% per annum, respectively (CSO, 1996). The area under sunflower production increased substantially between 1985 and 1990. The decline since 1990 was attributed to a decrease in the area under sunflower in the LSCF sector and the drought of the early 1990s.

Traditionally, public research into sunflower has been conducted by the DR&E. After independence in 1980, there was a government policy shift, directing DR&E to place more emphasis on the research needs of communal farmers, who had been left out by the research system. However, this directive was issued without additional manpower, transport and infrastructure.

Arguments have been put forward for financial support for sunflower research and the formulation of sound policies to guide the subsector (Johnson *et al.*, 1994). All this suggests the need for an ex-post evaluation to determine the costs and benefits of sunflower research and whether it is worthwhile to continue funding sunflower research.

Objectives

Impact assessment case studies were selected in collaboration with the National Agricultural Research System directors. The case study on edible oilseeds (groundnut, sunflower and soybean) was initiated in 1997. The general objective of this study was to assess the impact of sunflower research, extension and the enabling environment in Zimbabwe between 1976 and 2000. The specific objectives were: to describe and understand the existing sunflower production system in Zimbabwe, to trace the evolution of research and enabling environment with respect to sunflower technology development and transfer, to assess the impacts of sunflower research and enabling environment, to conduct preliminary analysis of the effects of the recent liberalization of the sunflower industry, and to assess policy implications for future investments in research and enabling environments.

Methodology

The comprehensive conceptual framework discussed in Part III of this book was used in this study. An attempt was made to measure the impact of R&D investments in terms of intermediate impact, economic impact, socio-cultural impact and environmental impact as well as the spill-over effect.

Economic impact measures the combined production and income effects associated with a set of R&D activities. In assessing the economic impacts, research is treated as an investment and rates of return (ROR) are estimated for this investment. Owing to lack of consistent time-series data, in this study, it was decided to use the surplus approach to estimate the average ROR for the R&D investments in the sunflower research programme. The average ROR takes the whole expenditure as given and calculates the ROR to the global set of expenditures. It indicates whether or not the entire investment package was successful, but not whether the allocation of resources between investment components was optimal.

In the social benefit–cost analysis, any costs or benefits accrued to individuals or societies outside Zimbabwe are not included. The prices and costs used in the analysis are opportunity costs and prices. The main feature of a benefit–cost analysis framework is the specification of the 'with' and 'without' scenarios. The 'with' R&D case corresponds to what has been observed over the years. In this study a 'trend' analysis of the yield data is used to define the 'without' scenario. In order to assess the economic impact of research, the technology that was made available to the farmers during the period of analysis should be clearly defined. In this particular case, several varieties and agronomic recommendations have been generated and disseminated to farmers during the period. What is being measured is the combined effect of these technologies.

The second component in the analysis was the estimation of the technology development and transfer costs. These costs were broken down into three categories: research, adoption and transfer costs. The costs of conducting research include staff salaries and benefits, recurrent administration and overhead expenditures, and provision for the depreciation of capital assets. Adoption costs were the costs that were associated with the uptake of the new technology by farmers over time. The technology transfer costs were mainly the extension costs, including the on-farm research costs.

In estimating the internal ROR (IRR), the incremental net benefit flow was calculated by subtracting the total cost flow from the incremental gross benefits. The incremental net benefit flow over the years was converted into values that could be compared by discounting and/or compounding. This took into account the time value of money. The discount rate used in this study was an approximation of the social time preference rate. This is normally taken as the rate of interest (real) on long-term governmental bonds. During the period 1983–1996, the lending rate, which was used to approximate the social time preference rate, averaged about 10% (World Bank, 1994). Rates of 5% and 15% were used as part of the sensitivity analysis.

The summation of the discounted incremental net benefits yielded the net present value (NPV) of sunflower R&D, which indicates positive impact when it is greater than zero.

Of interest is the discount rate that reduces the sum of discounted incremental net benefits to zero. This discount rate is the average IRR of sunflower R&D. When this is greater than the social time preference rate, then past investments in sunflower R&D have been socially profitable and justifiable.

Data collection

Both primary and secondary data were used in the analysis. Primary data on the current management practices and the adoption of various technological components over time were collected using a standard questionnaire. Personal interviews were conducted to collect the information from communal farmers, while a mail survey was used to collect data from the commercial farmers. Information on household and farm characteristics, details on sunflower cultivation practices over time, various aspects of marketing and support services were collected. In addition, individuals involved in the sunflower R&D activities were also interviewed to obtain specific information.

Secondary data, including cost information, evolution of the sunflower R&D programme and information on various technologies were collected from various published and unpublished reports. Information regarding the Consumer Price Index, prices, production and hectarage yields was obtained from the *Quarterly Digest of Statistics* from the Central Statistical Office. Various annual reports of the Agronomy Research Institute were used to estimate the proportion of sunflower trials to the total number of trials.

Different procedures were used to select the sample farmers for the survey in the two farming sectors. In the communal area sector, a multistage sampling technique was used for drawing up the sample of sunflower growers in the communal areas. The district-level production statistics, with figures provided by the Early Warning Unit of AGRITEX, were used in identifying districts producing sunflower. Ten districts were selected from the list of districts that produced at least 5% of the total communal area production. The average production figures from two cropping seasons, i.e. 1994/95 and 1995/96, were used for this purpose. In each selected district, two wards were randomly selected from the list of wards in the district. Randomly, ten sunflower growers were interviewed in each ward. The final sample was composed of 62, 62 and 100 farmers in NR II, III and IV, respectively.

For the large-scale commercial sector, a sampling frame of all registered large-scale commercial sunflower growers (640) was provided by the Commercial Oilseed Producers Association. Using a systematic sampling procedure, a total of 150 farmers were selected for the mail survey. Questionnaires were sent to the 150 selected farmers. Forty-seven farmers returned the completed questionnaires. This was 7.3% of the population and 31.3% of the sample selected for sending mail questionnaires.

Results and discussion

Sunflower management practices in the smallholder sector

Survey results showed that sunflower is a universally adopted crop across all the three natural regions. It ranked as the third most commonly grown crop by 50% of the farmers in NR II, 25% in NR III and 35% in NR IV. The land preparation and planting methods used by more than 80% of the farmers were in accordance with recommendations. Despite the campaign by extension to promote the use of hybrid sunflower seed in the communal areas, more than 70% of the farmers across the three NRs continue to use low-quality, outdated, open pollinated varieties, which were retained for generations. In fact, more than 70% of the communal farmers interviewed did not know the name of the cultivars they were growing. 'Peredovick' and its derivatives continue to be the most commonly grown traditional varieties. Furthermore, the seed rates used by the farmers were almost four times the recommended rate, suggesting either ignorance of the recommended rate or a deliberate attempt to improve sunflower plant stands on the sandy soils, where germination is a problem.

The survey results showed that more than 80% of the farmers made little effort to manage soil fertility in fields where sunflower was grown. This was probably a result of the farmers' strong belief that sunflower can perform well even in fields with low fertility soils. Pest and disease control measures were also limited across the NRs, with less than 20% of the farmers following the recommendations.

Sunflower kernel yields for the hybrids and open pollinated varieties were comparable, and this could be attributed to the poor management practices, which exerts a greater negative impact on hybrids than the open pollinated varieties, as the former are more sensitive to poor management than the latter. The average kernel yields observed in the survey were higher than the Central Statistical Office national averages of 500 kg/ha. This could be a result of an overestimation of yield by some farmers.

Prior to 1995, most sunflower seed was marketed through the Grain Marketing Board. Ninety-eight per cent of the sample farmers surveyed across the NRs did not use any credit for sunflower production because very few seasonal loans were provided by the Agricultural Finance Corporation for crops other than maize and cotton.

The survey results also indicated that there was a potential for improving sunflower yield in communal areas, particularly if farmers were encouraged to adopt the recommended cultural practices, such as improved varieties, timely weeding, control of pests and diseases, and applying adequate fertilizer. With good management, the production potential of the available varieties was much higher than the average yields obtained by the sample farmers. Therefore, adoption of improved varieties and better management can substantially improve productivity and overall production of sunflower in the communal farming sector.

Sunflower management practices in the commercial farming sector

The survey results revealed that the respondents had grown sunflower for up to 12 seasons, suggesting that a considerable percentage of farmers had found the crop viable enough to warrant growing over the years. About 60% of the sample farmers were growing improved varieties and, overall, sunflower management practices followed by commercial farmers were in accordance with the recommendations.

The average sunflower yields obtained by the sample of commercial farmers were above the national average, but below the potential realized at research stations. The average yield for the 1995/96 season was 1423 kg/ha, and 32% of the sample farmers realized a yield level of over 2 t/ha.

Impact assessment results

In estimating the NPV of the income stream, the social time preference rate of 10% was used. The estimated NPV for the investment at 10% discount rate was Z$5.33 million, as shown in Table 18.1. The associated IRR (ROR) for this investment over the years 1976–2000 was 12%. Given the social time preference rate of 10% (real), the investments in sunflower R&D were socially profitable. When the research and extension cost was increased by 25% in a sensitivity analysis, the estimated ROR declined to 6% (Table 18.1) indicating that the estimated ROR is

Table 18.1. Economic impacts of sunflower research and development for various scenarios.

Scenario	Net present value (10% discount rate) Z$ millions	Internal rate of return (%)
Base run	5.33	12
Research and extension costs increased by 25%	negative	6
Price based on oil content using current adoption rate		
Oil content ratio 2.7%	60.01	23
Oil content ratio 2.0%	37.86	20
Price based on oil content and entire area under improved varieties		
Oil content ratio 2.7%	119.97	29
Oil content ratio 2.0%	72.13	24

very sensitive to changes in research costs. Accordingly, the NPV of the benefit stream at the 10% discount rate was found to be negative.

Using current adoption rates of improved sunflower varieties, if the pricing is based on oil content, the ROR for the investment increased up to 20 (for the oil content ratio of 2%) and 23% (for the oil content ratio of 2.7%). Correspondingly, the estimated NPV at the 10% discount rate increased up to Z$60 million. If one assumes the entire areas under improved varieties and pricing based on oil content, then the ROR increased up to 29% and 24%, respectively, for the oil content ratios of 2.7% and 2.0%. The corresponding NPVs for the investment at the 10% discount rate are Z$120 million and Z$72 million (Table 18.1).

Conclusions

The major conclusions and implications of the impact assessment study were:

- Sunflower is grown in the communal areas across all three NRs. The area under sunflower has increased substantially over time. The production potential of the available improved sunflower varieties is much higher than the average yields obtained by the communal farmers. The farmers consider it to be a 'scavenger crop', thus use very little purchased inputs. The adoption of improved varieties and better management practices can substantially improve the productivity and overall production of sunflower in the communal farming sector. There is a need for on-farm testing of improved agronomic practices to assess their appropriateness to the socio-economic conditions of the specific target groups in the various NRs.
- Currently, the selection of varieties is based on oil content, but pricing is based on weight and not on oil content. Therefore, there is no incentive for farmers to use improved varieties. If the price is based on oil content, based on the current level of adoption, the ROR for the R&D investment is almost doubled, and the relative profitability of the crop at the farm level will improve significantly. Therefore, there is a critical need for reviewing the pricing policy of sunflower.
- Given current pricing policy of sunflower, future breeding work should target higher yield and oil content. Given the low level of input use 'Nitrogen', efficiency can also be a criterion in breeding and selection.

- With the emergence of the seed companies as alternative sources of improved varieties, there is also a need to critically examine the role of the public sector, private sector and NGOs in sunflower R&D in Zimbabwe.
- The main lesson from the various impact studies conducted so far is that an ex-post impact assessment is difficult and costly if no mechanism is in place to collect the relevant data on a continuous basis, i.e. a monitoring and evaluation system for R&D activities. It is also clear that the usefulness of the ex-post impact studies is limited, and the greatest benefit of impact assessment can only be realized if it is used as a planning tool, i.e. for ex-ante analysis.

Limitations of the study

The limitations of this study are related to the availability and quality of data used in this assessment. Given the data problems and techniques used, it was difficult to separate the effects of the different components of sunflower R&D.

- The yield levels without sunflower R&D, which are important in the calculation of benefits, had to be estimated. Extrapolations and opinions of key informants were used to calculate the 'without' technology yield levels.
- The response from the mail survey was much lower than expected; this reduces the level of confidence in interpreting the results for the commercial farming sector. There could be a bias resulting from self-selection.
- The absence of appropriate data made it impossible to adequately assess the impact of sunflower R&D on the environment, gender, employment and food security.

Despite these limitations, the study is useful in providing information on production practices of both large-scale and small-scale farming sectors. The results of this study clearly indicate that the ROR on past investments in sunflower R&D in Zimbabwe is attractive enough to warrant continued investment. Farm level and social profitability could be substantially improved if pricing is based on oil content.

Acknowledgements

The authors would like to express their thanks to the various individuals and institutions who contributed to making this study possible. Special thanks to the following: the Director and relevant institutes of the DR&E for granting the permission to conduct the study and making information available. The Commercial Oilseed Producers Association of the Commercial Farmers Union for providing the sampling frame for the large-scale commercial farmers. AGRITEX for providing the enumerators who assisted in the survey of smallholder farmers. The Southern African Center for Cooperation in Agricultural and Natural Resources Research for providing technical and financial assistance for conducting the study through a grant from the United States Agency for International Development. Last but not least, the farmers, both large-scale commercial and small-scale, who devoted a considerable amount of time and patience in completing the questionnaires. However, the authors are solely responsible for any errors or omissions in this chapter.

References

Central Statistical Office (CSO) (1996) *Quarterly Digest of Statistics*. CSO, Harare, Zimbabwe.

Commercial Oilseed Producers Association (COPA) (1986) *Oilseeds Handbook, 1981*. National Farmers Union, Harare.

Johnson, P., Zitsanza, N., Sukume, C. and Rukuni, M. (1994) *The Oilseed Sub-Sector in Zimbabwe.* World Bank, Washington, DC.

Mtindi, K. (1985) Crop Variety Release Notice. 'Msasa' hybrid sunflower. *Zimbabwe Agricultural Journal* 82, 164.

Mudzana, G. (1990) *Crop Variety Release Notice. 'Mopane' hybrid sunflower.* Department of Research and Specialist Services, Harare, Zimbabwe.

Weimann, H. (1972) *Agricultural research and development in Southern Rhodesia, 1890–1923.* Occasional paper Number 4. Department of Agriculture. University of Rhodesia, Salisbury.

World Bank (1994) *Infrastructure for Development.* World Bank, Washington, DC.

19 Impact Assessment of Cotton Research in Zimbabwe: 1970–1995

M. Mudhara, P. Anandajayasekeram, B. Kupfuma and E. Mazhangara

Abstract: This study analyses the impact of cotton research and development in Zimbabwe from 1970 to 1995. A benefit–cost analysis calculated a 47% rate of return to cotton R&D investments and showed that the investments in past cotton R&D were profitable. The assessment of the institutional impact showed that the success of the cotton industry was a result of carefully designed research, extension, marketing, credit and farmer training programmes. Overall, the study showed that integrated commodity-based research and development can generate positive dividends to public sector investment.

Background

Agriculture has always played an important part in the economy of Zimbabwe (CSO, 1989). It employs about 25% of the population in the formal labour force, but indirectly employs 70% of the population. In Zimbabwe, cotton research has produced a number of new varieties and accompanying technologies. However, these efforts were threatened by the reduction in resources by the government. In addition, the overhauling of the policy environment affected the productivity of the cotton industry. Therefore, there was a need to assess whether the impact of cotton research justified the current levels of investment.

Objectives

This study assesses the impact of cotton research, extension and the enabling environment in Zimbabwe over the period 1970–1995. The specific objectives were: to trace the evolution of cotton research, and the enabling environments; assess the impact of cotton research and the enabling environments; and to develop policy recommendations for future investments in research and the enabling environment.

History of cotton research

Cotton research was launched at the Cotton Research Institute (CRI) in Kadoma in 1925. The CRI developed a number of new cotton varieties and appropriate technologies.

Private agrochemical companies and universities also conducted some cotton research. In the 1960s cotton researchers developed varieties resistant to pests and diseases such as jassid and bacterial blight. The CRI has conducted trials on large-scale commercial farmers' (LSC) farms since the mid-1960s, and in the communal areas (CAs) from 1980. In CAs, research addressed moisture conservation, methods of scouting pests and selecting varieties for low input management. Recommendations on plant spacing, fertilizer use levels and moisture conservation techniques have been released.

Cotton extension and training

The field and technical divisions of AGRITEX were responsible for cotton extension (Reid, 1971). The crop specialists in the technical division trained extension workers who were in contact with farmers on a regular basis. However, with the extension worker to farmer ratio at 1 : 800, extension workers could not effectively meet the needs of individual farmers (Pazvakavambwa, 1994). The use of motorcycles has increased the ability of extension workers to contact farmers. Most extension workers contact farmers in groups. The Cotton Training Center (CTC) was established in 1973 by the Commercial Cotton Growers Association (Gledhill, 1979). In 1980 the CTC started to train both extension personnel and CA farmers.

Methodology

The rate of return calculations

Calculation of the rate of return (ROR) involves estimation of benefits, and the cost of research and transfer of technology. A simple benefit–cost analysis framework was used (Schuh and Tollini, 1979). This is the simplest version of the index-number approach that requires no elasticity estimates. For the 'without' technology scenario, it was assumed that cotton yields would have declined continuously. New cotton varieties and production technologies maintained average farm yields at 1684 kg/ha for LSC farming and 713 kg/ha for smallholders. Without the new cotton varieties, average farm yields would have fallen as old varieties become more susceptible to diseases and pests. The study, therefore, conservatively assumed an annual average rate of decline in average yields of 5%. A more conservative rate of 1% and a more optimistic rate of 10% were used in sensitivity analysis. Average farm yields of cotton increased by about 10% whenever a new variety was introduced. After the introduction of a new variety, average yields were initially maintained for a few years, but started to decline as disease and pest tolerance declined, triggering further releases of new varieties. The next new variety would push the yields back to the level before the decline. Hence, the first increase in average farm yields after 1970 was the only one that is considered in this study.

Export prices reflected the real value of cotton production in Zimbabwe. This is reasonable, as 70% of the cotton crop was exported during the 11-year period up to 1991. In addition, the domestic price was 70% of the export price over the same period. Prices were deflated using a CPI with 1990 as base year. The CPI was also used for deflating research and extension costs. This CPI was constructed for the lower income group. The yields without technology were estimated for both the LSC and CA farmers. The yield gain was then calculated. Multiplying the yield gain by the price and total area gave the gross benefits of cotton research and extension. Net benefits were obtained by deducting research and transfer costs from the gross benefits. Depreciation of capital assets was omitted, thus overestimating cotton research and extension benefits.

AGRITEX costs were lumped for all commodities. The ratio of cotton area to the total crop area was used for estimating the cotton extension budget. To estimate the annual budgets for cotton extension, total AGRITEX budget was divided into livestock, and crops budget, using the ratio of the value of livestock production, or crop production to total value of agricultural production. The crop budget was then multiplied by the ratio of the area under cotton, to the total crop area to obtain an estimate of the cotton extension budget. CTC extension costs for years after 1980 were obtained from its actual budgets. For the years before 1980, costs were estimated through backward extrapolation using exponential smoothing. The incremental net benefit flow was calculated by subtracting the total cost flow from the net benefits. The incremental net benefit flow over the years was converted into comparable values by discounting, to take account of the time value of money.

R&D expenditures were from research and extension. The research expenditures were lagged by 10 years. This period was considered long enough for the cotton R&D programme to produce an impact at the farm level. The impact of using longer and shorter lag lengths was examined as part of sensitivity analysis. The discount rate approximates the social time preference rate. This was taken as the rate of interest on long-term government bonds. During the period 1970 to 1995, the lending rate averaged 15% (World Bank, 1994). Sensitivity analysis was conducted using 10% and 20%.

Data collection

A sample survey was conducted to characterize cotton production practices in smallholder and LSC farming areas in Zimbabwe. One hundred and eighty smallholder farmers were interviewed. A mail questionnaire was administered to 66 LSC farmers. The number of farmers selected for each farm type depended on the area they planted to cotton. The proportions were determined from the average area under cotton, planted over the 5-year period 1985/86–1989/90. Multistage sampling was adopted for smallholder farmers. At the first stage, four of the country's nine depots were selected. Farmers were grouped into clusters and three clusters were randomly selected per depot. Finally, 15 farmers were selected randomly per cluster. One hundred and eighty smallholder farmers were selected and interviewed. Thirty-two LSC farmers returned completed questionnaires. For a detailed discussion of the cotton production systems in the CAs, LSC farms and the survey results, see Mudhara *et al.* (1995). The results of the surveys were used in studying the impact of cotton R&D.

Impact of Cotton Research, Extension and Enabling Environment

Past investments in cotton research and extension had intermediate and people-level impacts. Intermediate impacts affect the ability of the research system to efficiently produce the research product. They occur through changes in human resource development, availability of financial resources, the effectiveness of the research system and the strength of linkages that the research system establishes with other institutions. People-level impact looks at the socio-economic impact, measured using ROR, and/or benefit–cost ratios. People-level impact also includes changes in food security, gender and environmental implications. These impacts of past cotton research, extension and enabling environment are reported in the following sections.

Intermediate impacts

The long history of cotton research and extension in Zimbabwe made skilled manpower and financial resources available. In turn, this attracted scientists of high calibre into the research system leading to more productivity. In addition, cotton research and extension affected the organization of support services to cotton producers. Credit packages based on CRI research recommendations were available. CRI staff trained CA farmers and farm workers through a collaborative programme of the CTC.

Environmental impact

The assessment of the impacts of cotton R&D on the environment is qualitative. On the positive side, the research programme yielded positive environmental benefits through recommendations on land preparation and pest and disease management. In addition, the recommendations primarily aimed at reducing production costs to farmers. The environmental benefits were positive externalities.

People-level impact

The yields without technology were estimated for both the LSC and CA farmers. The yield gain was then calculated. The gross benefits of cotton research and extension are presented in Table 19.1. The gross benefits have increased throughout the years with fluctuations reflecting variability in weather patterns. Net benefits were obtained by deducting research and transfer costs from the gross benefits.

The payoff to investments in cotton research and extension are presented in Table 19.2. The net benefit streams were negative for the first 8 years. Thereafter, the net benefits were positive and increased with time. The variability in net benefits closely mirrored the weather pattern. However, the upward trend in net benefits was apparent. The net present value (NPV) of the net benefit stream was used for determining whether there was a payoff to the investments in cotton research and extension.

Table 19.1. Gross benefits of cotton research and extension, 1970–1995.

Year	Gross benefits (Z$ million)	Year	Gross benefits (Z$ million)
1970	0	1983	34.45
1971	0	1984	59.40
1972	0	1985	70.48
1973	2.40	1986	55.10
1974	3.32	1987	81.42
1975	15.43	1988	124.75
1976	27.03	1989	101.37
1977	24.71	1990	96.54
1978	28.14	1991	125.27
1979	25.67	1992	32.90
1980	28.49	1993	121.88
1981	31.52	1994	108.52
1982	28.81	1995	136.99

Source: Cotton case study estimates.

Table 19.2. Net benefits of cotton research and extension, 1965–1995 (constant 1990 Zimbabwe dollars).

Year	Net benefits (million Z$)	Year	Net benefits (million Z$)
1965	(0.47)	1980	27.32
1966	(0.55)	1981	30.18
1967	(0.60)	1982	27.58
1968	(0.60)	1983	33.38
1969	(0.64)	1984	58.34
1970	(0.64)	1985	69.34
1971	(0.89)	1986	53.94
1972	(1.04)	1987	80.14
1973	1.32	1988	123.50
1974	2.45	1989	100.05
1975	14.52	1990	95.37
1976	26.13	1991	124.11
1977	23.78	1992	31.90
1978	27.25	1993	121.08
1979	24.75	1994	107.69
		1995	135.43

Notes: Figures in parentheses are negative values.

Table 19.3. Rate of return at 1%, 5% and 10% yield declines.

Measure of return	5% Yield decline	1% Yield decline	10% Yield decline
IRR (%)	47.17	45.63	49.19
NPV at 10% (million Z$)	149.73	101.55	254.44
NPV at 15% (million Z$)	59.04	42.23	94.07

Source: Cotton case study estimates.

The internal RORs (IRRs) and the NPVs at 10% and 15% are shown in Table 19.3. Over the period 1965–1995, investments in cotton research and extension generated a ROR of about 47.17%. The rate was above the real interest rate in Zimbabwe. It compared favourably with figures from other parts of the world. Over the same period, and using a 10% discount rate, cotton research and extension investments generated Z$149.73 million, discounted back to 1965, an annual NPV of about Z$4.83 million. A discount rate of 15% still showed benefits of Z$59.05 million or Z$1.9 million annually.

Sensitivity analysis

The assess the extent to which changes in the variables that were assumed would affect the ROR estimates, sensitivity analysis was conducted by varying the yield differential, the period of benefits flow, the price of cotton and the costs of cotton R&D. Initially, a yield difference between the improved and the existing cotton varieties was assumed at 5%. The ROR was recalculated assuming 1% and 10% yield levels (see Table 19.3).

Changing the rate of decline in yields in the absence of cotton R&D did not significantly change the measures of return, especially the IRR. There were some significant changes in the NPV at both 10% and 15% levels. However, the focus was on the IRR, which at the 1% yield decline only fell by less than two percentage points to 45.63%. When the rate of yield decline in the absence of cotton R&D was assumed at 10%, the IRR increased by 2 percentage points to 49.19%. These changes were not significant.

Conclusions

The background to this study highlighted the pivotal role played by cotton in the agriculturally based Zimbabwean economy, in terms of economic growth, export earnings, employment, income generation and food security for rural people. Thus, the cotton research programme was one of the oldest, best supported and well-organized sectors in agriculture. This contributed to its success. The cotton survey and the impact assessment of cotton R&D programme highlight the prospects of cotton in Zimbabwe. First, the area under cotton in communal areas had grown substantially since 1980. The trend was continuing. These findings support the belief that with appropriate technologies and services, communal farmers are able to raise their incomes.

Secondly, the wide gap in cotton yields in CA and LSC farming sectors implied that there was scope for increasing cotton production by raising yields in CA. Low yields in CAs were due to limited use of fertilizer. Third, CA farmers faced transport problems in moving their cotton to intake points.

The ROR to cotton R&D of 47% showed that the investments were worthwhile. This ROR was higher than the returns from alternative investments of public funds. Sensitivity analysis, conducted by reducing the price of cotton, increasing the level of estimated R&D costs, and reducing the rate of decline in cotton yields without R&D did not significantly change the estimated ROR. This high ROR justified past investments, and indicated that there was under-investment in cotton R&D. The results of the assessment on the environmental impact of cotton R&D programmes were mixed. Recommendations on land preparation, moisture management and scouting had positive environmental effects, while the use of pesticides and the monocropping of cotton had negative externalities. Programmes that take account of environmental concerns were needed. Biophysical data to perform more complete environmental impact assessment were not available.

The assessment of the institutional impact revealed that the success of the cotton industry was a result of carefully designed research, extension, marketing, credit and farmer training programmes. These together have improved cotton production in the country. This study showed that an ex-post impact assessment was difficult when relevant time-series data are not available. Monitoring and evaluation systems to facilitate impact assessment, planning and priority setting were needed.

The limitations of this study relate to the availability and quality of data used in this assessment. In the absence of baseline data, the process of estimating yield level without cotton R&D was extremely difficult. The opinions of key informants were used to calculate the 'without' technology yield levels. The absence of appropriate data made it impossible to adequately assess the impact of cotton R&D on the environment, gender, employment and food security. As a result, the benefit and/or costs may have been underestimated or overestimated. Owing to wide ranging changes taking place in the cotton industry, it was difficult to predict the future impact of cotton R&D. Despite these limitations, the study provides information on production practices by both LSC farmers and communal farmers in Zimbabwe. The results indicate that the ROR on past investments in cotton R&D in Zimbabwe is attractive enough to warrant further investment.

References

CSO (1989) *Statistical Yearbook.* Central Statistical Office, Harare.

Gledhill, J.A. (1979) The Cotton Research Institute, Gatooma. *Rhodesia Agricultural Journal* 76, 103–118.

Mudhara, M., Anandajayasekeram, P., Kupfuma, B. and Mazhangara, E. (1995) *Impact Assessment of Cotton Research and the Enabling Environment in Zimbabwe, 1970–1995.* SACCAR and SADC, Gaborone.

Pazvakavambwa, S. (1994) Extension policy. In: Rukuni, M. and Eicher, C.K. (eds) *Zimbabwe's Agricultural Revolution.* University of Zimbabwe Publications, Harare, pp. 104–113.

Reid, M.G. (1971) An agricultural extension program at Gokwe. *Rhodesia Agricultural Journal* 68, 54–60.

Schuh, G.E. and Tollini, H. (1979) *Costs and Benefits of Agricultural Research: State of the Arts.* World Bank Staff Working Paper No. 360. World Bank, Washington, DC.

World Bank (1994) *World Development Report.* World Bank, Washington, DC.

20 The Socio-economic Impact of Proteaceae Research and Technology Development in South Africa

J.M.C. Esterhuizen (neé Wessels)

Abstract: This is a study of the socio-economic impact of the Proteaceae Technology Development and Transfer Programme. Farm-level data were collected from the major players in the industry, including cultivators, harvesters, agents and nurseries. The five major production regions were the Western Cape, South Western Cape, Southern Cape, Cape Peninsula and Eastern Cape, but other small regions in the Republic of South Africa (RSA) were also included in the study. Pre-tested mail questionnaires, supplemented with personal interviews were used to collect the primary data. This chapter reports the socio-economic impacts of the programme over the period 1974–2005. Results showed that the rate of return for the financial and economic analyses ranged between 7% and 12%, indicating that the research programme was a profitable investment to society. The Proteaceae R&D programme contributes to the conservation of biodiversity, has an institutional impact in terms of human capital development, technology transfer and employment generation.

Introduction

All countries, particularly developing countries, are faced with the basic economic problem of allocating limited resources, such as labour management and administrative, capital, land and other natural resources, and foreign exchange, for many different uses. Using limited resources in one enterprise reduces the resource availability in another. Thus, there are clearly trade-offs in resource allocation among the research and development programmes until there is no resource scarcity (Squire and van der Tak, 1975). Therefore, choices need to be made on the allocation of scarce resources for optimal use and to help a country in achieving its developmental objectives.

The basic rationale for research impact assessment (ex-ante) is to improve the efficiency of allocation of the research resources, as well as to improve the standard and effectiveness of decision making. Economic theory suggests, that to maximize the social output, resources should be allocated where their contribution is greatest at the margin. At the same time, research managers and other decision makers are interested in seeking feedback on what has been accomplished, in order to help them to direct the course of future work (ex-post).

 Impact of Science on African Agriculture and Food Security (eds P. Anandajayasekeram, M. Rukuni, S. Babu, F. Liebenberg and C.L. Keswani)

Impact assessment is a form of evaluation that deals with the effects of project output on the target beneficiaries. Impact assessment of technological activities and their transfer can be undertaken before initiating the research (ex-ante), or after the completion of the research activity (ex-post) including the technology transfer (Anandajayasekeram *et al.*, 1996). The purpose of conducting ex-ante impact assessment is to predict and compare outcomes, and to support efficient resource allocation through economically rational decisions. An impact assessment after the completion of the programme serves to evaluate the impact and to provide feedback for researchers, research managers, planners and policy makers. Thus, it is useful for accountability purposes, to establish the credibility of the research, and to justify allocations of scarce research resources.

Background

The Proteaceae flower industry is an integral part of horticulture in the Western Cape Province of the RSA. It originated in the harvesting and marketing of floral material, collected from the natural habitat in this area. These flowers were sold on the market and streets of Cape Town and dried flowers were exported to Europe as far back as the 19th century. This business flourished and has since become an integral part of the horticulture export industry of the RSA. The monetary value of the Fynbos industry is currently valued at R81.7 million. During 1995, 2860 t of fresh flowers were exported, which represents a 15.5% increase over the previous year. The industry also shows promising signs of growth potential. However, several aspects have been identified, which will need attention for this growth to attain its full potential. The limitations identified by the Agricultural Research Council (ARC) Roodeplaat Fynbos Unit include aspects of cultivation and marketability of flowers. This concern relates to the adding of time, form and place utility (Shepherd, 1968). These became the critical performance areas of the ARC-Fynbos Unit. The objectives of the breeding programme for the Proteaceae genera attempts to address limitations as they relate to time, form and place utility.

The Proteaceae research programme was initiated in 1974, as a programme of the Department of Agriculture. Currently, the programme is managed by the ARC-Fynbos Unit, which is a unit of the Vegetable and Ornamental Plant Research Institute in collaboration with the industry.

After the first flower consignments were successfully exported, the demand arose for a continuous supply of certain types or forms of products during the peak European marketing season. Also, during the years of political and economic isolation, opportunities arose for other countries in the Southern African Development Community region and elsewhere to initiate the cultivation of indigenous South African flora. The danger therefore exists for RSA producers that they have lost or will lose their market share to other competitors in the Southern Hemisphere, if the requirements of the international market cannot be successfully met. Even though the Fynbos industry originated from harvests in the wild, these products do not always adhere to international standards. The marketing of low-quality products negatively influences the price structure and could result in the RSA losing its market share permanently, against high-quality products in the international market. The quality of RSA products can, however, be upgraded by the use of improved cultivation practices. It has also been observed that new products, improved cultivars and plants flowering at the appropriate season (time) can be produced by utilizing the exceptionally diverse genetic resources available in the RSA. Through appropriate cultivation practices, RSA Fynbos producers could improve the quality and productivity of their material to successfully serve existing markets and develop new export opportunities. The overriding criterion in cultivar development is to produce unique products, which are different from the products harvested from the wild or obtained from seedling cultivation, and adhere to the requirements of the market demand, as it concerns time, form and place utility. These released cultivars are available as clonal material in the form of rooted or un-rooted cuttings (Wessels, 1998).

The role of production technology is a crucial element in this chain of activities, i.e. the release of clonal material. It was, therefore, visualized that successful research and development (R&D) will support the continued growth of the Fynbos cut flower industry, and will enhance the opportunity of turning a natural resource into valuable foreign currency earner. Apart from the monetary gains, other development consideration was the employment opportunities created by the labour-intensive nature of this industry, especially among the poorer population groups. This added a fourth utility dimension, namely social or equity utility. Given this scenario, the ARC-Fynbos Unit requested the impact assessment of technology R&D, not only to gauge the efficiency of this programme in terms of its value to society, but also to gain insight, into its future direction and investment potential. This study reports on the impact assessment analysis conducted on R&D on Proteaceae technology.

Objectives

Given its long history in developing and distributing cultivars, the Proteaceae technology R&D programme was chosen as a case study to assess the various technical and economic impacts. Therefore, the objectives of this study were to: describe the production and marketing system of Proteaceae in the RSA; establish the technology research and development priority areas, and assess the ARC's strategy in this context; trace the evolution and predict the future impact of technology R&D with respect to Proteaceae and propose a methodology for impact assessment analysis, apply this methodology and assess the impact of R&D initiatives with respect to Proteaceae.

Methodology

A comprehensive approach was taken to evaluate the impact of the Proteaceae R&D programme. The effectiveness, i.e. what was achieved vis-à-vis what was planned, was assessed through the application of a Logical Framework Analysis (ABOS, 1996; Farrington and Nelson, 1997). From this analysis, the main performance areas were identified.

The socio-economic impact of the Proteaceae R&D activities was estimated using the cost–benefit approach, a variation of the surplus approach. The advantage of the method is that it does not require elasticities (Anandajayasekeram *et al.*, 1996). These measures were not available for wild flowers at the time of the study. Only the costs and benefits accruing to domestic producers and consumers were included in the analysis. Primary data on the current production and marketing practices of the industry were collected through a mail questionnaire, supplemented with personal interviews. The industry is perceived to consist of veld harvesters, cultivators, wholesale agents and nurseries. All these stakeholders were considered in the study. The five main Proteaceae production regions included the Western Cape (WC), South Western Cape (SWC), Southern Cape (SC) and Cape Peninsula (CP) in the winter rainfall area, as well as the constant rainfall area of the Eastern Cape (EC). Proteas are also produced to a smaller extent in other areas of the RSA, including the summer rainfall area of Gauteng, Mpumalanga and the highlands of Kwazulu-Natal (Wessels *et al.*, 1997). Secondary data were collected from the ARC-Fynbos Unit and appropriate published sources.

Results

A typical agricultural research and development programme has several impacts on society. This not only relates to the direct product of research, but also considers institutional and social impacts. Institutional impact refers to changes in the institution's enabling environment, while social impact refers to economic, financial, social and environmental aspects (Anandajayasekeram *et al.*, 1996).

The Proteaceae production system

The industry consists of nurseries, cultivators, harvesters and agents (who can also be exporters). The ARC-Fynbos Unit produces potted plants and cuttings and also disseminates information on Proteaceae production to the industry. The South African Proteaceae Producers and Exporters Association (SAPPEX), also fulfils a further supporting role. The structure of the Fynbos industry and the various stakeholders involved as assessed at the time of the study (1997) are summarized in Fig. 20.1.

Most cultivators started producing Proteaceae during the period 1990–1997. At the time of the study, 19% of the sample farmers cultivated and also harvested products from the wild. Five different Proteaceae products (*Protea*, *Leucospermum*, *Leucadendron*, Cape Greens and dried products) were identified, which could be produced in three different ways, i.e. established by

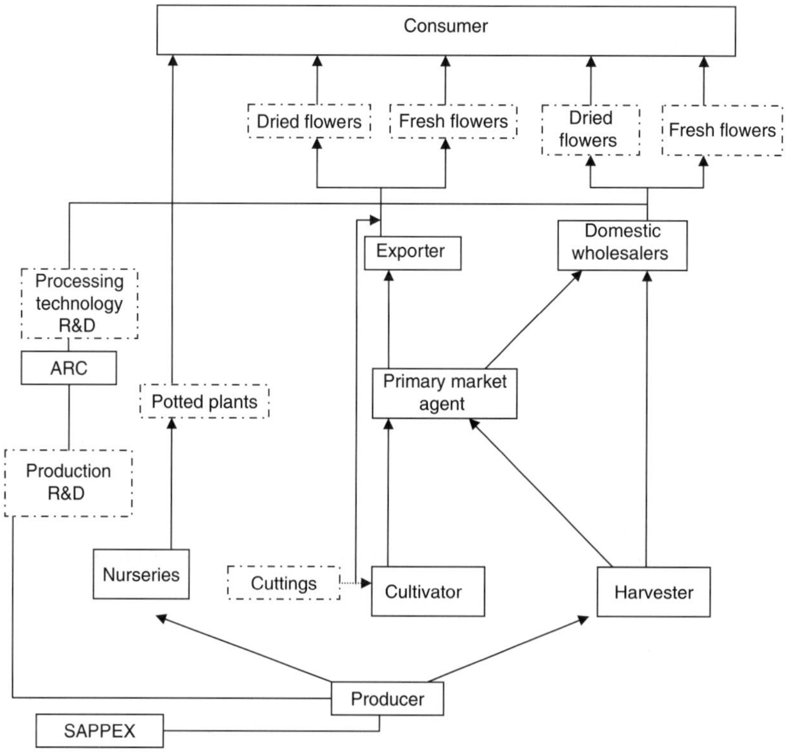

Fig. 20.1. Structure of the Fynbos industry.

broadcasting the seed or in plantations, using either seedlings or cuttings for establishment. According to the farmers, seedling and cutting establishment methods for Proteaceae were increasing (Wessels *et al.*, 1997). The veld harvesters are producers who utilize the natural veld as a source for their products. Ninety-two per cent of the sample farmers started harvesting after 1970, with the majority (38%) starting during the period 1990–1996.

Agents/exporters played the part of middle men and provided market outlets and marketing information to producers.

Effectiveness analysis

The wild flower industry in South Africa has enormous growth potential. However, growth of the industry will only take place if cultivation of commercially viable Fynbos plants is undertaken. Cultivation in plantations can only be successful if the industry is supported by research institutions, as research information is necessary to improve the cultivation practices of Proteaceae in orchards. The costs incurred by Proteaceae cultivation are fairly high, and the cost of on-farm research can also be very expensive; therefore, there is a need for a reliable source of research information on the cultivation of Proteaceae to improve the time, form and place utility characteristics of Proteaceae products. Improved access to production and value-added activities in the Proteaceae industry will enhance social utility and employment equity dimensions especially of the poorer groups. Therefore, the critical performance areas for research are to meet the challenges of time, form, place and social utility, through the activities of the Fynbos industry.

Effectiveness analysis measures the degree to which the project has achieved the intended objectives. First, the performance of the research programme was conceptualized through the application of a Logical Framework Analysis. The 'problem tree' was structured and entails the identification of relevant issues, and the logical positioning of these in terms of causes and effects in order to develop a clear and logical problem statement (Farrington and Nelson, 1997). The main problem, derived from the Logical Framework Analysis, was of a suboptimal income, due to the production and marketing system, which did not fully exploit time, form and place utilities. The ARC-Fynbos Unit strategy, and the actual objectives of the Proteaceae programme, largely confirmed this finding. The study suggested that the Proteaceae R&D programme was effective in addressing these problems.

Efficiency analysis

The industry started with harvesting from the wild, farmers then started to broadcast seed as a form of Proteaceae production and also to produce seedlings for planting Proteaceae. Without R&D efforts, the industry would have continued to depend on the wild harvest. With the research information available, farmers generally changed from harvesting to cultivation with vegetative propagation material.

In general, the findings and outputs of the Proteaceae programme had very limited effect on those producers who were broadcasting and/or using seedlings for cultivation. However, some of the agronomic recommendations, provided by the ARC-Fynbos programme, may have had a general effect on the productivity of cultivated orchards, using seed broadcasting and seedlings. Similarly, extension messages on pruning and post-harvest handling may have improved the quality of the products. No information was available to assess these generic gains. The only cultural practice that was a very certain result of the R&D programme was the shift to using new clonal materials. If a significant generic impact is expected, the project must include this cultural practice. Indications, however, are that many farmers entered the industry due to the new technology, while the impact of the new technology (cloning and cuttings) on broadcasting and seedling farming was relatively small. The position accepted for purposes of this study is that of increased efficiency due to the R&D and production of clonal material. Therefore, it was decided that the benefits attributable to organized cultivation using the clonal materials are the only clearly identified and quantifiable output of the Proteaceae technology R&D programme. This situation, however, resulted in underestimation of the benefits of the Proteaceae technology R&D programme, because the agronomic recommendations developed and the extension messages on pruning and post-harvest handling were also applicable, and had been applied to the other practices within the Proteaceae industry (broadcast and seedling establishments). However, these were not counted as an additional benefit. This needs to be kept in mind in interpreting the results of this study (Wessels *et al.*, 1997).

By using clonal material, the producers are in a position to produce better quality products. Clonal material also provides the opportunity to select for a flower type that is flowering at a specific time and has a specific appearance. It enables the farmers to cultivate wild flowers in established orchards.

A social cost–benefit framework was used in the analysis. The financial analysis was conducted for the individual farmers, and a separate financial and economic analysis for the Proteaceae industry. The different impacts of the Proteaceae technology R&D programme were further analysed, these were: environmental impact, institutional impact, socio-economic impact and spill-over effects. The financial, economic and different impact analysis of the Proteaceae technology R&D programme was conducted to determine the efficiency of the programme.

Financial analysis

The objective of a financial analysis is to assess the financial position of stakeholders involved in an enterprise, as well as the overall costs and affordability. A financial analysis was conducted for both individual farmers and for the industry. The financial rate of return (ROR) for individual farmers was estimated by including all costs and benefits, on a per hectare basis accruing to those farmers who adopted the new Proteaceae technologies. A real discount rate of 7% was applied. The estimated ROR for proteas was 29% and the net present value (NPV) was R1.73 million for the period 1991–2005 at a 7% real discount rate (1997 constant values). The average NPV per annum per hectare for Protea was R115,333. The ROR for *Leucospermum* was 45% and the NPV R1.66 million (average NPV per annum per hectare was R110,666). The ROR for *Leucadendron* was 33% and the NPV was R13 million at a 7% discount rate (average NPV per annum per hectare was R86,666). The analysis clearly indicated that the use of clonal material was profitable for individual farmers.

The industry analysis included the total area established under clonal material of Proteaceae products. The cost and benefits incurred by research activities were also included. The area under each of the Proteaceae production systems (i.e. broadcast seed, seedling or cutting establishment) was not available at the time of the study. It was, therefore, decided to use the distributed cuttings as the basis for estimating the area under improved cultivars. The discount rate (7%) was the same as that used for the financial analysis for farmers.

Apart from the distribution of cuttings the programme has also released several cultivars and agronomic recommendations. However, the only clearly measurable impact was that associated with the development and dissemination of clonal materials. Only the benefit associated with this effort was included in the analysis. However, the entire cost of the research programme was included. Appropriate data and records were not available to permit the joint cost allocation procedures usually needed in such a case (Gittinger, 1982; Wessels *et al.*, 1997). When interpreting the results from the analysis, it should therefore be kept in mind that the costs are overestimated and the benefits underestimated. The estimated financial ROR of the Proteaceae R&D investment to the industry was 7.7%. The NPV at a 7% (real) discount rate was R4 million in 1997 constant values. The estimated financial ROR was larger than the financial opportunity cost of capital (7%). The NPV was also positive, which indicated that the R&D investment is financially profitable (Wessels, 1998).

Several sensitivity tests were conducted, to incorporate the over and under estimation of costs and benefits, for example, increase in area under Proteaceae cultivation, reduction in overhead and administration costs, reduction in total research costs, an increase in gross revenue from protea, and an increase in area, and gross revenue simultaneously from Protea. These analyses clearly indicated positive investment figures for the more likely scenarios (Table 20.1). The sensitivity tests for overestimation of research cost clearly indicate that the project was profitable with an internal rate of return of 8.7 and a NPV of R9.8 million at a 7% discount rate. The sensitivity tests clearly indicate that the most likely scenario (increase in Protea area and gross revenue) delivered positive investment figures, indicating that the project should proceed.

Table 20.1. Estimated NPV and IRR for the various financial scenarios.

Scenario	Net present value: 5% discount rate	Net present value: 7% discount rate	Net present value: 8% discount rate	Internal rate of return
Base scenario	24,924,187	4,333,223	(1,041,237)	7.70
Sensitivity tests				
Increase in area Protea under cultivation				
S 5% increase from year 2000	26,386,040	5,035,812	(551,464)	7.90
S 10% increase from year 2000	27,847,893	5,738,400	(61,691)	8.00
S 50% increase in Protea and the rest increased at 10%	33,708,948	8,457,615	1,800,568	8.30
Reduction in overhead and administration				
S 25% reduction	26,106,757	5,086,121	(437,293)	7.90
S 50% reduction	27,289,327	5,839,019	166,651	8.00
Reduction in total research cost				
S 25% reduction	29,253,761	7,080,654	1,159,042	8.25
S 50% reduction	33,583,334	9,828,084	3,359,322	8.70
Increase in gross revenue from Protea				
S 25% increase	36,281,746	10,146,953	3,153,415	8.60
S 50% increase	47,639,305	15,960,684	7,348,069	9.40
S 100% increase	70,354,422	27,588,145	15,737,376	10.50
S 150% increase	93,069,540	39,215,605	24,126,683	11.50
Increase in Protea area and gross revenue per hectare from Protea				
S 50% increase in area and 100% increase in gross revenue	123,110,977	54,242,085	34,843,816	12.00
Increase in 'without' value				
S 10% increase	24,918,679	4,330,209	(1,043,486)	7.70
S 50% increase	24,896,644	4,318,154	(1,052,482)	7.70
S 100% increase	24,869,100	4,303,084	(1,063,727)	7.70

Economic impact

The economic impact measures the effects on economic efficiency associated with a set of R&D interventions. The real costs are systematically compared with the real benefits of the project, valued at economic shadow prices. Certain adjustments are required when the market price does not reflect the true economic value of a commodity or service. In the case of Proteaceae, the costs of cultivation inputs and irrigation water, as well as the value of land, were adjusted for distortions to reflect the true economic value. The cost of cultivation inputs were distorted due to the tariff charged on these products and were adjusted with the tariff protection rate (Bradfield, 1993). The economic value of land should be expressed in terms of the opportunity cost of producing alternative crops. This could be translated to the rental value of the land to produce these alternatives. The value of grazing land was found to be the most realistic measure of alternative land use. This value also represented the 'without' project value scenario, as most farmers introduced Proteaceae on grazing land. The market price of irrigation water also does not reflect the opportunity cost of irrigation water due to current and historical levels of subsidies. Three different approaches were considered to determine the economic value for irrigation water, and included the production cost value, the alternative land use and the marginal value product approaches. These shadow prices were not found to have a significant influence on the investment decision.

The long-term bond rate in the RSA, adjusted for inflation, represents society's time preference for money. This was estimated at 5% and was used as the social discount rate in the economic analysis. The estimated economic ROR was 7.5% and the NPV, at a 5% real discount rate, amounted to R22.5 million. The real bond rate of 5% is lower than the ROR of 7.5%, which indicated that the public investment in the Proteaceae R&D was profitable. In the economic analysis, sensitivity tests were conducted to test for changes in the shadow values of irrigation water and land. The ROR was found to be insensitive to a projected substantial change (100%) in both irrigation water and land values (Table 20.2).

Environmental impact

At the time when Fynbos was recognized as a commercial enterprise there was concern that these practices would, in the long-term, affect the remaining natural Fynbos vegetation on private land. Some of the issues raised included: (i) that the shift from veld harvesting to orchard

Table 20.2. Estimated NPV and internal rate of return for the various economic scenarios.

	Net present value				
Scenario	3% discount rate	4% discount rate	5% discount rate	7% discount rate	Internal rate of return
Base scenario	66,349,671.00	40,117,242.00	22,545,895.00	3,104,726.00	7.5
Sensitivity tests					
Increase in shadow value of irrigation water					
50% increase	65,301,523.00	39,384,597.00	22,033,590.00	2,855,153.00	7.50
100% increase	63,980,417.00	38,422,522.00	21,328,558.00	2,469,592.00	7.40
Increase in the 'without' value					
10% increase	66,342,610.00	40,112,104.00	22,542,133.00	3,102,671.00	7.50
50% increase	66,314,369.00	40,091,554.00	22,527,084.00	3,094,451.00	7.50
100% increase	66,279,067.00	40,065,866.00	22,508,273.00	3,084,176.00	7.50

Source: Wessels *et al.* (1997).

cultivation would result in more natural vegetation and Fynbos vegetation being removed for cultivation; (ii) that state support to Fynbos farming in orchards would stimulate interest in the establishment of new orchards, thereby changing the present land use pattern; (iii) that the establishment of Proteaceae cultivar material would result in the genetic contamination of the natural vegetation; and (iv) that the use of pesticides in Fynbos cultivation would pose a threat to biodiversity. Several mitigation proposals were put forward by Cape Nature Conservation to minimize these concerns and to conserve the natural ecology and biodiversity (Wessels *et al.*, 1997).

It was interesting to note that the majority of harvesters sampled (59%) did not notice any decrease in the number of flowers harvested per year from the same unit of land, neither had any specific plant types disappeared over the years. The average quantity of fertilizer applied was relatively low for Proteaceae species, whereas the amount of pesticides used in the control of pests and diseases was considered to be relatively higher. Given this, the cultivation of Proteaceae may not have any significant incremental environmental impact compared with enterprises such as horticulture and viticulture. However, if one is moving from broadcast seed orchards to seedling/cutting orchards, this will have a negative effect on the environment through increased contamination.

An important environmental contribution of the Proteaceae research programme related to the conservation of biodiversity in terms of gene bank accessions. The number of these accessions increased from 139 in 1974 to 1774 in 1996. This is an almost 800% increase, which is remarkable over a 22-year period. About 44 of these species have been classified as rare, endangered or vulnerable (Wessels *et al.*, 1997). Thus, there was strong evidence that the Proteaceae technology R&D programme was playing a crucial part in maintaining and conserving the biodiversity.

Institutional impact

This impact was considered in terms of aspects impacting on institutional and organizational arrangements. These included training provided by the Fynbos Unit as part of the programme, budget allocation, changes in the number and composition of the research team, and the contribution to technology transfer through the Proteaceae technology R&D programme.

A total of 65 participants were trained through the 2-day courses, and another 152 through the 4-day course over the period 1994–1996. The entire group of cultivators indicated that the short-term training should continue. This was an indication that the training was contributing positively to the industry. In addition, the programme assisted in producing eight MSc, four PhD and three BTech degrees during this period.

Sixty-six per cent of the sampled cultivators attended field days, 38% of the harvesters received some form of extension advice, and 85% of the sampled cultivators indicated they received consultancy service from the ARC-Fynbos Unit. Almost all the sampled cultivators indicated that these services should continue.

Socio-economic impact

The socio-economic impact of the research was measured in terms of contribution to employment, income distribution and other indirect impacts.

In 1974, the sampled agents/exporters received Proteaceae products from eight harvesters and two cultivators. By 1985, this number had increased to 42 cultivators and 12 harvesters. During 1996, 44 cultivators and 343 harvesters supplied Fynbos products to the sampled agents. The total mass of fresh flowers marketed increased from 1.6 million kg in 1976 to

4.8 million kg in 1996. During the same period the dried flower market increased from 1.49 million kg to 1.54 million kg (Wessels *et al.*, 1997). This represents a modest increase within the dried flower industry, but a threefold increase in fresh flowers, thereby clearly indicating that the industry has contributed significantly to job creation. During 1996, the additional employment created through the Proteaceae technology R&D activities was estimated at 858 full-time and 309 part-time jobs. This number was expected to increase to a total of 1006 full-time employees and 457 part-time employees by the year 2000. Again, these were very conservative estimates (Wessels *et al.*, 1997).

All indications were that the Proteaceae R&D activities contribute substantially to the economy of the Western Cape. From a strategic micro- and macro-modelling project analysis conducted by Eckert *et al.* (1997), the Fynbos industry was ranked ninth out of 48 sectors in the Western Cape. Whereas the total impact should not be ascribed to Proteaceae R&D activities alone, the steady beneficial influence of the R&D operations described in this study on the Fynbos industry in the Western Cape should not be discounted. A significant and positive impact could at least be claimed. This is so because the Fynbos industry creates jobs, which are generally taken up especially by the poorer population and this leads to a redistribution of income through the employment of poorer households.

The wide applicability of research over a range of agricultural production conditions or environments, often cut across geographical and national boundaries, and across commodities, is generally referred to as 'spill-over' effects. The Proteaceae materials are being used by other countries both within and outside the Southern African Development Community region. The number of flower cuttings exported increased by 182% during the period 1994–1996. These planting materials will be used in the breeding and development programmes of the importing countries, thereby contributing to the welfare gains of the producers and consumers in those countries.

Conclusions

The Proteaceae R&D programme was shown to have a positive financial, social, economic, institutional and environmental impact on society. The ROR of over 7% for all considered scenarios indicated that the Proteaceae R&D programme was socially profitable. The ARC-Fynbos Unit has also contributed significantly to human capital development, technology transfer through the multiplication and distribution of propagation materials, formal and informal training, publications and consultation services.

Though the economic ROR of the Proteaceae research programme indicated a profitable investment to society, it was not as high as the estimated returns to public sector research (44%) (Khatri *et al.*, 1996). The lower ROR for the Proteaceae can be ascribed to the perennial nature of the crop and the time needed (19–33 years) to develop a cultivar from the initial seed collection to cultivar acceptance. The benefits were further underestimated due to lack of data, where only the benefits associated with clonal materials were included, while the entire cost of the R&D programme was, however, accounted for. The cost–benefit variation of the economic surplus approach further tends to underestimate the benefits of R&D activities (Anandajayasekeram *et al.*, 1996). In the presence of more dependable data, the ROR and other benefits would undoubtedly improve positively.

References

ABOS (1996) *Handboek voor de toepassing van de Doelgerichte Interventieplanning (DIP) op het ABOS.* Algemeen Bestuur voor Ontwikkelingssamewerking, Ministerie van Buitelandse Zaken, Brussels, 2nd edition.

Anandajayasekeram, P., Martella, D.R. and Rukuni, M. (1996) *A Training Manual on Research and Development Evaluation and Impact Assessment of Investments in Agricultural and Natural Resources Research.* SADC-SACCAR, Gaborone.

Eckert, J.B., Liebenberg, G.F. and Troskie, D.P. (1997) *The Western Cape Agricultural Social Accounting Matrix (WCAGRSAM): Strategic Micro and Macro Modelling Project (SM^3).* A joint project between the Department of Agriculture, Western Cape, and The Development Impact Analysis Unit of the Agricultural Research Council: Progress Report: Vol. 1. Elsenburg, August.

Farrington, J. and Nelson, J. (1997) *Using Logframes to Monitor and Review Farmer Participatory Research.* ODI, Agricultural Research & Extension Network, No. 73. ODI, London.

Gittinger, J.P. (1982) *Economic Analysis of Agricultural Projects.* Johns Hopkins University Press, Baltimore, Maryland.

Khatri, Y., Thirtle, C. and van Zyl, J. (1996) Public research and development as a source of productivity change in South African agriculture. *South African Journal of Science* 92, 143–150.

Niederwieser, F., Anandajayasekeram, P., Coetzee, M., Martella, D., Pieterse, B. and Marasas, C. (1997) *Socio-economic Impact of the Lachenalia Research Program.* ARC/SACCAR Report. SACCAR, Gabarone, Botswana.

Shepherd, G.S. (1968) *Agricultural Price Analysis.* Iowa State University Press, Ames, Iowa.

Squire, J. and van der Tak, H.G. (1975) *Economic Analysis of Projects.* Johns Hopkins University Press, Baltimore, Maryland.

Wessels, J. (1998) The socio-economic impact of the Proteaceae technology development and transfer programme. MSc thesis, University of Pretoria, Pretoria.

Wessels, J., Anandajayasekeram, P., Littlejohn, G., Martella, D., Marasas, C. and Coetzee, C. (1997) *Socio-economic Impact of the Proteaceae Development and Transfer Program.* ARC Report. Agricultural Research Council, Pretoria.

21 Impact Assessment of the Biological Control of *Prosopis* Species in the Britstown–De Aar District, South Africa

J.M.C. Esterhuizen (neé Wessels)

Abstract: This chapter reports on the results of an impact assessment of the biological control programme against invasive *Prosopis* species or mesquite in South Africa. The study developed a framework for structuring an effectiveness and efficiency analysis of public expenditure as related to agricultural technology research and development. A logical framework analysis was applied to facilitate the assessment of the programme's effectiveness and a cost–benefit approach to facilitate the assessment of the programme's efficiency. Primary data were collected from the Britstown–De Aar study area (Northern Cape Province) by means of a questionnaire and also a pod survey to collect data on the success of the biological control agents. It was found that since 1986, the biological control programme has had a positive impact by reducing the rate of *Prosopis* invasion. The analysis found that the biological control programme only has to be 2.5% effective, in terms of area not invaded but with the potential to be invaded, in order to be a financially profitable investment and only 1.3% effective for it to be an economically beneficial investment to society.

Introduction

Thousands of plant species, from around the world, have been introduced into South Africa, for a range of purposes – as crop species, for timber and firewood, as garden ornamentals and for sand dune stabilization, as well as barrier and hedge plants. Many of these exotic plant species have become naturalized, surviving in the South African landscape without needing to be tended, and some of these naturalized species have become invasive. Invasive alien species are able to survive, reproduce and spread unaided and sometimes have established at alarming rates across the landscape (Department of Water Affairs and Forestry, 1999).

South Africa with a land mass of 1,221,040 km^2 with its conducive environment is particularly prone to invasions – from giant eucalyptus to scrubby cactus. Invasions by exotic woody weeds displace the indigenous vegetation, increase the biomass in catchment areas and substantially decrease the surface runoff (Verseveld *et al.*, 1998). This phenomenon has a major impact on natural resources, particularly on water resources of the country.

According to Verseveld *et al.* (1998), *Prosopis* L. (Fabaceae) is classified as one of the top ten plant invaders in South Africa. The importance of different species varies between provinces

 Impact of Science on African Agriculture and Food Security (eds P. Anandajayasekeram, M. Rukuni, S. Babu, F. Liebenberg and C.L. Keswani)

with *Prosopis* and *Opuntia* occurring mainly in the arid interior (Karoo and Kalahari) and the remaining species from the top ten occurring in the region between the coast and the escarpment.

The current recognized expenditure on invading alien plant clearing projects exceeds R100 million (US$125 million) per year, having been stepped up to this level in 1995 in response to growing concerns. The cost incurred by farmers, foresters, landowners, local authorities, utilities and other institutions in clearing plant invasions must be added to this amount. Preliminary investigations have led to the generally accepted view that, if water alone is considered, the expense of clearing programmes can be financially justified. It is accepted that the clearing of invasions at an early stage and prevention of invasion is the most cost-effective option, in view of the increasing exponential cost of clearing and the impact of their incremental water-use if invasion is allowed to continue unchecked (Verseveld *et al.*, 1998).

South Africa is faced with the problem of limited public resources, and a situation of trade-off in achieving the country's fundamental objectives. Economic theory suggests that to maximize the social output, resources should be allocated where their contribution is greatest at the margin. To achieve maximization of social output, the debate on the allocation of scarce public resources should be aided by evidence of net gain from research to society. It was realized that impact assessment procedures should be developed to provide evidence of the performance of a research programme, in order to negotiate continued public sector investment (Van Rooyen *et al.*, 1996).

Background

Several species of the genus *Prosopis* L. (Fabaceae), commonly known as mesquite or *Prosopis*, were introduced into the arid western parts of South Africa and Namibia around the beginning of the 19th century, as agroforestry trees to provide shade, fuel wood and fodder from the nutritious seed pods. The *Prosopis* species are large, thorny shrubs or trees that can grow up to 10 m tall. The thickest infestations of *Prosopis* spp. occur in the northwestern Cape of South Africa. In view of the uncertainty on the taxonomy of the *Prosopis* spp. in South Africa, the invasive taxa and any possible hybrids are referred to under the generic name *Prosopis* in the rest of this study.

Prosopis species are thicket-forming trees or multi-stemmed shrubs. Because *Prosopis* can utilize deep groundwater it is able to form dense canopies with high productivity which overshadow low growing natural vegetation that usually relies on other methods of surviving drought. Trees grow rapidly and soon dominate the natural vegetation and produce large numbers of hard-coated seeds with almost 100% germination under optional conditions. These factors, along with active encouragement to farmers in the past from the Department of Agriculture to propagate *Prosopis* as a dry-land fodder plant, resulted in a considerable increase in the area and density of these trees (Harding, 1988). The ability of *Prosopis* to utilize deep groundwater often causes dense stands to occur on river banks (riparian zones), in dry riverbeds or on old water courses (alluvial plains and landscapes) (Harding, 1988). These areas generally have better soils, i.e. deep soils with accessible groundwater or stagnant water, which may then become encircled by the trees. These prime production areas are subsequently lost to agriculture.

Prosopis species are the most aggressive water-using invaders of the semi-arid and arid interior of South Africa. The Northern Cape has the greatest estimated percentage water-use by invaders, primarily because of the low marginal annual runoff in this largely semi-arid to arid province. This is particularly important, because the province has very limited water resources and a growing population (Verseveld *et al.*, 1998). Thus the rate and danger of invasion, the decrease in value of agricultural land due to invasion, the utilization of groundwater and the negative impact on biodiversity due to *Prosopis* invasion are factors causing reasons for action against the further spread of *Prosopis*. Presently all *Prosopis* species have been declared invader plants under the Agricultural Resource Act, 43 of 1983 (Henderson *et al.*, 1987), with an estimated area of 1.8 million ha invaded by *Prosopis* in South Africa (Verseveld *et al.*, 1998).

Biological control of *Prosopis* in South Africa was initiated in 1986 to reduce the impact and spread of this invader. This method of control was preferred and implemented due to the conflicting interests over the continued existence of *Prosopis* at that time, and the lower cost implications compared with mechanical and chemical control methods. Seed-feeding beetles were chosen as a method of biological control, because the major enhancing factor for *Prosopis* invasion is its ability to produce large quantities of viable seeds. The second reason was that the seed- feeding biocontrol agents would not reduce the number of seeds available for stock feed, but only impact on seed germination. Thirdly, the objective was not to eradicate *Prosopis*, but only to control further spread, which relates to the situation of conflicting interests over *Prosopis*. Therefore, the objectives of the biological control of *Prosopis* were to reduce the invasion rate of *Prosopis* (further spread and densification) without removing the beneficial attributes of the plant.

Farmers, who were greatly affected by invasions, were using chemical and mechanical control to some extent, with limited success, due to high cost and the labour-intensive requirements of these methods. The cost of mechanical and chemical control usually exceeds the value of the land reclaimed by farmers (Harding, 1988). Various factors had to be taken into account when the biological control of *Prosopis* was planned. In the first place, the beneficial attributes of the plants had to be taken into account. Conflicting interests with regard to the continued existence of the trees were apparent due to the beneficial attributes of *Prosopis*, where it was still in its useful form. The large quantities of seed produced annually by *Prosopis*, attributed largely to the invasiveness of *Prosopis*, were also considered.

The objective of the programme was to reduce the invasiveness and spread of *Prosopis* in South Africa, while preserving beneficial attributes of the trees. Based on these objectives seed-feeding insects were chosen as biological control agents to reduce the reproductive potential of *Prosopis*, curb their dispersal and reduce the amount of follow-up work needed after complementary clearing methods have been implemented, while still allowing for the continued utilization of the beneficial attributes. For this reason, research focused on seed-feeding insects of which three species, namely *Algarobius prosopis*, *Algarobius bottimeri* and *Neltumius arizonensis*, were introduced into South Africa and released at various sites of invasion (Zimmerman, 1991; Impson *et al.*, 1999). The control of *Prosopis* can only be achieved at the expense of time, energy and/or money, which are all scarce items in an increasingly competitive society. Owing to the general trend towards reduced funding for agricultural research, the *Prosopis* biological control programme is competing for the use of scarce resources. Thus, it was important to evaluate the impact of the *Prosopis* biological control programme from a financial and economic viewpoint to assist in the economic management of scarce resources and contribute towards research policy formulation. This study can help establish credibility of the research programme and provides feedback to researchers, research managers and policy makers, on what has been accomplished.

Objectives

The main purpose of this study was to establish and apply an analytical procedure to determine the impact of the *Prosopis* biological control programme at the Plant Protection Research Institute of the Agricultural Research Council of South Africa. Therefore, the objectives of this study were to: (i) establish an impact assessment for a biological control programme of the *Prosopis* species – an invader plant; (ii) describe the problems related to the effective biological control of *Prosopis* invasion; and (iii) assess the impact of the biological control programme through a case study in the Britstown–De Aar District.

Analytical Framework and Methodology

In this study, a framework was developed to structure an effectiveness and efficiency analysis on public expenditures related to agricultural technology research and development (Wessels, 2000). Effectiveness analysis measures the degree to which the programme or activity has achieved its desired objectives, i.e. 'what is' in contrast to 'what should be' (Farrington and Nelson, 1997). A logical framework analysis was applied to facilitate the assessment of main performance areas (Wessels, 2000).

An efficiency analysis was conducted to determine the economic efficiency of the programme in terms of costs and benefits. The technique applied was a simple cost–benefit framework, with financial and economic analysis methodology using shadow pricing and rate of return (ROR) estimates. The efficiency analysis refers to the comparison of costs and benefits in a 'with' and 'without' scenario to calculate the incremental net benefits from which net present value and ROR estimates are derived (Table 21.1). The cost items included in the analysis were the cost of research activities and livestock production. The financial returns included the returns from sales of livestock products.

Table 21.1. Framework of analysis for the *Prosopis* biological control programme.

Programme		'With' scenario	'Without' scenario
Biological control of *Prosopis* as invader plant	• Technology development and transfer cost of biological control programme • Reduction in the rate of invasion and densification, which translates into agricultural activity related benefits • The benefit in reduced total water use of *Prosopis* • The reduction in the danger of invasion • The benefit of no increased disturbance of biodiversity and ecosystems	Technology development and transfer activities related to biological control of *Prosopis*. Biological agents were released in *Prosopis* invaded areas to control further spread and densification of the invader in 1986. With the reduction in rate of invasion the productivity of the land is improved due to the reduced water use, which reflects in the increased quality of grazing land and the reduced rate of reduction in grazing land. The positive impact on biodiversity and the reduction in possible further degradation and erosion of land in watercourses and alluvial plains	The invasion and spread of *Prosopis* without any control. Therefore, increased invasion and densification, which leads to an increased rate of reduction in grazing land and biodiversity, increased utilization of ground water, increased erosion and degradation of land along water-courses and alluvial plains

Source: Wessels (2000).

The impact of the *Prosopis* biological control programme was assessed in Britstown and De Aar magisterial districts of the Northern Cape Province. This was chosen as the study area because it represents one of the main centres of invasion and historical data on invaded areas were available. Primary data were collected from the study area using a personally administrated questionnaire, which was mainly to determine the magnitude of invasion, and secondly, a pod survey collecting data on the success of the introduced biological control agents (Stoltsz, 2000).

Results

Problems related to the biological control of *Prosopis* invasion

Various factors play a crucial part in the success of exotic introduced biocontrol agents. The *Prosopis* seed-feeding beetle was introduced from the USA. The following factors could play a crucial part in the success of biocontrol of *Prosopis*:

- Indigenous parasitoids attacking the eggs or larvae of the beetles may reduce the efficiency of the beetles. Many beetle parasitoids in Southern Africa attack a wide range of host species, and therefore pose a threat to the effectiveness of the beetles (Zimmerman, 1991).
- Host-plant preferences of seed-feeding beetles may diminish their overall effectiveness if some of the taxa within the *Prosopis* species complex in South Africa are unacceptable or are only marginal hosts (Zimmerman, 1991).
- Survival of adult beetles in the winter when pods are scarce, or even absent, will determine how quickly the beetle populations will turn to the new pods produced during the summer (Zimmerman, 1991).
- Utilization of pods by livestock and other mammalian herbivores. The utilization of pods as fodder may affect the biological control programme because population increase of the beetles will be hindered when eggs and larvae in ingested pods are killed. The effectiveness of beetles, preferring mature pods to lay eggs, is reduced if pods are removed by livestock before the beetles can destroy sufficient seeds (Impson *et al.*, 1999).
- The factor in the success of the beetles is interspecific competitive interactions (Impson *et al.*, 1999). All the beetle species released on to *Prosopis* in South Africa utilize the same food source during the larval stages and interspecific competition is anticipated in the field if the insects become abundant (Impson *et al.*, 1999).

The effectiveness of the *Prosopis* biological control programme in South Africa

In assessing the actual objectives of the programme, logical framework analysis was used to construct a logical problem statement. The main problem derived from this analysis is that of an increased rate of invasion and densification by *Prosopis*, due to the ability of *Prosopis* to produce large quantities of seeds and the dispersal of seeds by livestock, mammals and other vertebrate herbivores (Harding, 1988; Verseveld *et al.*, 1998). This in turn leads to the consequence of reduced farm income, due to the loss of productive farmland.

The core strategy of the biological control programme was to reduce the invasiveness and spread of *Prosopis* by reducing the number of viable seeds produced. The effectiveness of the programme, as it relates to the specific objective of reducing viable seeds, was assessed from a pod survey in the study area (Stoltsz, 2000). The results of the pod survey showed that at least some of the bio-agents have become widely established in the Britstown–De Aar area, and are currently neutralizing on average 41% (range 20–70%) of the annual seed crop, which is an improvement over previous estimates for free-range areas (Impson *et al.*, 1999). The success of the biocontrol agents is jeopardized due to the availability of *Prosopis* pods for livestock forage. Pods utilized by livestock are excluded from the pods being attacked by the seed-feeding beetles. Fencing off *Prosopis* infestation to exclude livestock is not a common practice in the area. It is therefore assumed that the biocontrol agents have not yet reached their full potential, but never may, as long as livestock are allowed to forage pods before the seed-feeding beetles have had sufficient opportunity to damage the seed. However, the exact impact of biological control towards reducing the rate of spread could not be quantified, due to the lack of information on the population dynamics of *Prosopis* in South Africa. This was one of the limitations of the study. Therefore, various scenarios were applied in the cost–benefit analysis, to make a provision for this lack of data.

In order to assess the effectiveness of the programme in achieving its overall objective of reducing the rate of *Prosopis* spread, the actual invaded areas ('with' scenario) were compared with a projected area ('without' scenario) over time. The initial reference point was taken to be the extent of invasion at the time of implementation of the biological control programme in 1986. Survey records of the Department of Agriculture (1992) for the study area were used as the source of this information. Taking into account the maximum potential area suitable for invasion in the study site (Lloyd, 2000), the projected area of invasion was estimated from a 'logistic' model, developed from spatial maps of invasion over time, for an area where no control measures were in place (Verseveld *et al.*, 1998). The model applies the following formula:

$$N_t = N_{t-1} + r(1 - N_{t-1}/K)\,N_{t-1}$$

where N = the hectares that are occupied, t = the current time step, $t-1$ = the previous time step, r = the rate of increase or expansion coefficient and K = the hectares that potentially could have been occupied.

Based on the information of areas currently invaded in comparison to the areas projected by the logistic model, the rate of spread was reduced. In the 'with' and 'without' scenarios, the expansion coefficient (r) was 0.13 and 0.18, respectively (Verseveld *et al.*, 1998). K was estimated to be 36,667 ha (on a condensed area basis) and N_{t-1} at 2331 ha (on a condensed area basis), where $t-1$ was 1985. Data on actual area invaded were available from the Department of Agriculture (1992) for specific years and together with more recent survey data (Wessels 2000), which provided verification of the expansion model for the 'with' scenario. The output for the 'with' scenario was always less than the 'without' scenario, implying a reduced rate of invasion in the case of the former. As it was discovered that mechanical and chemical control measures were not implemented to a large scale during this period (Wessels, 2000), the reduction in rate of spread was attributed to biological control. Based on these findings, it could be concluded that the biological control programme does have an impact on the rate of invasion, although it is not quantified, thus addressing the main objective of the programme.

Efficiency analysis of the *Prosopis* biological control programme in South Africa

A cost–benefit framework was used in the analysis. Financial and economic analysis was conducted to determine the efficiency of the programme as it impacts on the Britstown–De Aar district. The cost–benefit analysis was conducted for the period 1986 up to the year 2010, thus a 25-year period. Biological control is a long-term activity; therefore, periods of 25 years for the analysis were chosen.

Estimation of areas as related to the impact of the biological control programme for *Prosopis*

According to a study conducted by Harding (1988), the total possible area of invasion for *Prosopis* was estimated to be 935,000 ha. This total possible area of invasion, minus the area which was invaded in 1985 (180,406 ha) was used as the area (754,594 ha) at which the biological control programme activities were aimed. This area was converted to an area of 100% canopy cover. The total potential area of invasion was calculated as 584,375 ha with 100% canopy cover minus 27,060 ha 100% cover, which is the area already invaded in 1985. This equals 557,315 ha of 100% canopy cover at which biological control was targeted.

The objective of a reduced rate of invasion from biological control directly translates into areas not invaded due to the control of *Prosopis*. Therefore, the area to be used in cost and benefits calculations as they relate to the programme is crucial. This directly relates with the 'with' and 'without' scenarios. The costs and benefits as applicable in this study are given in a Rand per hectare unit. The area, used in the analysis, is all converted to 100% canopy cover, which is equal to total dense invasion. Therefore, the area invaded is lost to any agricultural activities.

The total area of Britstown and De Aar was assessed to determine which areas were subjected to possible invasion. A computer model developed by the Agricultural Research Council, Institute for Soil, Climate and Water, was used to estimate the areas of possible invasion by *Prosopis*. A total area for the Britstown–De Aar district was estimated to be 58,668 ha, and was adjusted to 36,667 ha of 100% canopy cover (Wessels, 2000).

According to a survey conducted to estimate the area of *Prosopis* invasion for the Northern Cape in 1985, the area invaded in the Britstown and De Aar district was 14,970 ha. The survey was again repeated in 1991, and the total area invaded was found to be 28,415 ha (Department of Agriculture, Nature Conservation and Land Reform: Northern Cape Province, 1992). When these total areas with invasion were converted to 100% canopy cover, the corresponding area of invasion (100% canopy cover) was 2331 ha (1985) and 4420 ha (1991), respectively (Wessels, 2000).

Defining the benefits and costs

The benefits included in this analysis are derived from sheep farming in the Britstown–De Aar district, as it corresponds with the areas not yet invaded but with the potential to be invaded as derived for both the 'with' and 'without' biological programme scenarios. The costs included in this analysis are the costs incurred by sheep farming in the Britstown–De Aar district, again for the 'with' and 'without' biological programme scenarios. In the 'with' programme situation the costs of research activities and release of the technology are included in this analysis. All costs and benefits were included in the analysis as a Rand per hectare value. Therefore, the total research and beetle release costs were included as a value per hectare by using the calculated area at which biological control was aimed.

The financial analysis

In this analysis, a simple cost–benefit approach was used to calculate the incremental net benefits derived from the biological control programme. To estimate the ROR, the annual research and production costs in real values were subtracted from the total benefits (income from sheep farming), which gave the annual net benefits (real) from the biological control programme activities. The net benefit (real values) from the 'without' scenario was subtracted from the net benefits from the 'with' scenario to obtain the incremental net benefit from the *Prosopis* biological control programme activities. This was done for the period 1986–2010. In this study, a real interest rate of 7% (market rate adjusted for inflation) was used as the discount rate. The estimated ROR for the financial base scenario was 104%. This estimated ROR is far greater than the opportunity cost of capital (7%). The above analysis showed that investment in the biological control of *Prosopis* is very profitable from the financial point of view. The ROR was high because of the low cost implication of research due to the per hectare allocation of research cost for the study area. It was therefore decided to include the total research cost of the programme, as if the biological control programme was aimed at the study area only. In this analysis, a ROR of 18% was obtained, which is still greater than the opportunity cost of capital (7%).

The biggest shortcoming in estimating the benefits derived due from the biological control programme was the lack of quantified data on the reduction in the rate of *Prosopis* invasion. Owing to this lack of data, a sensitivity analysis was conducted to estimate the minimum required rate of effectiveness of biological control as it was reflected in the areas with the potential to be invaded but not yet invaded for the investment to be profitable. When the biological control of *Prosopis* is 80% effective, the ROR was 88% and with 50% effectiveness the ROR was 64%. Even if the biological control accounts for only 10% of reduced invasion the ROR is 25%, which is still higher than the opportunity cost of capital. When the effectiveness was reduced to 2.5%, the ROR dropped to 9%, thus indicating that at least 2.5% of the area not invaded due to some control should be ascribed to biological control for the programme to be financially profitable. A second aspect addressed with sensitivity analysis was the cost of research for the biological control programme. Combinations of these two aspects were also tested (Table 21.2).

The economic analysis

In the economic analysis a simple cost–benefit approach was used. The benefits and the costs occurring due to farm production activities are presented according to the estimated area not yet invaded but with the potential to be invaded if the project was not introduced. The assumption was made that this area was mainly used for sheep farming, as this is the major farming activity practised as found from the *Prosopis* invasion survey for the Britstown–De Aar district (2000). The cost of the research activities for biological control of *Prosopis* was also included.

The market price may not always be a good reflection of the shadow value of a commodity or service. Certain adjustments were, therefore, made where necessary to the market price to reflect the economic value of a commodity or service. In this case study, the outputs from the farming activities were found to be distorted due to tariff charges. These were adjusted to reflect the economic value by applying the concept of outputs classified as tradable items. This was applied in both the 'with' and 'without' biological programme situations. The inputs used in farming-related activities were not distorted significantly and therefore no adjustments were needed. The base scenario for the financial analysis was adjusted to economic values by applying shadow values to programme outputs classified as tradable.

The ROR for the base scenario was 119%. This is much higher than the opportunity cost of capital, which is 5% in the case of economic analysis. From a society viewpoint, the opportunity cost of capital was reflected by the real value for long-term government bond rates. When comparing the ROR estimates with the real bond rate, the investment in the *Prosopis* biological control programme was a profitable investment to society.

Various scenarios were developed for the different levels of effectiveness of biological control and inclusion of research cost as well as various combinations of these. When, the biological control of *Prosopis* was 80% effective, the ROR was 100% and with 50% effectiveness, the ROR was 71%. Even if biological control accounted for only 10% of reduced invasion, the ROR was

Table 21.2. Estimated ROR for the various scenarios in the financial analysis.

Scenario	Rate of return (%)
Base scenario	104
Levels of effectiveness and research cost scenarios	
• Various levels of biological control effectiveness	
80% effective	88
50% effective	64
10% effective	25
2.5% effective	9
Increase in research cost	
20% increase	91
50% increase	77
100% increase	64
Inclusion of total research cost	18
• Various levels of biological control effectiveness	
80% effective	16
50% effective	11
40% effective	9
Combinations of increased research cost per hectare, and various levels of biological control effectiveness	
• 50% increase in research cost	
80% effective	67
50% effective	49
10% effective	19
• 100% increase	
80% effective	55
50% effective	42
10% effective	16

Source: Wessels (2000).

27%, which is still higher than the opportunity cost of capital. When the effectiveness was reduced to 1.3%, the ROR dropped to 5%, indicating that at least 1.3% of the area not invaded due to some control measure should be ascribed to biological control activities for the programme to be economically profitable (Table 21.3). It was decided to develop different scenarios with regard to the inclusion of research cost. The total research cost of the programme was included in the analysis as if the programme activities were for the benefit of the study area only. This resulted in an 18% ROR for the base scenario. In this scenario, the effectiveness of the biological control was reduced to incorporate the different levels of success from biological control. With an 80% effectiveness the ROR was estimated at 16%, 50% effectiveness derived an ROR of 11%. In this scenario of total research cost inclusion, the biological control programme only has to be 20% effective to be economically profitable (ROR 5%) (Table 21.3).

Environmental impact

The environmental impact of the biological control programme was assessed in relation to the negative impact of *Prosopis* invasion on the environment, in terms of land degradation, loss of

Table 21.3. Estimated ROR for the various scenarios in the economic analysis.

Scenario	Rate of return (%)
Base scenario	119
Levels of effectiveness and research cost scenarios	
• Various levels of biological control effectiveness	
80% effective	100
50% effective	71
10% effective	27
1.3% effective	5
Increase in research cost	
20% increase	103
50% increase	87
100% increase	71
Inclusion of total research cost	20
• Various levels of biological control effectiveness	
80% effective	18
50% effective	13
20% effective	5
Combinations of increased research cost per hectare and various levels of biological control effectiveness	
• 50% increase in research cost	
80% effective	75
50% effective	55
10% effective	21
• 100% increase	
80% effective	62
50% effective	46
10% effective	18

Source: Wessels (2000).

biodiversity and water consumption (Harding, 1987; Verseveld *et al.*, 1998; Le Maitre, 1999). Any reduction in the rate of invasion by *Prosopis* due to biological control therefore has a positive environmental impact, by reduction of the negative impact of *Prosopis* invasions on the environment.

The biological control agents introduced in the programme were host specific, and therefore have no negative impact on other indigenous plant species. Biological control does not cause pollution, as is the case in chemical control, or soil disturbance and sudden vacant niches, as from mechanical control. These characteristics of biological control have a positive impact on the environment.

Conclusions

A major constraint on assessing the effectiveness and efficiency of the biological control programme – by reducing the rate of *Prosopis* invasion – was the lack of quantitative data on the 'real' impact of the biological control agents. Their effectiveness in reducing the number of seeds available for dispersal could be ascertained, but not the degree to which this reduces the actual rates of invasion. For this, detailed investigations into seed bank and dispersal dynamics of the seed are required. However, in the absence of quantitative data, an impact analysis was performed on the biological control programme in the Britstown–De Aar area, based on various scenarios, related to different levels of reduction in invasion rate, cost of research activities and various combinations thereof. This study showed that the programme was financially and economically justifiable, even at very low levels of effectiveness in reducing the invaded area. This strengthens the debate for increased investment of public funds into the biological control of *Prosopis*, while providing long-term benefits to society. The analysis also provided a decision-making tool to predict and compare outcomes of different biological control activities, which supports efficient resource allocation through economically rational decisions.

References

Department of Agriculture, Nature Conservation and Land Reform, Northern Cape (1992) *Ondersoek Na Prosopis-Indringing in The Britstown En De Aar Landdrosdistrkte.* Department of Agriculture, Nature Conservation and Land Reform Extension Office, De Aar, South Africa.

Department of Water Affairs and Forestry (1999) *The Environmental Impacts of Invading Alien Plants in South Africa.* Department of Water Affairs and Forestry, Pretoria.

Farrington, J. and Nelson, J. (1997) *Using Logframes to Monitor and Review Farmer Participatory Research.* ODI, Agricultural Research & Extension Network, No. 73. ODI, London.

Harding, G.B. (1987) The status of prosopis as a weed. *Applied Plant Science* 1, 43–48.

Harding, G.B. (1988) The genus *Prosopis* spp. as an invasive alien plant in South Africa. PhD thesis, University of Port Elizabeth, South Africa.

Henderson, M., Fourie, D.M.D., Wells, M.J. and Henderson, L. (1987) *Declared Weeds and Alien Invader Plants in South Africa.* Department of Agriculture and Water Supply, Government Printer, Pretoria.

Impson, F.A.C., Moran, V.C. and Hoffman, J.H. (1999) A review of the effectiveness of seed feeding beetles in the biological control of mesquite, *Prosopis* species (Fabeaceae), in South Africa. In: Olckers, T. and Hill, M.P. (1999) *Biological Control of Weeds in South Africa: 1990–1998.* Entomology Society of South Africa, Pretoria, South Africa, pp. 81–88.

Le Maitre, D.C. (1999) *Working for Water Programme: Prosopis and Groundwater: a Literature Review and Bibliography.* Environmentek Csir, Stellenbosch, South Africa.

Lloyd, W. (2000) *Potential Areas of Prosopis Invasion in the Britstown–De Aar District.* Agricultural Research Council, Institute for Soil, Climate and Water, Stellenbosch, South Africa.

Stoltsz, C.W. (2000) *Prosopis Invasion in Britstown–De Aar District.* Agricultural Research Council, Plant Protection Research Institute, Addo, South Africa.

Van Rooyen, C.J., Carstens, J.P. and Nortjé, A. (1996) *The Role of Agriculture in South Africa's Economy – a Research and Technology Challenge.* Agricultural Research Council, Pretoria.

Verseveld, D.B., Le Maitre, D.C. and Chapman, R.A. (1998) *Alien Invading Plants and Water Resources in South Africa: a Preliminary Assessment.* Council for Scientific and Industrial Research, Division of Water, Environment and Forestry Technology, Stellenbosch.

Wessels, J. (2000) Impact analysis of the biological control program of *Prosopis* species in the Britstown–De Aar District, South Africa. MSc thesis, University of Ghent, South Africa.

Zimmerman, H.G. (1991) Biological control of Mesquite, *Prosopis* spp. (Fabeaceae), in South Africa. *Agriculture, Ecosystems and Environment*, 37.

22 Socio-economic Impact of the Control of Ticks and Tick-borne Diseases in South Africa

R. RANDELA

Abstract: This study assesses the socio-economic impact of the dipping programme as well as determining whether there is any economic justification for the government to control ticks and tick-borne diseases in the Northern Province of South Africa. The study is based on a cross sectional survey of 125 small-scale farms in the Venda region of the Northern Province. At 4% mortality rate, the cost–benefit analysis revealed a ratio of 0.8 and it is sensitive to changes in the mortality rate. This ratio indicates that the control of ticks and tick-borne diseases by government is not economically justified. However, because of the economic nature of the service it provides (public good) and the significant socio-economic benefits to poor households, it could be argued that cattle dipping still deserves government support. The currently envisioned tick control strategy is based on an integrated tick management system where acaricides are strategically applied.

Introduction

In many African nations, virtually all of the large animals are raised on small-scale family farms (Scholtz *et al.*, 1991; Umali *et al.*, 1994). For many farmers, animals are a form of stored wealth, a cushion against starvation, a source of fertilizer, a means of transportation and a source of traction. However, many smallholder farmers in developing countries, including South Africa, face problems with ticks and tick-borne diseases. The control of ticks is of importance in maintaining the health and productivity of cattle. As a result, various tick control programmes have been put in place in the rural areas of South Africa to alleviate the problem. Chemical control is currently the most common tick control measure. The plunge dip system is the most widely used and it forms part of a comprehensive strategy of disease surveillance and extension in South Africa.

Background

After the eradication of East Coast fever in the 1960s, it was found out that the most efficient and economical manner for tick control was to target treatment at the parasitic stage. Short interval

 Impact of Science on African Agriculture and Food Security (eds P. Anandajayasekeram, M. Rukuni, S. Babu, F. Liebenberg and C.L. Keswani)

dipping of livestock became the standard method of tick control and it was compulsory. In South Africa compulsory dipping was abolished and the choice was left to individual farmers (Norval *et al.*, 1994). However, in some of the autonomous South African homelands, compulsory dipping carried out by the government on communal grazing continued to be enforced through legislation. Although indirectly stated, this was also the case in Venda as was stipulated in the Venda government gazette which states that, by law a stockowner is now compelled to produce his cattle for inspection and tally recording by a stock inspector at the stock inspection point (diptanks) at least every 14 days. This practice is still in operation today. Compulsory dipping may have been thought to lead towards eventual tick eradication, which would have provided a possible rationale for public funding. However, it is worrying that resistance by ticks to acaricides may be on the increase. The increase of tick resistance to acaricides and the rapidly rising costs of running dipping services raised doubts as to whether the Northern Province will be able to sustain effective tick control.

Cattle are either dipped on a weekly or fortnightly cycle depending on the tick challenge in a particular area. There is a large variety of acaricides to choose from, each having its own application and management system (Dipping Policy Survey, 1997). The commonly used acaricides are Grenade, Triatix and Clout (pour-on).

In South Africa there is an ongoing debate over which type of institution (public or private) can provide an optimal delivery of a particular service. Currently, the Directorate of Animal Health is reluctant to fund animal health services. This is probably due to the government failure to realize the value to the country of the veterinary service or failure to foresee the consequences to society if such a service is not provided.

Objectives

The objective of this study was to determine whether it is economically justified to control ticks and tick-borne diseases as well assessing the socio-economic impact of the dipping programme at the farm level. The study examined the costs and benefits of tick control. The 'with' and 'without' dipping scenario was then incorporated into the analysis. The study then focused on the socio-economic impact of the tick control programme.

Methodology

Sampling procedure, sample size and study design

One of the important issues in sampling is to determine the size of the sample. The major criterion to use when deciding on the sample size is the extent to which the sample size is representative of the population. Two-stage sampling was used in this study. First, three diptanks were chosen within each area. Secondly, cattle within each diptank were grouped based on the number of cattle, namely 1–10 head of cattle, 11–20, 21–30 and more than 30. Before sampling, a list of cattle numbers for all cattle owners was made available by the extension officers in charge of the respective diptanks. Then, within each stratum, a simple random sampling was performed using random number tables. From each diptank 25 respondents were chosen.

Owing to time limitations and financial constraints, a standard questionnaire was used to collect the primary data. Prior to the survey, it was learned that children or herd boys bring some of the cattle herds to the diptank. Respondents were then notified about the planned interviews a week in advance of the dipping day. The actual number of farmers who participated in the surveys is, however, less than the targeted number of 150. This can be ascribed to poor dipping attendance (caused by rain and other urgent commitments) and uncooperative behaviour by some cattle owners. This reduced the number of actual respondents to 125. This sample is about 80% of the targeted sample, and is considered large enough to be representative of the cattle farmers in the study area.

Socio-economic impact assessment technique

The study is conducted using a cost–benefit analysis approach because it provides a sound theoretical framework for analysing the economic and social impacts of research and development activities. The study aimed at establishing whether or not various tick and tick-borne disease control interventions are producing their intended effects. Data from the veterinary service dip record were used to complement the primary data collected through the survey. However, certain factors cannot be easily assessed quantitatively. For instance, the role of cattle in marriage contracts, prestige and ceremonial activities cannot be assessed in terms of a quantitative comparison, but should not be ignored. The 'with' and 'without' programme was included in the analysis to provide a basis for cost–benefit analysis.

Study area

There is always a problem when the researcher has to make a decision as to whether a single site or several sites will be sufficient to give a better picture of the situation. Owing to resource constraints, the former Venda homeland was chosen as a study area. Venda now forms part of the Northern Province.

Venda is divided into three veterinary zones, namely the Red Line, Yellow Line and Open area. On the advice of the Venda Department of Agriculture (Veterinary Services Division), it was decided that the survey be conducted in all the veterinary zones. However, because of socio-political reasons, surveys could not be conducted in the Red Line area. The two areas surveyed (Yellow Line and the Open area) are generally known as non-foot-and-mouth disease areas. In comparison, however, the Open area is known as a low tick prevalent area and the Yellow Line as a high tick prevalent area. Within the Yellow Line area the following diptanks were chosen: Tshifudi, Malongana and Matshena and within the Open area Vyeboom, Guyuni and Dzanani.

A quantitative assessment of the value of livestock to the rural community: a micro perspective

This section discusses the importance of cattle in terms of milk production, cattle sales, manure and draught power. A quantitative valuation of cattle is done because it forms the basis for the cost–benefit analysis. In addition, the rationale for undertaking this analysis demonstrates the magnitude of the socio-economic impact, which mortalities could have, particularly to those who make decisions on the financial allocations for disease control programmes. The role of cattle in developing areas is well documented in Doran *et al.* (1979), Scoones (1992) and Randela (2000). On average the value of cattle in the surveyed regions is R1152 (see Table 22.1). Milk contributes the greatest percentage (46%) followed by sales (31%), work of cattle (23%) and manure (1%).

Table 22.1. The value of cattle to a rural household in Venda (Rand/animal).

Area	Milk	Sales	Manure	Work of cattle	Total value per animal
Vyeboom	1022	335	10	–	1367
Malongana	793	446	12	–	1251
Guyuni	365	376	15	182	938
Matshena	532	374	14	–	920
Dzanani	240	316	17	–	573
Tshifudi	205	273	17	336	831
Total	3157	2120	85	518	5880
Average	526	353	14	259	1152
Average % contribution	46	31	1	23	100

Source: Author's estimate.

The financial burden of dipping

In several countries in southern Africa, government veterinary services have special responsiblity for funding and implementation of extensive dipping programmes for farming communities (Norval *et al.*, 1992). The Venda government's network of some 127 cattle diptanks served 104,778 head of cattle owned by communal farmers in 1997. The control of ticks and tick-borne diseases is economical only if ticks are effectively controlled. However, it becomes expensive to farmers if tick-borne diseases emerge and farmers have to buy their own medicines, usually from farmers' co-operatives. The cost of acaricides alone in the control of ticks amounted to R80 million in 1994 (Onderstepoort Veterinary Institute, 1996). The government also spends considerable amounts of money on research, training and extension services related to the control of ticks and tick-borne diseases. Parastatals (e.g. Agricultural Research Council–Onderstepoort Veterinary Institute) and international organizations, the Food and Agriculture Organization, also spend large sums of money on research aimed at developing improved control methods. Ticks cost South African agriculture more than R500 million per annum in direct and indirect losses (*Farmers Weekly* 1998, p. 47). The dipping service for cattle in Venda is provided at heavily subsidized rates to the communal cattle owners (Mukhebi and Perry, 1992). It currently costs communal farmers in Venda 1 Rand and the government 12 Rand to dip a head of cattle each year. However, there is a group of commercial farmers who bear the full cost of controlling ticks and tick-borne diseases in their own herds. Depending on the frequency of application, annual costs of acaricides to farmers who bear the full cost of control can range from US$2 to US$20 per animal (Mukhebi, 1992). In order to choose the optimal level of treatment, Norval *et al.* (1992) argue that one must be able to model the impact of different scenarios of intervention upon tick infestation, tick population dynamics, breed susceptibility to tick infestation and damage coefficients (Dipping Policy Survey, 1997).

Costs and benefits of alternative control programmes

The outbreak of East Coast fever in the early 19th century presented the Venda government (then the South African Government) with only two basic options, namely to control or not to control. The assessment of an animal health project takes into account the incidence of disease outbreaks before and after the project, as well as the losses that would result from failure to implement the project (Sidibe, 1981). This section examines those two options/scenarios herein referred to as:

- A 'do nothing' strategy ('without' dipping scenario)
- A control strategy ('with' dipping scenario).

Emphasis will be given to the comparison of the costs and benefits of each strategy at the household level.

Estimations of economic costs and benefits of disease control programmes

Knowledge of factors relating to costs incurred and benefits accrued is important when dealing with the appraisal of alternative control strategies. Establishing standard methods to measure both costs and benefits in order to determine the cost–benefit ratio is not possible, as this differs in terms of the disease, the species affected and their levels of production (Paniagua Arellano and Yubero, 1983). Perry and Mukhebi (1995) also admit that the quantification of output losses (costs) is not a simple task. They argue that there are often problems in separating these losses from the effects of confounding factors, such as other diseases and production constraints. Accurate measurement of the level of diseases and their effects on productivity requires intensive record keeping, which are often non-existent or unavailable, or requires time-consuming and expensive field surveys and monitoring studies to be conducted. The determination of the cost–benefit ratio requires the following information: the mortality and morbidity rate; losses in meat, milk, etc.; infertility and abortion; treatment (acaricides) cost; and other potential losses (expenses resulting from the closing of domestic and export markets) (Paniagua Arellano and Yubero, 1983).

The following benefits will accrue from the control of the disease: a reduction in the mortality and morbidity rate; increase in the production of meat, milk, hides, etc.; increase in the fertility and birth rates, other potential benefits (i.e. the opening up of markets, the improved sanitary quality of the products and thus a higher sales price); and the relief from socio-psychological embarrassment borne by the animal owner and his neighbours as a result of the disease. There are costs that must be incurred in order to achieve these benefits. These costs can be grouped into: direct costs, including investment and salaries; and indirect costs, including improved standard of hygiene in the animal husbandry facilities. However, it is not always possible to evaluate the above-mentioned losses as some cannot be quantitatively assessed, for others a system for valuation is non-existent. The most common methods used to calculate losses or benefits are: estimates by experts, verification based on research or survey data, observations made while programmes are put into practice and findings from comparative studies. Once these physical output losses and/or benefits have been quantified, they must be valued in monetary terms to reduce them to a common unit of measurement before they can be aggregated. This requires the use of appropriate prices for valuation, which are not often obvious.

A 'do nothing' strategy

Owing to heavy stock losses in the face of rinderpest disease, the introduction of exotic breeds made Africa vulnerable to tick-borne diseases. Owing to breed susceptibility, it would probably have been irrational at the time to adopt this strategy for economic reasons. If this strategy had been adopted, tick-borne diseases would have been an epidemic on the continent, but as time progresses cattle would have adapted in a move to enzootic stability (A.M. Spickett, 1998, personal communication). The existence of enzootic stability implies that control could have been selective, strategic and focused only on susceptible target cattle populations.

A 'do nothing' strategy would have incurred no additional expenditure to the government and farmers would have maintained the use of their cattle, although the cattle would not have performed as efficiently as before the introduction of the disease. In addition, the income foregone from cattle products would have been higher relative to the income foregone under the control strategy to be discussed below. This is because of the high morbidity and mortality rate estimated to be about 4%. Although this strategy was not adopted, an effort should be made to estimate the losses in the absence of control and then estimate the extent to which these losses will be reduced by the strategy adopted by the society.

A control strategy

One of the most significant consequences of the East Coast fever control programmes was the introduction of compulsory dipping in acaricides throughout the infected areas. Whether eradication of East Coast fever can be attributed entirely to the dipping programme and the other control measures that were applied, or whether climatic changes brought this about or at least facilitated the elimination of a specific population of ticks responsible for transmission of East Coast fever, is open to speculation (Norval, 1992). The benefits derived under this strategy are higher than the benefits derived under the 'do nothing' strategy. This is because of a relatively low morbidity and mortality rate of 3% in the sampled region. Most developing regions were left with a tradition of short interval dipping, which has persisted to the present day. If properly applied, dipping provides very good control of all tick-borne diseases.

Estimates of Costs and Benefits

The 'without' dipping scenario

One of the most important components in a cost–benefit analysis of the disease control programmes is the estimation of losses per animal from disease. The estimation of physical losses for individual animals is always a problem, especially in a situation where the 'without' situation cannot be modelled because dipping has been in operation since the turn of the 20th century. This lack of data created problems in the accurate determination of the effects to be assessed, the costs and benefits and the exact method to be used to measure these effects. As a consequence, it was decided to assess only direct benefits that appear at the farm level in terms of milk and meat production (gains/losses), low productivity due to longer calving interval and impacts on draught power.

The derivation of benefits for the 'without' scenario was performed on the basis of various estimates obtained using the Delphi technique. Experimental data comparing the effect of ticks and tick-borne diseases on livestock productivity between tick-free and tick-infested herds are unavailable for South Africa. Estimates of impact on productivity done by the Food and Agriculture Organization in Zambia were used with minor adjustments (Pegram *et al.*, 1993).

Table 22.2 presents various figures used for the estimation of costs and benefits for both milk and meat (beef). Based on the income from milk production, the impact on morbidity and mortality was calculated for the 'without' situation as follows (i.e. income gained from milk production):

Morbidity = 326 (sampled average herd size) × 38% (cows percentage) × 405 (value per year)

Mortality = 326 (sampled average herd size) × 4% (mortality rate) × 38% (cows percentage) × 405 (value per year)

Table 22.2. Values for the possible calculation of milk and meat benefits.

Dipping scenarios	Average live body weight (kg)	Beef price/kg* (R)	Average herd size	Cows herd composition (%)	Mortality rate (%)	Value of milk/year (R)	Value of a cow/year (R)
'With'	241	5	329	39	3	526	1152
'Without'	234	5	326	38	4	405	1036

*Source: Own survey data.

It was estimated that on average the mortality rate would have stabilized at 4% without the control programme (A.M. Spickett, 1998, personal communication). However, the initial mortality rate would have been 10% without dipping at the beginning of the 20th century. This possible decline in mortality is based on the assumption that breeds with a strong immune system (predominantly the Nguni breed) would have survived better and would have dominated the national herd.

Income gained from beef production is calculated as follows:

Morbidity = 57 (average number of cattle sold) × 234 kg (weight) × R5 (price/kg)

Mortality = 326 (sampled average herd size) × 4% (mortality rate) × 234 kg (weight) × R5 (price/kg).

For a small-scale farmer the selling price of a live cow is not a function of weight. However, it is assumed that price is a function of weight in this analysis so that the price of beef per kilogram can be multiplied by the total weight of an animal. Moreover, the traditional farmer sometimes salvages a sick animal for meat. In this analysis all reported deaths are considered to be unsalvaged.

With regard to draught power, two significant assumptions had to be made to perform the analysis, namely, only oxen are used for ploughing and infected oxen are incapable of ploughing. Oxen were grouped into two breed categories, indigenous and exotic breed, so as to attach the infection probability for each group. Grouping of breeds was done using the sampled breed percentage composition. Nguni constitute 65% of the sampled herd and it was assumed that Nguni is the only indigenous breed. Again it was assumed that both breeds have an equal infection probability of 25% without dipping. The value of oxen indicated in Table 22.3 is for an average herd of 18 oxen after taking into consideration the infection probability. The average number of oxen for the sampled herd is 24.

Ticks may be responsible for loss in udder quarters and may increase the calving interval. The latter is the most significant parameter in determining herd productivity and, hence, profitability. This type of loss is better estimated on a survey basis (Pegram *et al.*, 1993). Low productivity in milk production due to a longer calving interval for the infected cow is estimated as follows (i.e. foregone income due to longer calving interval):

Productivity = 61 (number of calves in a year) × 12.5% (probability of losing a calf per cow in a year) 1036 rand (average value of cattle).

The 'with' dipping scenario

The derivation of benefits for this scenario is similar to the 'without' scenario. Unlike the 'without' scenario, this scenario relies on the survey results. A survey result reveals the mortality rate as 3%. Mortality rates exhibit a strong seasonal pattern that closely parallels the rainy season

Table 22.3. Comparison of costs and benefits for the with and without dipping scenarios (1997). Based on a 4% mortality rate.

Costs (R)		Income (R)		
'With' dipping scenario				
Government costs*		Milk: sold		67,854
Motor transport allowance, maintenance and insurance	11,538	Mortality	−2,025	65,829
Heavy machinery (water provision)	661	Meat: sold		69,890
Stationery	236	Mortality	−11,893	
Livestock maintenance (acaricides)	15,535			57,997
Diptank maintenance	21,000	Draught power		6,216
Total	**48,970**			
Farmers costs				
Tickicides	1,164			
Purchased drugs	682			
Total	**846**			
Salary	6,341			
Total cost	**57,157**	**Total income**		**130,042**
		Net income	**72,885**	
'Without' dipping scenario				
		Milk: sold		50,171
		Mortality	−2,007	48,165
		Meat: sold		66,690
		Mortality	−15,257	51,433
		Draught power		4,662
		Total income		**104,260**
		Low calving productivity		−7,900
		Total income		**96,360**
		Benefit–cost ratio		0.8

*Information obtained from secondary sources (e.g. veterinary budget).

when ticks are most active. No attempt was made here to differentiate the various mortality rates for the different seasons of the year. To measure the mortality rate, farmers were asked about the number of cattle deaths caused by tick-borne diseases. This relies only on the farmers' diagnosis and it is acknowledged that the reliance on farmer diagnosis has a potential for error depending on, *inter alia*, familiarity with the disease, memory and the desire to attract veterinary attention. In communal areas, mortalities are not always reported, or if reported, seldom investigated. Moreover, disease reporting in general in South Africa is unsatisfactory, and results in underreporting of most diseases and causing unreliable statistics (National Department of Agriculture, 1996/7). Although these limitations render the estimations less reliable and accurate, they serve as the only basis on which the analysis can be built. Both milk and meat production incomes can be obtained using the values in Table 22.2.

For a small-scale farmer the costs of tick control can be divided into two categories, namely: those that the government incurs; and those incurred by stockowners. Data on government expenditure for ticks and tick-borne diseases control were obtained from the government veterinary budget, reports and records in the Department of Veterinary services. Costs incurred by the government include costs of vehicle maintenance, insurance and allowances, stationery, dip maintenance and heavy machinery for water provision. The cost of water provision is low because the location of many diptanks was determined by the available water sources. Costs of tick and tick-borne disease control to individual farmers are fairly straightforward and are based upon expenses for animal treatment and acaricides application. Treatment costs largely include cost of purchased drugs, etc. These costs were obtained from sampled farmers' interviews and are shown in Table 22.3. Treatment of the disease is regarded as an ex post use of resources to restore animal performance to its previous level of health before the disease occurred. Whether it is worth doing, as opposed to accepting the reduced productivity or even culling the animal, is an economic decision.

A comparison of the 'with' and 'without' scenarios

Project impact analysis attempts to value costs and benefits that arise with the project and compare them with the situation as it would have been without the project (Gittinger, 1982). Both costs and benefits are clearly shown in Table 22.3 with milk and beef calculated in terms of mortality and a decrease in the productivity of the herd (i.e. morbidity). The foregone income for the 'without' scenario due to mortality is higher than the foregone income for the 'with' scenario. The difference between the 'with' dipping scenario net income and the net income balance for the 'without' scenario gives an incremental net loss of R23475 per sampled herd. This amount represents the loss in income occurred in spite of the existence of the dipping programme. Disease control is economically justified only if the estimated benefits outweigh the costs incurred, i.e. the benefit–cost ratio must be greater than or equal to 1.

Table 22.3 shows a benefit–cost ratio that is almost equal to one (0.8). According to Morris and Meek (1980, p. 165) 'monetary values must be used with caution that the numerical values obtained in an economic analysis should be seen principally as a basis for ranking strategies, not as representing the actual benefit which will be achieved under all circumstances'. This benefit–cost ratio was estimated by dividing the increase in total income from both scenarios by the increase in total costs for both scenarios. This ratio suggests that the control of ticks and tick-borne diseases by the government is not economically justified and probably needs to be a private sector responsibility. However, in establishing the appropriate roles for the public and private sector in the livestock services industry, it is necessary to obtain a clear understanding of the nature of the tick control service. Not only will the economic nature of the service determine whether private delivery will be feasible, but also whether private provision will result in a socially optimal level of supply (Kirsten and Randela, 1998).

The control of ticks is a private good with externality. This implies that the control of ticks by an individual farmer also benefits other farmers by reducing tick population and the chances of other farmers from getting tick-borne diseases. Owing to this free rider problem associated with tick control service delivery, there will be a tendency towards underprovision or no provision of this service when the production decision is profit motivated. Thus, private firms will have no incentive to provide this service because it will not be in the interest of any individual to pay for it.

Furthermore, although the benefit–cost ratio is slightly less than 1, this ratio can be improved by reducing the costs involved. This can be done by moving from intensive dipping to a strategic acaricides application. Such a decrease in costs might possibly result in two consequences, namely, either the same output can be produced more cheaply or the savings can be converted into increases in output. The strategic control of ticks is a strategy based on the fact that ticks exhibit a seasonal cycle and that concentration of acaricide application during the peak month of tick activity will effectively interrupt the tick feeding cycle, reduce the tick population thereby reducing the number of engorged females which lay eggs to perpetuate the generation. This dipping regime is possible taking into consideration that livestock owners are more particular about regular immersion of their cattle during the spring/summer period (September/October–March/April) when the presence of ticks and the damage they cause are obvious, than in winter when only a few or no ticks are seen. The presence of ticks on cattle in winter varies from region to region, from few to none, depending on the degree of dryness and severity of winter cold. Thus, in some areas, there may in fact be no need to dip cattle in winter instead of dipping cattle throughout the year. It appears, therefore, that if the strategic dipping regime can be adopted, there can be a significant reduction in the dipping costs and an increase in the benefit–cost ratio. The magnitude of the reduction in dipping costs depends on the knowledge of the optimal tick control. In addition, the range and magnitude of physical losses avoided, *inter alia*, depends on the control system used, and is determined by the technique of control and the success with which it is implemented.

Sensitivity analysis

The mortality rate for the 'without' dipping scenario is, however, subject to debate and is based on optimistic assumptions. Therefore, sensitivity analysis is conducted to determine the robustness of benefit–cost ratio against possible changes in the mortality rates. The sensitivity analysis presented in this section addresses the effects of changes in the mortality rate to 10%, and the results in Table 22.4 with a 10% mortality rate without the control programme show the benefit–cost ratio increases from 0.8 to 1.2. Thus, it is evident that the benefit–cost ratio is sensitive to changes in the mortality rate. The 10% mortality rate was chosen assuming that it would have been probably the second lowest mortality rate to prevail in the absence of the control programme.

The sensitivity analysis reveals an incremental net benefit of R10,658 per sampled herd. It is important to note that the incremental net benefit derived represents the loss in income that would have resulted from failure to implement the dipping programme. Sensitivity analysis based on various mortality rates (i.e. worst case scenario) was also performed. In addition to the 4 and 10% mortality rates already analysed, Table 22.5 further shows the probability of occurrence for various mortality rates together with their respective benefit–cost ratios. In the 'without' dipping scenario, the mortality rate is inversely related to the probability of occurrence. This is because the higher mortality rate, e.g. 21%, is relatively a worst case scenario that is unlikely to occur, hence the low probability of 2%. At 15% and 21% mortality rate the benefit–cost ratio is 1.5 and 2, respectively.

Attempts to supplement this analysis with other measures of project value, internal rate of return and net present value, were considered. However, lack of both benefit and costs series of data as well as data on initial investment costs precluded the performance of the analysis. This lack of data surely underscores the importance of institutionalizing monitoring and evaluation within the South African agricultural research system. However, wherever programmes involve public expenditure the authorities responsible now require, *inter alia*, clear demonstration that programmes will show a net benefit. In performing analyses of ongoing programmes, there has to be a comparison of projected future progress with what would happen in the future if the programme is discontinued. Therefore, what is of relevance today is the analysis of what will happen in future if dipping can be completely discontinued. It is estimated that in the first, second and third year farmers are likely to face an average mortality rate of 10, 6 and 4%, respectively (A.M. Spickett, 1998, personal communication). Thus, it will only take 3 years for the herd to develop an acceptable immune system to the extent that the mortality rate can stabilize at 4%.

The reason for the high mortality rate in the first year is that dipping to some extent destroyed the immune system of especially the indigenous African breeds. In addition, it resulted in an interruption in the transmission of tick-borne pathogens and led to the establishment of a susceptible cattle population which was followed by the loss of the existing situation of enzootic stability.

Table 22.4. Sensitivity analysis of a change in the mortality rate to 10%.

Costs (R)		Income (R)		
'With' dipping scenario				
Government costs*		Milk: sold		67,854
Motor transport allowance, maintenance and insurance.	11,538	Mortality	–2,025	
Heavy machinery (water provision)	661			65,829
Stationery	236	Meat: sold		69,890
Livestock maintenance (acaricides)	15,535	Mortality	–11,893	
Diptank maintenance	21,000			57,997
Total	**48,970**	Draught power		6,216
Farmers costs				
Tickicides	1,164			
Purchased drugs	682			
Total	**1,846**			
Salary	6,341			
Total cost	57,157	**Total income**		**130,042**
		Net income		**72,885**
'Without' dipping scenario				
		Milk: sold		38,418
		Mortality	–3,842	
				34,576
		Meat: sold		66,690
		Mortality	–35,802	
				30,888
		Draught power 4662		
		Total income		70,126
		Low calving productivity		–7,900
		Total income		62,226
		Benefit–cost ratio		1.2

*Information obtained from secondary sources (e.g. veterinary budget).

Table 22.5. Estimates of mortality rates for the 'without' dipping scenario.

Mortality rate (%)*	Probability of occurrence (%)*	Benefit–cost ratio
4	80	0.8
10	40	1.2
15	10	1.5
21	2	2

Source: A.M. Spickett (1998, personal communication).

Considering a 10% mortality rate, of the total sampled herd of cattle (1976) 198 cattle are expected to die of tick-borne diseases resulting in a R228,096 (198 × R1152) loss to the sampled cattle owners. Obviously, dipping should continue but the manner in which it should continue would require further investigation involving all stakeholders within the livestock industry.

The direct economic impact of tick-borne disease mortalities at farm level

Tick-borne diseases cause great economic losses to the individual farmer. Mukhebi and Perry (1992) categorized losses caused by tick-borne diseases into direct and indirect production losses, losses through costs incurred for controlling the disease and costs for providing research training and extension services pertaining to the diseases. Economic losses caused by tick-borne diseases are at present difficult to estimate accurately. Direct production losses can be attributed to the presence of the diseases in the cattle herd largely through mortality. Cattle that become severely infected die unless treated. Often it is the mortality rate and the cost of control that appear to receive the greatest attention and concern from those interested in controlling the diseases. This is due to the fact that mortality is more discernible than any tick-borne disease effects. Moreover, losses from morbidity are negligible compared with losses from mortality and may be ignored. It is for this reason that this section solely focuses on the effect that the mortality caused by tick-borne diseases has for the sampled small-scale cattle farmers.

A total number of 64 cattle died from tick-borne diseases as reported by 22% of the sampled farmers. Total cattle losses range from two in Dzanani to 16 in Guyuni. The most commonly found tick-borne diseases in the surveyed areas are redwater, heartwater and gall-sickness. The majority of the respondents (71%) are farming with the Nguni breed, which represents about 65% of the total sampled herd size. However, it constitutes 80% and 50% of the total sampled herd in the Open and the Yellow Line areas, respectively. With the exception of Tshifudi, group discussion indicates that the respondents within the Yellow Line area have always preferred cattle breeds that grow fast for commercial purposes.

The Nguni breed is resistant to diseases to such an extent that little or no losses from the infection may be experienced, especially in areas where tick-borne diseases are endemically stable (Norval, 1994). However, contrary to expectation the Nguni breed exhibited a relatively higher mortality of 2.8%, followed by the Nguni–Brahman mixed breed mortality of 0.3% in Matshena. The relatively high mortality due to tick-borne diseases in the Nguni breed can probably be ascribed to the loss of immunity caused by indiscriminatory acaricide application. More importantly, a possible explanation for this paradox could relate to poor management practices of small-scale farmers. Evidently, tick-borne diseases hardly make it possible for a resource-poor farmer to maintain the desired rate of genetic gain through selection as tick-borne diseases affect the demographic structure of the herd population. Animals that recover from tick-borne diseases may suffer from weight loss, produce low milk yields, provide less draught power and could possibly suffer from reduced fertility and delays in reaching maturity. These animals according to A.M. Spickett (1998, personal communication) remain carriers and can spread the infection, particularly heartwater.

The cost of the loss of cattle as a result of tick-borne diseases in Vyeboom ranges from R1146 for Dzanani to R19,138 for Vyeboom. The financial loss caused by animal diseases is determined, *inter alia*, by the form of diseases. A disease can either be in an enzootic or epizootic state. The former refers to those diseases whose incidences vary from farm to farm and which the individual livestock owner can control (Dijkhuizen *et al.*, 1991, p. 263). Whereas epizootic diseases on the other hand are those that are usually contagious and rarely occur in certain areas and require regional and (inter-) national control measures.

The total loss of cattle value to the surveyed areas is estimated at R73,728. The total loss was calculated by the number of cattle which died by the average value per cattle. The estimate indicates that the total direct loss in milk, traction, manure and sales caused by tick-borne diseases in the surveyed areas is R73,728. As the number of cattle deaths increases, the total cattle value lost also increases. Cattle death represents the greatest financial loss followed by meat sales, work of cattle and manure. The loss of milk diminishes the supply of protein for impoverished households. The monetary value of livestock is heavily influenced by how one values them and such a value in turn depends upon market forces. This is one of the reasons for the differences in cattle values amongst and within countries.

Under the 'without' dipping scenario, the mortality rate would have been 4% on average. This means that of the total 5094 head of cattle in the surveyed areas, 204 head of cattle would have died resulting in the potential loss in income of R211,344. This value is the potential estimated direct mortality impact (loss) of tick-borne diseases without the control programme. And it is apparent that the higher farmers value their cattle and the higher the death rate, the higher the losses experienced by farmers, *ceteris paribus*.

Without the control programme, tick-borne diseases induced mortalities have a relatively greater impact on the livelihood of the rural households. The socio-economic impact of diseases is determined by the loss to the community of benefits and products derived from cattle. Thus, the determination of the socio-economic impact of diseases is related to the reasons for cattle farming and the value of cattle. The benefits derived from cattle farming by surveyed households include food (e.g. milk, meat), monetary (e.g. capital wealth, investment), manure (e.g. fertilizer) and work (e.g. draught power). Taking into consideration that the income received from cattle or cattle products is largely used to meet some basic needs, the impacts of cattle mortalities to rural livelihood are both direct and indirect. For instance, cattle mortality means fewer cattle or cattle products available for sale, resulting in low income. This could possibly result in reduced expenditure on education and training, and this consequently would give rise to a higher illiteracy rate leading to lower productivity, thus unemployment, thereby putting pressure on the government budget for social services. In essence cattle mortalities reduce the per capita consumption of both livestock and other products, *ceteris paribus*.

The 'with' scenario results in increased agricultural production through reduced morbidity and mortality rates. The mortality savings (deaths prevented) directly affect total off-take per time horizon and herd population structure. If mortality savings are high and the herd expands, higher off-take rates are possible. This contributes towards an increase in human well-being through increased farm income and food availability. Under these circumstances of increased production, the price of the product is expected to decline thereby enhancing food affordability, *ceteris paribus*.

A reduction in cattle deaths by any means for tick control reduces the loss, *ceteris paribus*. Apparently the socio-economic value for the 'with' scenario, in particular, significantly contributes towards food security. It is therefore important to adopt strategies aimed at avoiding the potential occurrence of a catastrophe in less developed regions. Similarly, in poor rural areas the provision of dipping services should continue to be the responsibility of the governments to ensure food security, alleviate poverty and provide other social benefits.

Conclusions

The cost–benefit analysis of the ticks and tick-borne diseases control programme was done in two scenarios, namely the 'with' and 'without' control scenario. Emphasis was given to the comparison of the costs and benefits of each strategy at the household level, had both strategies been adopted. The income foregone in terms of milk, meat, draught power and low productivity under the 'without' scenario is larger than the income foregone under the 'with' scenario. The 'with' scenario has significant socio-economic impact to the rural households and such impact is related or linked to the different products, benefits and services obtained as a result of the control of ticks and tick-borne diseases.

The results of this study reveal a benefit–cost ratio, which is less than 1 (0.8) using a 4% mortality rate. This ratio can possibly be improved by reducing the current dipping costs through the adoption of a strategic tick control. Under this strategy, acaricides are applied when it is necessary, especially during period of high tick challenge. It appears that over time the current benefits obtained, more especially from the Nguni breed, can be obtained with less cost.

This benefit–cost ratio seems to be sensitive to changes in the mortality rate yielding a benefit–cost ratio of 1.2 when the mortality rate is changed from 4% to 10%. Although the benefit–cost ratio is less than 1, for the 4% mortality scenario, the dipping programme still deserves government support because of the economic (public) nature of the service it provides. More importantly, in less developed areas dipping forms part of the whole veterinary surveillance programme, which in essence is a purely public good. The information from a surveillance programme benefits the whole sector and cannot be appropriated by any livestock farmer or private sector.

References

Dijkhuizen, A.A., Renkema, J.A. and Stelwagen, J. (1991) Modeling to support animal health control. *Agricultural Economics* 5, 263–277.

Dipping Policy Survey (1997) *Report on the Implementation of a Dipping Policy for Cattle in the Northern Transvaal Region*, Petersburg, South Africa, pp 6.

Doran, M.H., Low, A.R.C. and Kemp, R.L. (1979) Cattle as a store of wealth in Swaziland. *American Journal of Agricultural Economics* 61, 41–47.

Farmers Weekly (1998) Combat Ticks and Win a Rugby Ticket to the World Cup. 30 October.

Gittinger, J.P. (1982) *Economic Analysis of Agricultural Projects.* Johns Hopkins University Press, London, p. 503.

Kirsten, J.F. and Randela, R. (1998) The Value of the Diagnostic Programme of OVI and OIED to South Africa: a case for Government Intervention. Unpublished Report, p. 12.

Morris, R.S. and Meek, A.H. (1980) Measurement and evaluation of the economic effects of parasitic diseases. *Veterinary Parasitology* 6, 165–184.

Mukhebi, A.W. (1992) Economic impact of *Theileriosis* and its control in Africa. In: Norval, R.A.I., Perry, B.D. and Young, A.S. (eds) *The Epidemiology of* Theileriosis *in Africa.* Academic Press, London, pp. 107–112.

Mukhebi, A.W. and Perry, B.D. (1992) Economic implications of the control of East Coast fever in Eastern, Central and Southern Africa. In: *Future of Livestock Industries in East and Southern Africa.* Proceedings of a workshop held in Nairobi, 20–23 July 1992, eds Kategile, J.A. and Mubi, S. Kenya International Livestock Research Institute, pp. 107–112.

National Department of Agriculture (1996/97) *Annual Report: Veterinary Services and Livestock Improvement.* Pretoria, South Africa.

Norval, R.A.I. (1994) Vectors. In: Coetzer, J.A.W., Thomson, G.R., Kriek, R.J. and Tustin, R.C. (eds) *Infectious Disease of Livestock with Special Reference to Southern Africa*, Vol. 1. Oxford University Press, Cape Town.

Norval, R.A.I., Barrett, J.C., Perry, B.D. and Mukhebi, A.W. (1992) Economics, epidemiology and ecology: a multidisciplinary approach to the planning and appraisal of tick and tick borne disease control in Southern Africa. In: Fivaz, B., Petney, T. and Horak, I.G. (eds) *Tick Vector Biology: Medical and Veterinary Aspects.* Springer-Verlag, Berlin, pp. 35–54.

Onderstepoort Veterinary Institute (1996) *Project Description and Progress Report.* Onderstepoort Veterinary Institute, Pretoria, South Africa.

Paniagua Arellano, F.J. and Yubero, D. (1983) *Animal Health and Economics*, Technical Series No. 3. Office International Des Epizootic, France.

Pegram, R.G., Tatchell, R.J., De Castro, J.J., Chizyuka, H.G.B., Creek, M.J., McCosker, P.J., Moran, M.C. and Nigarura, G. (1993) Tick Control: New Concepts. *Medical and Veterinary Entomology* 13(1):33–40.

Perry, B.D. and Mukhebi, A.W. (1995) The socioeconomic factors influencing the occurrence of diseases in animals. Workshop on Socioeconomic Aspects of Animal Diseases in Southern Africa, 3 May 1995. Foundation for Research Development Auditorium, Pretoria, South Africa.

Randela, R. (2000) Socio-economic impact analysis of livestock disease control programme with special references to ticks and tick-borne diseases. MSc thesis. University of Pretoria, South Africa.

Scholtz, M.M., Spickett, A.M., Lombard, P.E. and Enslin, B. (1991) The effects of tick infestation on the productivity of cows of three breeds of cattle. *Onderstepoort Journal of Veterinary Research* 58, 71–74.

Scoones, I. (1992) The economic value of livestock in the communal areas of southern Zimbabwe. *Agricultural Systems* 39, 339–359.

Sidibe, S. (1981) Animal health and economics in Mali. *Bulletin of International Epizootics* 93(5–6), 849–854.

Umali, D.L., Feder, G. and De Haan, C. (1994) Animal health services: finding the balance between public and private delivery. *The World Bank Research Observer* 9, 71–96.

23 Impact of Investments in Livestock Research and Development Programmes in South Africa

M.R. Mokoena

Abstract: This study assesses the economic impact of livestock production research and development (R&D) programmes in South Africa from 1970 to 1996. The objectives of the analysis are to estimate the costs and returns to investments in livestock improvement schemes. The net present values (NPVs) and the rates of return (RORs) to investments in the livestock improvement R&D programmes were estimated using the Akino–Hayami model. The estimated RORs to the dairy cattle, beef cattle, mutton sheep and pig improvement schemes were 51%, 29–44%, 25–54% and 2–14%, respectively. The estimated NPVs for all the schemes at all discount rates (i.e. 5%, 10% and 15%) were positive, and substantial, except for the pig scheme at 15% discount rate. These RORs are, except for the pig scheme, higher than the opportunity cost of investment in South Africa. The high RORs to livestock improvement R&D programmes do not only justify past investments, but also indicate possible under-investments in some livestock R&D programmes. The NPVs show that the payoffs to livestock production research can more than pay for the investment costs. These livestock improvement schemes depend on public funding. In the climate of increased competition for government resources, future alternative cost recovery measures should be pursued with an increasing share of private sector investment.

Introduction

Before the discovery of important mineral deposits in South Africa in the second half of the 19th century, agriculture provided the base for economic growth and development (Laubscher and Kotze, 1982; Van Zyl *et al.*, 1988; Van Rooyen *et al.*, 1996). Agricultural growth is important to the Reconstruction and Development Programme and for the economic well-being of South Africans. Agriculture also plays a vital role in supporting the rural areas in maintaining infrastructure, housing and other services. Some 5400 primary farm schools owe their continued existence directly or indirectly to agriculture (Van Rooyen *et al.*, 1996). Thus an investment in research and development (R&D) must be viewed as being central to the economic transformation of the country.

 Impact of Science on African Agriculture and Food Security (eds P. Anandajayasekeram, M. Rukuni, S. Babu, F. Liebenberg and C.L. Keswani)

Background

The livestock sector is an integral component of the agricultural sector in South Africa (Mokoena, 1998). Over the past several decades, livestock products accounted for about 40% of the total value of agricultural output, which is not surprising as 80% of the agricultural land in South Africa is not suitable for crop production. Livestock also performs a number of important roles or functions in the lives of rural communities in developing areas. These functions can be categorized into social, economic and cultural roles. Recent research has shown that economic/commercial objectives are becoming increasingly more important (Nkosi and Kirsten, 1993). Livestock also provide an important source of foreign exchange earnings. As new production technologies are developed, and adopted in other livestock exporting countries of the world, the competitiveness of South African livestock products on world markets will only be preserved if there is a concurrent adoption of new technologies in South Africa. Similarly, livestock productivity will have to improve to preserve the profitability of commercial livestock enterprises on domestic markets.

Livestock R&D is an important component of the livestock production sector. Five research centres were established to meet the demands for new technologies, including the Animal Improvement Institute, the Animal Products and Animal Nutrition Institute, the Range and Forage Institute, the Onderstepoort Veterinary Institute and the Onderstepoort Institute for Exotic Diseases. The activities of the first three of these institutions are geared towards improving livestock productivity while the latter two are charged with preventing a decline in productivity.

The Animal Improvement Institute is comprised of a number of livestock improvement schemes, namely: the National Beef Cattle Performance and Progeny Testing Scheme (NBCPPTS), the National Dairy Cattle Performance and Progeny Testing Scheme (NDCPPTS), the National Pig Performance and Progeny Testing Scheme (NPPPTS), the National Mutton Sheep Performance and Progeny Testing Scheme (NMSPPTS), the National Wooled Sheep Performance and Progeny Testing Scheme (NWSPPTS) and the National Poultry Performance Testing Scheme (NPPTS). The objective of these improvement schemes is to develop technologies to improve livestock productivity.

Despite the importance of the livestock industry in South Africa, and the extensive livestock research activities undertaken, there have been concerns that the rate of return (ROR) to livestock research has been very low. Initial estimates suggested that the ROR to livestock research was about 5% (Van Zyl, 1996), which is much lower than the estimated ROR to public sector research of 44%, with benefits concentrated in the field crop and horticulture subsectors (Khatri *et al.*, 1996). This chapter reports on a study of the NDCPPTS, NBCPPTS, NMSPPTS and NPPPTS. These livestock improvement schemes were initially introduced in 1919, 1959, 1965 and 1956, respectively. The need for more financial support for livestock research, and the formulation of sound research policies to guide the livestock industry requires evidence that past investments in R&D efforts in livestock technology development and transfer have been competitive with other attractive investment alternatives. Thus, there was a need for the economic assessment of livestock research.

Objectives

The overall objective of this study was to assess the economic impact of livestock production research in South Africa. The specific objectives were as follows:

- To estimate the returns to the investments in livestock production research.

- To discuss the debate on public versus private funding for livestock research and alternative cost recovery methods.
- To develop policy recommendations for future investments and cost recovery measures for livestock research.

Methodology

Several methods have been used to estimate the returns on research investments, which usually require the estimation of the supply or production functions, as initiated by Griliches (1958). Similarly, this study used the above methods as extended by Akino and Hayami (1975).

Data

The data used in this study included annual R&D expenditures, annual production data of different livestock products and their sale prices, annual adoption rates of technologies, etc. The data requirements differed with respect to the methods of analysis (Mokoena, 1998). Therefore, this aspect will be further discussed under each method, with its specific data requirements.

The data used in this study were collected from different sources, namely: the Agricultural Abstracts from the Department of Agriculture, Government Expenditures from the State Library, South African Statistics from Central Statistical Services, Annual Records of the Livestock Improvement Schemes from Agricultural Research Council-Animal Improvement Institute (ARC-AII), as well as from different studies conducted in this country. When some of the relevant information was missing, meetings were arranged with the personnel of the ARC-AII.

Research and development expenditures

The gathering of R&D expenditures data was one of the greatest challenges of this study. The historical financial record keeping at the Research Institute has been poor, with no or very little historical data. This may be some reflection on the lack of accountability and planning of spending government funds in the past. Only recently, with the establishment of the Agricultural Research Council (and the institute now called Animal Improvement Institute) in 1992, a systematic record of R&D expenditures data has been kept. The only recorded R&D expenditures in South Africa appear in the Government Budget or Parliamentary Grants to each improvement scheme within the institute. Although these data are not ideal (as they changed slightly over time), they represent a good approximation of what each scheme was spending. Government budgets from 1970 to present were used to derive the R&D expenditure series.

The costs included in the estimation of the R&D series are staff salaries and benefits, recruitment, administration and overhead expenditures, input costs and provision for the depreciation of capital assets. These costs were compiled on an annual basis to derive annual R&D expenditures in nominal terms. However, the series were inconsistent due to the fact that during certain years the R&D expenditure components, e.g. staff salaries and benefits, were missing in the records. As a result, the derived inconsistent nominal R&D series was used to estimate an R&D index. The estimated index was then used to derive a consistent series by using 1996 R&D costs to interpolate a series backwards to 1970. The interpolation procedure was performed in order to account for the missing information. The nominal R&D expenditures were deflated by the consumer price index to derive real values with 1996 as a base year.

Prices and price elasticities

The study used the producer prices of direct animal products (i.e. fresh milk, beef, mutton and pork) that were obtained mainly from the *Abstract of Agricultural Statistics* (Department of Agriculture, various years). The nominal prices were deflated to real prices using a consumer price index with 1996 as a base year. The price elasticities used in the study were obtained from previous studies. The long-term demand elasticities of fresh milk, beef, mutton and pork were –0.51 (Mckenzie and Nieuwoudt, 1985), –0.96, –1.93 and –1.86, respectively (Hancock *et al.*, 1984). The demand elasticities of these livestock products show that a 10% increase in the prices of these products will reduce the quantity demanded for fresh milk, beef, mutton and pork by 5.1%, 9.6%, 19.3% and 18.6%, respectively.

Studies on the supply and demand of agricultural products were less readily available. Consequently, the long run price elasticity of supply of aggregate livestock output was used in this study. The estimated long-term supply elasticity of livestock output was 0.32 (Townsend, 1998). Thus a 10% increase in the price of livestock increased livestock output by 3.2%.

Yield and yield differentials

The gathering of yield data for performance-tested animals was another challenge. The historical data for yields of beef cattle, mutton sheep and pigs in the scheme were not available. However, the historical fresh milk production data for the dairy cattle scheme have been recorded annually as well as the yield advantage percentages, ranging from 50% to 58% (Department of Agriculture, 1979–94; ARC-AII, 1995/1996). For the estimation of yields for beef, mutton and pork, the national production averages derived from the *Abstract of Agricultural Statistics* (Department of Agriculture, various years) were used. The study assumed that yield of performance-tested animals was 50% above the national average. For the estimation of yield differentials between, with and without scheme animals, a 50% yield advantage was assumed for all schemes.

Adoption rates and adoption behaviour

Rogers (1962), as quoted by Feder *et al.* (1985), define the adoption process as 'the mental process an individual passes from first hearing about an innovation to final adoption'. However, for rigorous theoretical and empirical analysis, a precise quantitative definition of adoption is needed. Such a definition must distinguish between individual (farm-level) adoption and aggregate adoption. Final farmer-level adoption is defined as the degree of use of a new technology, when the farmer has full information about the new technology and its potential. This definition corresponds to Schultz's (1975) contention that the introduction of new technologies results in a period of disequilibrium behaviour, where resources are not utilized efficiently by the individual farm, and learning and experimenting lead the farmer toward new equilibrium levels.

In the context of aggregate adoption behaviour, the diffusion process is defined as 'the process of spread of a new technology within a region'. Aggregate adoption, is defined as 'the degree of use of a new technology as a quantitative measure of the extent of adoption'. It is measured by the aggregate level of use of a specific technology, within a given geographical area or a given population (e.g. the percentage of farmers using the technology). In most cases, agricultural technologies are introduced in packages, which include several components, for example, high yielding crop varieties, fertilizers, land preparation practices, etc. While the components of a package may complement each other, some of them can be adopted independently, thus farmers may face several distinct technological options. They may adopt the complete package of innovations introduced in the region, or subsets of the package. In most cases, several adoption and diffusion processes may occur simultaneously. However, such adoption processes may follow specific (and predictable) sequential patterns (Feder *et al.*, 1985).

This study assumed the aggregate adoption in all the animal improvement schemes, and that farmers have adopted a complete package of innovations (breeding technologies, feeding technologies, etc.) introduced in the schemes. It was further assumed that this adoption followed a linear pattern. As the study has assumed 1970 as the initial year for all the technology developments as well as a lead period of 10 years, the adoption rates were calculated from 1980 (i.e. the initial year of adoption). The rates of adoption were estimated as a percentage of the number of animals under the improvement scheme over a total number of animals registered in the South African Studbook. The rates were then linearized according to the assumed length period of 11 years, when the adoption was increasing to reach the ceiling and declined to zero during the following 6 years. Briefly, the adoption rates of the technologies in all the schemes started in 1980 and increased until they reached the ceiling in 1990, and then start declining from 1991 until they are completely taken out of the production system in 1996. The technological adoption ceilings for NDCPPTS, NBCPPTS, NMSPPTS and NPPPTS were 40.1%, 11.6%, 36.9% and 3.9%, respectively.

Results

Estimated benefits to livestock improvement research

The total benefits to animal improvement schemes were estimated by adding the consumer and producer surpluses. The R&D expenditures over the period 1970–1979 were also estimated and deflated to 1996 real values. The study assumed that the R&D costs occurred during the first 10 years (1970–1979) of the programme only, and there were no benefits during that period. Gross benefits and R&D expenditures, for the study period, of NDCPPTS and NBCPPTS, and NMSPPTS and NPPPTS are shown in Tables 23.1 and 23.2, respectively. These tables reveal that the benefits from animal improvement schemes were not realized until 10 years of research had elapsed, because of the assumption that it took an average of 10 years of research to generate benefits thereafter. On an average the annual gross benefits of all the schemes, except the pig scheme, significantly increased from 1980 to 1990 (peak year), and then declined to zero in 1996 (when the technology was out of the production system). However, the gross benefits from the pig scheme increased from 1980 to 1989, and thereafter started declining from 1990 to 1996 and were fluctuating.

Payoffs to investments in livestock improvement R&D programmes

The two estimates, i.e. gross benefits and R&D costs, allow the estimation of the payoff to investment in livestock production research. This is presented in the form of incremental net benefits (INB) obtained after subtracting the R&D costs from the gross benefits. The net benefits for each scheme are also shown in Tables 23.1 and 23.2. The INB stream for all the schemes was negative for the first 10 years. Thereafter, the net benefits were positive and increased over time, when adoption rates were increasing, until the adoption ceiling was reached, then declined until the technology was no longer adopted. However, the positive and negative streams of INB do not determine whether there was a payoff to the investment in livestock production research or not. The best way of determining whether there was a payoff is by calculating the

Table 23.1. Gross annual research benefits (GARB), research and development (R&D) costs and incremental net benefits (INB) for the NDCPPTS and NBCPPTS: 1970–1996.

	NDCPPTS			NBCPPTS		
Year	GARB (Rm)	R&D (Rm)	INB (Rm)	GARB (Rm)	R&D (Rm)	INB (Rm)
1970	0.0	5.0	(5.0)*	0.0	4.8	(4.8)
1971	0.0	5.7	(5.7)	0.0	5.4	(5.4)
1972	0.0	5.5	(5.5)	0.0	5.2	(5.2)
1973	0.0	5.0	(5.0)	0.0	4.8	(4.8)
1974	0.0	6.0	(6.0)	0.0	5.7	(5.7)
1975	0.0	5.9	(5.9)	0.0	5.6	(5.6)
1976	0.0	5.9	(5.9)	0.0	5.6	(5.6)
1977	0.0	5.7	(5.7)	0.0	5.4	(5.4)
1978	0.0	5.6	(5.6)	0.0	5.3	(5.3)
1979	0.0	5.1	(5.1)	0.0	4.9	(4.9)
1980	123.2	0.0	123.2	70.0	0.0	70.0
1981	263.7	0.0	263.7	138.3	0.0	138.3
1982	382.9	0.0	382.9	196.8	0.0	196.8
1983	467.8	0.0	467.8	251.3	0.0	251.3
1984	570.5	0.0	570.5	284.8	0.0	284.8
1985	584.8	0.0	584.8	307.4	0.0	307.4
1986	593.2	0.0	593.2	294.1	0.0	294.1
1987	615.2	0.0	615.2	471.5	0.0	471.5
1988	770.2	0.0	770.2	500.1	0.0	500.1
1989	921.8	0.0	921.8	534.2	0.0	534.2
1990	948.9	0.0	948.9	553.1	0.0	553.1
1991	687.3	0.0	687.3	466.9	0.0	466.9
1992	569.4	0.0	569.4	404.7	0.0	404.7
1993	364.5	0.0	364.5	222.6	0.0	222.6
1994	246.8	0.0	246.8	159.1	0.0	159.1
1995	120.0	0.0	120.0	67.1	0.0	67.1
1996	0.0	0.0	0.0	0.0	0.0	0.0

Figures in parentheses are negative values.
Rm = million rand.

net present value (NPV) and the ROR of the INB stream. These NPVs and RORs are the measures of the return to and efficiency of investment in livestock production research.

The NPVs for all the animal improvement schemes were calculated at three levels of the real discount rate, i.e. 5%, 10% and 15%. Equating the NPV to 0 gave the ROR. The estimated RORs and NPVs at 5, 10 and 15% are shown in Table 23.3. During 1970–1996, investments in the NDCPPTS, NBCPPTS, NMSPPTS and NPPPTS generated the RORs of about 51%, 44%, 54% and 14%, respectively. These ROR figures indicated that for every R1.00 spent in these animal improvement schemes during this period, the returns were R1.51, R1.44, R1.54 and R1.14, respectively, or annual profits of 51, 44, 54 and 14 cents, respectively. These ROR to investments in all animal improvement schemes are high compared with the opportunity costs of investment in South Africa during the period 1970–1996. This was because the real interest rate was about 8.5% in 1996 (Fényes and Meyer, 1998).

During the past 27 years, using a 5% discount rate, investments in NDCPPTS, NBCPPTS, NMSPPTS and NPPPTS generated 3305.0, 1934.9, 1143.1 and 51.7 million Rand,

Table 23.2. Gross annual research benefits (GARB), research and development (R&D) costs and incremental net benefits (INB) for the NMSPPTS and NPPPTS: 1970–1996.

	NMSPPTS			NPPPTS		
Year	GARB (Rm)	R&D (Rm)	INB (Rm)	GARB (Rm)	R&D (Rm)	INB (Rm)
1970	0.0	1.4	(1.4)*	0.0	3.2	(3.2)
1971	0.0	1.6	(1.6)	0.0	3.7	(3.7)
1972	0.0	1.5	(1.5)	0.0	3.5	(3.5)
1973	0.0	1.4	(1.4)	0.0	3.2	(3.2)
1974	0.0	1.6	(1.6)	0.0	3.8	(3.8)
1975	0.0	1.6	(1.6)	0.0	3.8	(3.8)
1976	0.0	1.6	(1.6)	0.0	3.8	(3.8)
1977	0.0	1.6	(1.6)	0.0	3.6	(3.6)
1978	0.0	1.5	(1.5)	0.0	3.6	(3.6)
1979	0.0	1.4	(1.4)	0.0	3.3	(3.3)
1980	41.7	0.0	41.7	2.4	0.0	2.4
1981	85.9	0.0	85.9	4.9	0.0	4.9
1982	125.1	0.0	125.1	6.7	0.0	6.7
1983	160.8	0.0	160.8	10.3	0.0	10.3
1984	204.5	0.0	204.5	12.7	0.0	12.7
1985	228.8	0.0	228.8	13.2	0.0	13.2
1986	268.8	0.0	268.8	16.2	0.0	16.2
1987	289.3	0.0	289.3	18.7	0.0	18.7
1988	301.7	0.0	301.7	22.3	0.0	22.3
1989	291.1	0.0	291.1	22.3	0.0	22.3
1990	292.0	0.0	292.0	14.8	0.0	14.8
1991	215.5	0.0	215.5	16.1	0.0	16.1
1992	137.7	0.0	137.7	14.7	0.0	14.7
1993	82.0	0.0	82.0	10.0	0.0	10.0
1994	39.7	0.0	39.7	7.8	0.0	7.8
1995	17.6	0.0	17.6	3.2	0.0	3.2
1996	0.0	0.0	0.0	0.0	0.0	0.0

Figures in parentheses are negative values.
Rm = million rand.

Table 23.3. Returns to investments in the animal improvement schemes: 1970–1996.

Measure of return	NDCPPTS	NBCPPTS	NMSPPTS	NPPPTS
Rate of return (%)	51	44	54	14
NPV at 5% (Rm)	3305.0	1934.9	1143.1	51.7
NPV at 10% (Rm)	1428.2	819.1	502.5	12.4
NPV at 15% (Rm)	652.8	364.9	232.9	(2.0)

NPV, net present value.
Figures in parenthesis are negative values.
Rm = million rand.

respectively. In other words, the annual NPVs for these schemes are about R122.4, R71.7, R42.3 and R1.9 millions, respectively. Changing the discount rate to 10% the NPVs for all the schemes were still substantial and positive. However, on changing the discount rate to 15% the NPV for NPPPTS became negative, whereas those of others were still positive. The NPVs for NDCPPTS were significantly higher than for all other schemes at all discount rates. However, these animal improvement schemes are not mutually exclusive.

People-level impacts of livestock improvement R&D programmes

The RORs to the animal improvement schemes, as shown in Table 23.3, quantify the economic impact of investment in livestock production research. These RORs clearly indicate that the investments in these schemes represent a high return on public funds during the period 1970–1996. However, the ROR alone does not tell us who are the beneficiaries or losers from public investment in research. In order to determine the beneficiaries and losers due to investment in livestock production research, the Akino–Hayami model enabled us to perform welfare analysis by estimating the annual consumers' and producers' surpluses and estimate how the benefits from investments in livestock production research programmes are distributed among consumers and producers.

Table 23.4 shows the annual welfare gains and losses due to investments in livestock production research programmes from 1980 to 1996. Table 23.4 shows that the consumers gain more than the producers in all the schemes. The consumer gains due to NDCPPTS during 1980–1995 were more than the total gains (economic surplus), implying that the milk quotas

Table 23.4. Welfare gains and/or losses due to the animal improvement schemes: 1980–1996.

	NDCPPTS		NBCPPTS		NMSPPTS		NPPPTS	
Year	CS (Rm)	PS (Rm)	CS (Rm)	PS (Rm)	CS (Rm)	PS (Rm)	CS (Rm)	PS (Rm)
1980	156.8	(33.6)	71.1	(1.0)	28.9	12.8	1.7	0.7
1981	331.7	(68.0)	139.9	(1.5)	58.8	27.1	3.5	1.4
1982	475.9	(93.0)	198.3	(1.5)	84.5	40.5	4.8	1.9
1983	574.3	(106.6)	252.2	(0.9)	107.3	53.5	7.4	2.9
1984	691.9	(121.4)	284.7	0.1	134.7	69.8	9.1	3.6
1985	700.4	(115.6)	306.3	1.2	148.7	80.1	9.4	3.8
1986	701.5	(108.3)	291.9	2.2	172.3	96.4	11.5	4.7
1987	718.3	(103.1)	466.2	5.3	183.0	106.3	13.3	5.4
1988	887.6	(117.4)	492.7	7.5	188.2	113.5	15.8	6.5
1989	1048.5	(126.7)	524.3	10.0	179.1	112.0	15.8	6.5
1990	1065.0	(116.1)	540.7	12.4	177.1	114.9	10.5	4.3
1991	790.4	(103.1)	459.7	7.3	134.1	81.4	11.4	4.7
1992	670.5	(101.1)	401.2	3.5	87.9	49.8	10.4	4.2
1993	439.3	(74.8)	222.1	0.4	53.6	28.4	7.1	2.9
1994	304.3	(57.5)	159.9	(0.8)	26.6	13.1	5.6	2.2
1995	151.3	(31.3)	67.9	(0.8)	12.1	5.5	2.3	0.9
1996	0.0	0.0	0.0	0.0	0.0	0.0	0.0	0.0

CS stands for consumer surplus, while PS stands for producer surplus.
Figures in parenthesis are negative values.
Rm = million rand.

were imposed during this period (Mckenzie and Nieuwoudt, 1985), i.e. the production shift had caused prices to decline, but this decline may have been offset by the quotas. However, the quota system treats the effect and not the cause of overproduction. In summary, if the supply function remained static and farmers receive lower prices in excess of quota, there would be no surplus produced as the marginal costs would exceed the marginal revenue.

The producer's surplus due to NBCPPTS was also negative during the periods 1980–1984 as well as 1994–1995, implying that during those periods the consumer surpluses were more than the economic surpluses. This was especially true when the supply and demand functions are inelastic. The price elasticities of demand and supply for fresh milk and beef, used for the estimation of benefits to investments in NDCPPTS and NBCPPTS are inelastic, thus the consumer benefits due to these schemes were more than total benefits resulting in negative producer benefits. The same results were found by Fox *et al.* (1990) regarding Canadian dairy cattle research, but were in contrast to the findings by Widmer *et al.* (1988) regarding Canadian beef cattle research.

The gains to consumers and producers, due to investments in both NMSPPTS and NPPPTS, were all positive during the period 1980–1995, but the consumers' gains were always more than the gains to producers. This is attributed to the fact that the long-term price elasticities of demand for mutton and pork used for the estimation of research benefits due to these schemes are elastic, whereas the price elasticities of supply for mutton and pork are inelastic. Despite the lower and negative figures of the producers' gains, they also gain, because generally the producers are also the consumers, as they retain some of their produce for home consumption.

Sensitivity analysis

A simple benefit–cost analysis (BCA) approach was used to test the sensitivity or robustness of the results to the assumptions of the Akino–Hayami model, i.e. a pivotal divergent supply shift and constant elasticities of demand and supply curves. The BCA approach is a simplified version of the surplus approach or index number method. In contrast to the Akino–Hayami model, the BCA model assumes perfectly inelastic supply and perfectly elastic demand curves. In addition, the BCA does not require the information of price elasticities of demand and supply when estimating the benefits. The gross benefits were estimated by multiplying the yield gain due to technological change by the constant price. Similar R&D expenditures as well as lead period, as in the Akino–Hayami model, were used when estimating the INB due to livestock production research programmes. The results from this analysis are presented in Table 23.5.

The results indicate that the ROR to NDCPPTS was the same as that estimated by Akino–Hayami model, but the NPVs were lower at all discount rates. However, the RORs and the NPVs of the other animal improvement schemes were significantly lower than those

Table 23.5. Returns to investments using the BCA approach: 1970–1996.

Measure of return	NDCPPTS	NBCPPTS	NMSPPTS	NPPPTS
Rate of return (%)	51	29	25	(2)
NPV at 5% (Rm)	2573.8	774.9	79.9	(17.6)
NPV at 10% (Rm)	1102.9	267.5	32.0	(17.7)
NPV at 15% (Rm)	507.9	98.3	12.2	(15.9)

NPV, net present value.
Figures in parenthesis are negative values.
Rm = million rand.

estimated by the Akino–Hayami model, thus indicating a larger assumptions variation. Therefore, the larger the variation of the assumptions between the Akino–Hayami and BCA models, the larger the difference between the estimated RORs. In addition, the ROR and the NPVs (at all discount rates) for the NPPPTS were negative, indicating that there was no payoff in this scheme during the period 1970–1996. The ROR was negative 2%, indicating that for every R1.00 invested in NPPPTS during the period 1970–1996 the return was 98 cents. In other words, for every R1.00 spent in NPPPTS there was a loss of 2 cents.

Aggregate returns to livestock research in South Africa

Aggregate returns were estimated by calculating the marginal internal RORs at three levels of the alternative research expenditures and two levels of output. The first level of research expenditures included the sum of animal improvements, range and forage and veterinary research expenditures. The second level consisted of the sum of animal improvements, and range and forage research expenditures. The third level included the animal improvements research expenditures only, whereas the first output level includes the total livestock output, and the second level was made up of the total livestock output less poultry output (Townsend, 1998). The results are presented in Table 23.6.

The ROR to livestock research using all the expenditures, including animal health was between 5% and 7%, which was consistent with the value derived by Van Zyl (1996). This method attributes that all livestock research expenditures were used to increase livestock productivity. This was clearly a mis-specification, as veterinary research is primarily focused on preventing livestock diseases. If the veterinary research expenditures were removed, the ROR increases to between 10% and 14%. This was expected, as the cost used to derive the returns has been reduced. The increasing ROR, due to the reduction of veterinary research expenditures, seemed more realistic. However, one should bear in mind that if there was no veterinary research, the ROR could have been lower, because the livestock productivity would have declined due to disease.

If the expenditures to range and forage research are removed, i.e. using research expenditures to animal improvements only, the ROR increased to 16–22%. This represents the aggregate ROR to livestock production research, i.e. animal improvements. The disaggregate ROR to animal improvement schemes (i.e. NDCPPTS, NBCPPTS, NMSPPTS and NPPPTS) as mentioned above are significantly higher, compared with the aggregate ROR for livestock production research, except that the ROR for the pig scheme was slightly lower, i.e. 14%. The lower ROR to the pig scheme, as compared with the aggregate ROR, suggests the reason why

Table 23.6. Marginal internal rates of return to R&D expenditures on livestock research: 1927–1994.

	Marginal internal rates of return (%)	
Alternative research expenditures used for the rate of return calculation	Using total output	Using total output net of poultry
Animal improvements + range and forage research + veterinary research	5–7	5
Animal improvements + range and forage research	11–14	10
Animal improvements	18–22	16

Source: Townsend (1998).

the aggregate ROR to animal improvement research was less than the disaggregate RORs to dairy cattle, beef cattle and mutton sheep schemes. In general, these results showed that the investments in the animal improvement schemes contributed significantly to the livestock productivity, because both aggregate and disaggregate RORs for livestock production research were still substantial in absolute terms.

Conclusions

The high RORs to livestock improvement research programmes justify past investments, but also indicate a high return to livestock R&D expenditures. They also confirm the importance of allocating resources to R&D. The NPVs showed that the payoffs to livestock production research can more than pay for the investment costs. The distributions of benefits are such that they are mostly captured by consumers. If the consumers benefit more, the benefit is distributed across a larger population, i.e. all households that consume livestock products. The benefit would decline more on those households who spend a higher proportion of their income on livestock products. Benefits to producers, however, are only realized by definition, because producers represent a small proportion of the population.

Presently, the burden of funding livestock production research is borne by taxpayers, who, according to this study, could be seen as subsidizing consumers and producers who are the beneficiaries of research. These livestock production research programmes are primarily dependent on public funding, and in the climate of increased competition for government resources, alternative cost recovery approaches need to be pursued. Therefore, the private sector should provide an increasing proportion of the funding, and the high estimated returns should be sufficient enough to attract private investment (whether from farmers or industry).

Pressure on government expenditures in the interests of deficit reduction, as well as pressures for productivity-enhancing research, had led to the consideration of other means of financing agricultural research. Producer levies represent one possible source of funds. In South Africa, this is possible for the large-scale commercial farmers, who have been the chief beneficiaries of public R&D in the past. The adoption of the technology developed by the animal improvement schemes remains confined to large-scale commercial producers. Therefore, this bias towards commercial production and lack of emphasis on technology development for the communal areas needs to change and the appropriate technology to suit the wider South African clientele should be developed. At the same time, research policy must maintain a balance between the need for assisting the communal areas and maintaining the higher productivity of the commercial sector.

The main lesson from impact assessment is that an ex-post impact assessment was difficult if no mechanism is in place to collect the relevant data on a continuous basis, i.e. a monitoring and evaluation system for R&D activities. In general, the Agricultural Research Council-Animal Improvement Institute as well as other Research Institutes should consider the institutionalization of a monitoring and evaluation system for their programmes to facilitate impact assessment, planning and priority setting.

Limitations of the study

This study has several limitations that should be considered when interpreting the results and the recommendations. The major limitations were:

- The annual R&D expenditures collected from various records were not consistent in nominal terms. As a result, the derived inconsistent nominal R&D series was used to estimate an R&D index, which was then used to derive a consistent series by using 1996 R&D costs and was interpolated back to 1970, using agricultural R&D expenditure series index.
- The interpolation procedure was performed in order to account for the missing information during certain years, so that the series should be consistent to represent true nominal R&D expenditures.
- The yield levels of 'with' and 'without' livestock improvement scheme, which are an important component in the calculation of benefits, had to be estimated for all schemes, except the dairy scheme. In the absence of baseline data, the process was extremely difficult. The opinions of key informants (mainly researchers) were used in the estimation process.
- The absence of appropriate data made it impossible to adequately assess the impact of livestock improvement R&D on the environment, gender, employment and food security. As a result, the benefits and costs may have been under- or overestimated.

Acknowledgements

Thanks are due to many individuals who provided inputs in various aspects of this study, several of whom I need to mention by name. These are Dr Robert Townsend and Professor Johan Kirsten for their efforts and encouragement, as well as their valuable comments to this study; and Dr Ponniah Anandajayasekeram who taught me impact assessment methodologies. My sincere appreciation goes to the staff members of the Agricultural Research Council, for their helpful discussions and provision of data. I also thank my family for their patience, tolerance, encouragement and support. Finally, I owe my thanks to the Centre for Science Development for their financial support. May God bless all!!

References

Akino, M. and Hayami, Y. (1975) Efficiency and equity in public research: rice breeding in Japan's economic development. *American Journal of Agricultural Economics* 57, 1–10.

ARC-Animal Improvement Institute (1995/1996) *Annual and Technical Reports of the Livestock Improvement Schemes*. ARC-Animal Improvement Institute, Irene, South Africa.

Department of Agriculture (1979–94) *Biennial Reports of the Animal and Dairy Science Research Institute*. Department of Agriculture, Irene, South Africa.

Department of Agriculture (various years) *Abstracts of Agricultural Statistics*. Republic of South Africa.

Feder, G., Just, R.E. and Zilberman, D. (1985) Adoption of agricultural innovations in developing countries: a survey. *Economic Development and Cultural Change* 33(2), 254–298.

Fènyes, T.I. and Meyer, N.G. (1998) Agricultural policy and macroeconomic issues in South Africa in perspective. *Agrekon* 37, 63–82.

Fox, G., Roberts, B. and Brinkman, G.L. (1990) Canadian dairy policy and the returns to federal dairy cattle research. *Agricultural Economics* 6, 267–285.

Griliches, Z. (1958) Research costs and social returns: hybrid corn and related innovations. *Journal of Political Economy* 66, 419–431.

Hancock, P.J., Nieuwoudt, W.L. and Lyne, M.C. (1984) Demand analysis of meat in South Africa. *Agrekon* 23(2), 26–29.

Khatri, Y., Thirtle, C. and van Zyl, J. (1996) Public research and development as a source of productivity change in South African Agriculture. *South African Journal of Science* 92, 143–151.

Laubscher, J.M. and Kotze, H.A. (1982) *The South African Beef Cattle Industry: its Relative Importance to Agriculture, as a Whole and Recent Developments in Supply and Demand*. University of Orange Free State, Bloemfontein, South Africa.

Mckenzie, C.C. and Nieuwoudt, W.L. (1985) Estimation of demand and supply functions for fresh and industrial milk in South Africa. *Agrekon* 24(2), 27–33.

Mokoena, M.R. (1998) Assessment of livestock research in South Africa. MSc thesis, University of Pretoria, South Africa.

Nkosi, S.A. and Kirsten, J.F. (1993) The marketing of livestock in South Africa's developing areas: a case study of the role of speculators, auctioneers, butchers and private buyers in Lebowa. *Agrekon* 32(4), 230–237.

Rogers, E. (1962) *Diffusion of Innovations*. The Free Press, New York.

Schultz, T.W. (1975) The value of ability to deal with disequilibrium. *Journal of Economic Literature* 13, 827–846.

Townsend, R.F. (1998) *Returns to Livestock Research in South African Agriculture: a Re-Examination*. Agricultural Research Council-Unit of Development Impact Analysis, Pretoria, South Africa.

Van Rooyen, C.J., Carstens, J.P. and Nortje, A. (1996) *The Role of Agriculture in South Africa's Economy: a Research and Technology Challenge*. Agricultural Research Council, Pretoria, South Africa.

Van Zyl, J. (1996) *Returns to Research on Crops, Horticulture and Livestock in South Africa*. Agricultural Research Council, Pretoria, South Africa.

Van Zyl, J., Nel, H.J.G. and Groenewald, J.A. (1988) Agriculture's contribution to the South African economy. *Agrekon* 27(2), 1–9.

Widmer, L., Fox, G. and Brinkman, G.L. (1988) The rate of return to agricultural research in a small country: the case of beef cattle research in Canada. *Canadian Journal of Agricultural Economics* 36, 23–35.

24 Overview of Impact Assessment Methodologies

P. ANANDAJAYASEKERAM AND S. BABU

Abstract: The ultimate interest of investors in agricultural research and development (R&D) is the extent to which their investment eventually bears a positive outcome on the lives of the poor and hungry people of developing nations without seriously harming the natural resource base. Consequently, most donors, governments and financiers of agricultural and natural resource research focus on outcome and impact assessment as a major step in their strategy. In impact assessment of R&D investments, one needs to differentiate between the research results and the contribution of research to development (i.e. the people-level impact). Moreover, both of these aspects should be addressed simultaneously.

This chapter summarizes the various methods and techniques used in R&D using a comprehensive framework that addresses intermediate product, direct product and people-level impact simultaneously. Owing to the wide-ranging implications of agricultural research results to society, no single technique or method is sufficient to adequately address the impact. However, there is consensus that the most appropriate approaches to impact assessment should involve a mixture of both qualitative and quantitative methods and active participation of the beneficiaries. Case studies are also essential for further refining the approaches and lessons learned. The chapter concludes by outlining suggested best practices in understanding impact assessment of agricultural R&D investments.

Introduction

Today, it is a well-known fact that investors in public research and development (R&D) are no longer satisfied with activity-based progress reports. They in fact expect outcome and impact evaluation, i.e. an objective assessment. The continuous decline in funding on the other hand is exerting pressure on R&D managers to carefully select a portfolio of investments in order to maximize the social impacts of scarce research resources. Therefore, the ultimate interest of those who invest in agricultural R&D is the extent to which their investment eventually bears a positive outcome on the lives of the poor and hungry people of developing nations without seriously harming the natural resource base. Thus, impact assessment has become a major step in the strategies of most donors, governments and financiers of agricultural and natural resource research. There are at least four rationales for conducting impact studies: (i) accountability and

 Impact of Science on African Agriculture and Food Security (eds P. Anandajayasekeram, M. Rukuni, S. Babu, F. Liebenberg and C.L. Keswani)

credibility; (ii) quality and relevance; (iii) programme/project design and implementation; and (iv) future planning and priority setting.

In assessing the impact of R&D investment, there is a need to differentiate between the research results and the contribution of research to development (i.e. the people-level impact) and both aspects should be addressed simultaneously. Research produces both 'hard' technologies such as finished variety or a breed which diffuse among farm population as well as 'soft' results such as information or advice that other actors use as inputs in a broad innovation process. Owing to the wide-ranging implications of agricultural research results to society, no single technique or method is sufficient to adequately address the impact. In this chapter, attempt is made to summarize the various methods and techniques used in R&D impact assessment. A comprehensive framework has been used for this purpose. This framework addresses intermediate product, direct product as well as people-level impact simultaneously.

The most appropriate approaches to impact assessment should involve a mixture of both qualitative and quantitative methods. Active participation of the beneficiaries is crucial in the process. A retrospective narrative is an essential component of the qualitative assessment and indeed provides the basis for quantitative estimates and the related issues of attribution. Case studies are essential for further refining approaches and methodologies and learn lessons, which can enhance the future impacts related to impact estimates. Suggested best practices are also outlined in this section of the book.

Impacts and Impact Types

Impact assessment is a special form of evaluation. In the literature the term 'impact' is used in many different ways. It is sometimes taken to mean 'any effects' (both intended and unintended) that can be attributed to a specific action. Impact analysis can also be defined as a mechanism that determines 'whether the programme made a difference compared to either no programme or an alternative programme'. Rossi and Freeman (1993) define the purpose of impact assessment as 'establishing whether or not an intervention is producing its intended effects'. In other cases, the concept of impact is used in a more restrictive manner and refers only to the 'long-term outcomes of results of development programme on the people, economy, society or environment' (Kumar, 1995) or the 'ultimate effects on the country or organization' (DANIDA, 1994). In recent years, the concept has been extended to look at the impact of research on their ultimate development goals: food security, protection of the environment and poverty alleviation (Cracknell, 1996). In a recent review, Pingali (2001) concluded that there is a logical evolution from the relatively narrow focus in the 1970s and 1980s on assessment of impacts of germplasm adoption and crop management research, to formal rate of return (ROR) and benefit distribution studies starting in the 1980s. The next major broadening in the 1980s was the work on spill-over and inter-sectoral impacts. Finally in the 1990s the activities have broadened further into gender, environmental impact assessment (EIA), research work and poverty-related work. It has been argued that impact begins to occur when there is a behavioural change.

In discussing the agricultural R&D programme, one can identify two broad categories of interpretation (Anderson and Herdt, 1990). Some people (mostly biological scientists) look at the direct output of the research activity, i.e. a variety, a breed or a set of recommendations, and call this an impact. The other group (mostly donors, planners and policy makers) goes beyond the direct product and tries to study the effects of these direct products on the ultimate users, i.e. the so-called 'people-level' impact. This second type of impact deals with the actual adoption of the research output and subsequent effects on production, income, environment, poverty and/or whatever the development objective may be. The people-level impact of any R&D activity cannot be assessed without information about the (extent) number of users and the degree (intensity) of adoption of improved techniques and the effects of these techniques on the production costs and outputs. However, the adoption of any technology is determined by several factors that are not part of the original research activity in many cases. The people-level impact can be economic (including production), socio-cultural (including effects on gender, income distribution, nutrition, poverty, etc.) and environmental.

Horton (1990) classified technologies broadly into two categories, namely, production technology and R&D technology. Production technology refers to all methods that farmers, market agents and consumers use to cultivate, harvest, store, process, handle, transport and prepare food crops, industrial crops and livestock or any other enterprises for ultimate consumption. R&D technology, on the other hand, refers to organizational strategies and methods used by research and extension programmes in conducting their work. This may include scientific procedures, organizational models, institutional strategies (for programme and project planning, evaluation, training, networking, etc.) and interdisciplinary team research. The production technology leads to production impacts, whereas the R&D technology leads to institutional impact. Institutional impact refers to the effect of new R&D technology on the capacity of research and extension programmes to generate and disseminate new production technologies. R&D technologies are a necessary pre-condition to enhance the development of production technologies, and both are complementary. Therefore, invariably most R&D efforts can simultaneously lead to both production impacts as well as institutional impacts.

The people-level impact cannot be achieved without achieving the intended direct product of research. Therefore, in any comprehensive impact assessment, there is a need to differentiate between the research results and the contribution of research to developmental efforts (i.e. the people-level impact) and both aspects should be addressed simultaneously. A programme that has an impact is one that achieves some movement or change towards the desired direction. This implies the need for a set of operationally defined goals and criteria of success. It is important to ensure that the changes observed are a function of the specific intervention and cannot be accounted for in any other way. In assessing the impact of any R&D programme attention should be paid to causality, attribution and incrementality.

All these observations are used in developing a conceptual framework for assessing the R&D programme and are presented in Fig. 24.1. The different types of impacts associated with a typical R&D investment are summarized in Box 24.1.

The direct products of a research programme can be a technology, a set of specialized information, reports and even journal articles. All these may have impact on the ultimate users, which may be totally different. In general, for problem-focused research, the R&D

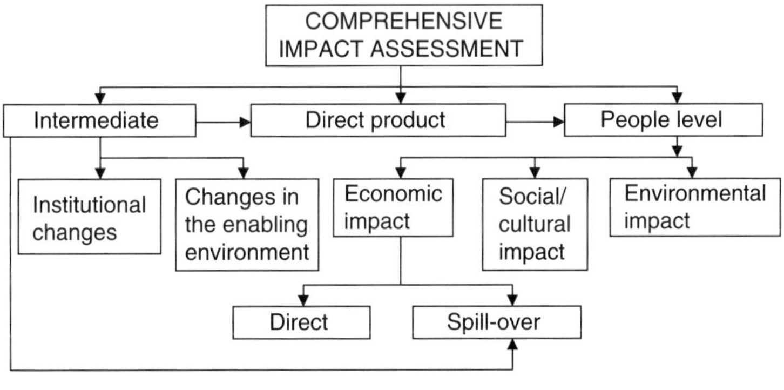

Fig. 24.1. Comprehensive assessment of R&D investments. Source: Anandajayasekeram *et al.* (1996).

Box 24.1. Types of impact associated with R&D investment.

Production impact

- Yield/productivity gains
- Hectarage

Economic impact – comparison of benefits and costs

- Income
- Rate of returns
- Reduced risk
- Number and type of jobs created or reduction in employment rates per type
- Distribution of benefits by:
  - gender
  - income group
  - location
- Changes in resource allocation, e.g. labour patterns
- Nutritional implications

Social/cultural impact (can be positive or negative)

- Changes in the status of women
- Changes in the knowledge and skill level of people
- Changes in the health of various groups of people

Environmental impact (can be positive or negative)

- Air and water pollution
- Soil erosion and sedimentation
- Contamination of soil and water by herbicide or pesticide residues
- Effects on the long-term functioning of biosphere, potential climate change, etc.
- Effects on biodiversity

Institutional impact

- Changes in intermediate organizational structures of methods and plans
- Changes in the number and composition of scientists
- Changes in the proportion of funds allocated to research
- Changes in the mix of public and private sector participation
- Improvement in interdisciplinary involvement

activities will result in three broad categories of impacts that form an integral part of a comprehensive impact assessment.

The impact of any R&D initiative can include production effect, income effect, institutional impact, socio-cultural impact and environmental impact (see Box 24.1). The production effect is measured in terms of increased yield and hectarage under production. The economic impact considers research as an investment activity and attempts to estimate the ROR to research investment. The ROR is the most commonly used evaluative measure of investments in technology development and transfer (TDT). This measure summarizes the benefits, costs and income from the activity in a single number. This number is easily compared with the cost of obtaining funds or ROR obtained from alternative investments. This can also include the additional income generated to the farmer as well as risk considerations.

The third form of impact is the institutional impact, also called the intermediate impact because it is the effect of agricultural research on the so-called 'enabling environment' within the organization that conducts research. The R&D technologies include scientific procedures, organizational models and institutional strategies for programme planning, evaluation, networking and other models (Horton, 1990). For example, the International Maize and Wheat Improvement Center (CIMMYT) has attributed the fall in the coefficient of variation of its yearly international maize trials from 25% in 1974 to 16.5% in 1982 to learning through collaborative interaction between CIMMYT and national scientists (Byerlee and Moya, 1993). Although the linkages between such intermediate effects and ultimate benefit to the public may not be clear, ignoring them is likely to result in underestimating the value of research.

The socio-cultural impact of agricultural technologies is also important especially if the technology improves the food security, status of women and the nutritional status of the rural infants.

The adoption of modern agricultural technologies has often resulted in external benefits and costs largely through its effects on the environment. For example, the use of fertilizers or pesticides may lead to surface and groundwater contamination by toxic chemicals and algae, resulting in significant environmental costs. On the other hand, adoption of minimum tillage technology and herbicides by farmers has probably had environmental benefits in the form of reduced soil erosion and nutrient loss.

Though environmental impacts are important and at times very significant (integrated pest management programmes), due to the lack of biophysical data and complications involved in valuing some of these externalities, most studies do not include a detailed environmental impact of technologies. As a result, in empirical studies, very often one is forced to use some qualitative EIA.

Purpose of Impact Assessment

The purpose of undertaking impact assessment of agricultural R&D activities depends on when the assessment is done. Impact assessment can be undertaken before initiating the research (ex-ante) or after the completion of the research activity (ex-post), including the technology transfer.

Ex ante impact studies (pro-active) can indicate the potential benefits to be gained from research and, therefore, assist the research manager in planning, priority setting and resource allocation. The ex-ante impact assessment will enable one to:

- Study the likely socio-economic impact of the proposed research activity/project;
- Formulate research priorities by examining the relative benefits of different research programmes;
- Identify the optimal combination of research programmes;
- Provide a framework for gathering information to carry out an effective ex-post evaluation.

Ex-post impact studies (reactive) can demonstrate the benefits of past investments in achieving the broader social and economic benefits. The various purposes of conducting ex-post impact assessments include:

- To study the impact and provide feedback for researchers, research managers, planners and policy makers;
- Lessons learned can be used to improve the management and decision-making process with respect to priority setting, implementation, and management of research activities as well as technology transfer;
- For accountability[1] purposes;
- To establish the credibility of the public sector research;
- To justify increased allocations of research resource.

Given the resource constraints confronting the research manager in the region, in future ex-ante impact assessment will become a powerful planning tool in agricultural and natural resource research management.

Impact assessment can be carried out at different levels of aggregation based on the objectives of the exercise. On the one hand, the impact assessment can move from individual research activities to a research programme plus complementary services within a country. On the other hand, technology impact can be measured at individual farm households, on a target population, at the sectoral level, at the national level or even at the regional level. At times, the impact can spread to a secondary sector also. So far, most impact assessment studies have focused on individual commodities at the programme level, looking at the impact on the society as a whole. Given the current emphasis and growth in regional research networks in most developing countries, there is a growing need to assess the impacts of these research projects at the regional level.

Methods of Assessing Impact of R&D Technologies

Agricultural research generates several indirect economic effects, including labour market effects and growth linkages of technical change, that go beyond the standard measure of consumers' and producers' surpluses that is used in obtaining the RORs to R&D investment. The overall effects of technical change as a result of research ultimately depend on general growth and income effects induced throughout the economy by consumption, input and output linkages. A full accounting of these effects requires a general equilibrium model.

Although conceptually it is not difficult to formulate a general equilibrium model the practical application will necessarily be limited by the information, time and analytical capacity required. The incorporation of the various effects may require regional desegregation of product and factor markets and inter-regional trade flows. However, the modelling of the entire economy is an exercise that is likely to be beyond the scope of most research evaluation and priority setting studies and the capacity of most agricultural research and development institutions in the developing countries to implement. In addition, there is as yet no evidence that including indirect effects of research would lead to a difference in ranking of research programme impacts, or in return to research related to public investments.

In many countries general equilibrium models do not exist at the moment and it may take considerable efforts and time to construct such models. Hence, the commonly used empirical models fall under the partial equilibrium analysis category.

A comprehensive impact assessment should simultaneously assess the various impacts, of the TDT. The various techniques and methods used to assess the different types of impact are summarized in Table 24.1 and discussed in the subsequent sections.

Direct product of research: effectiveness analysis

The most commonly used approach for assessing the direct product of research is known as effectiveness analysis. A useful starting point for effectiveness analysis is the logical framework of the project. The logical framework permits the assessment of the degree to which the research activities have made changes in the desired directions. The logical framework itself is a simple matrix that provides a structure for one to specify the components of a programme/activity and the logical linkages between the set of means (inputs and activities) and the set of ends (outputs). This logical framework makes the impact assessment process transparent by explicitly stating the underlying assumptions of the analysis.

Table 24.1. Impact types, techniques, and methods used in a comprehensive assessment.

Impact type	Technique	Method
Intermediate impact Institutional changes Changes in the enabling environment	Simple comparison/trend analysis	Survey, monitoring of selected variables
Direct product of research	Simple comparison – target versus actual	Effectiveness analysis using logical framework
Economic impact	Various	ROR estimates
Socio-cultural impact	Comparison over time	Socio-economic survey/ adoption survey, monitoring of selected variables
Environmental impact	Various (need biophysical information) Qualitative and quantitative	Environmental impact assessment Qualitative assessment

Source: Anandajayasekeram *et al.* (1996).

The effectiveness analysis is a simple comparison of the targets with actual or observed performance of the project. Three sets of comparisons are identified in the literature: 'before' and 'after' comparison (also called historical comparison); 'with' and 'without' comparison; and 'target' versus 'achievement' comparison. The most useful comparison is target versus achievement. This means clearly identifiable quantified targets should be set during the planning process. The targets need not be completely achieved for the project to be deemed effective, but movement in the direction of the desired target is evidence of project effectiveness.

Evaluating the impact of intermediate product(s)

The link between the intermediate product and the ultimate economic benefit is often not clear and, therefore, tends to be ignored in most impact assessment studies. The evaluation of the intermediate product is made difficult by the fact that the benefits of these products are not easy to quantify. Thus, most studies acknowledge the fact that having the institutional capacity to conduct agricultural TDT is of paramount importance. These studies, however, do not include all the benefits in the assessment of the impact. The costs that are easy to quantify are usually included. Thus, the assessment of the intermediate product has been a tricky issue. The practice has been to trace the changes in institutional capacity over time using either simple trend analysis or comparisons. This requires baseline information on these indicators and careful monitoring. The results from these analyses can be incorporated in the quantitative analysis through a multi-criteria analysis.

People-level impact

As pointed out earlier, the people-level impact can be economic, socio-cultural and environmental.

The economic impact

The economic impact of TDT initiatives can be traced through its effect on production and income. The approach used is called efficiency analysis. Efficiency analysis assesses the people-level impact by comparing the benefits that society gets from TDT and the costs incurred in conducting TDT programmes. The benefits and costs are normally collapsed into a single number, the ROR. There are two broad ways of calculating the ROR to TDT: ex-ante and ex-post. The ex-ante methods are useful as research planning tools as they aid in the selection of the research portfolio, priority setting and resource allocation. The ex-post studies are useful for justifying past TDT investments and demonstrating the payoff of such investments.

The ex-ante methods for estimating ROR include benefit–cost analysis, simulation models and mathematical programming models. The last two methods are data- and skill-intensive and, therefore, rarely used.

Commonly used techniques to estimate the ROR are discussed individually in the subsequent sections.

Socio-cultural impact

Socio-cultural impacts include the effects of research on the attitude, beliefs, resource distribution, status of women, income distribution, nutritional implications, etc., of the community. These can be assessed through socio-economic surveys and careful monitoring. To be cost-effective, appropriate socio-cultural questions can be included in adoption survey questionnaires.

Environmental impact

The full assessment of environmental quality issues requires complex analysis of physical, biological, social and economic processes. This also leads into some measurement problems. Such a breadth of analysis is likely to be beyond the scope of most agricultural research assessment activities. Nevertheless, some assessment of environmental impact is necessary when evaluating agricultural research, especially where the environmental impact of the application of the research is likely to be significant. In the absence of data required for a thorough analysis, it may still be possible to identify qualitatively the nature of the social benefits and costs, together with the likely gainers and losers.

Spill-over effects

At times research findings may be potentially applicable to a range of agricultural production conditions or environments cutting across geographical and national boundaries. The wide applicability of research results over a range of agricultural production conditions or environments often cutting across geographical and national boundaries are generally referred to as 'spill-over effects'. Evenson (1987) identified four classes of spill-overs, namely inter-locational, inter-foci, inter-commodity and inter-sectoral.

INTER-LOCATIONAL SPILL-OVER. Inter-locational spill-overs deal with the spill-over of technology from one location to another that is important in regional commodity networking programmes. Locational spill-over will be greater between two locations with similar geo-climatic characteristics than between locations with dissimilar geo-climatic conditions.

INTER-FOCI SPILL-OVER. According to Evenson (1987), most agricultural research programmes invest in the following specializations:

- Pre-technology science (basic research);
- Technology invention and development (strategic and applied research);
- Technology development and sub-invention.

The development invention of technology is the primary objective of most research organizations. However, this technology development is dependent on the science base and the technology base that together define the invention potential of the research programme. Most organizations invest in pre-technology science programmes to facilitate the development of invention potential.

The pre-technology science research, e.g. gene transfer, in one country system has spill-over effects on its own country programme as well as on other countries via enhancement of invention potential. The technology developed in one country may also enhance the invention potential of another country even if the technology is not directly transferable to the other country. Germplasm in a breeding programme is a good example of this type.

Substations within a larger country and some small national agricultural research systems fit with this hierarchy in that they concentrate on adaptive development and sub-inventions targeted to relatively small regions. They are typically technologically dependent on the main station or larger national agricultural research systems. According to Evenson (1987), in the USA, the chief inter-foci spill-over is between pre-technology and applied invention- oriented research.

INTER-COMMODITY SPILL-OVER. The indirect spill-over mechanism will not be confined to a single commodity. Research on input-based technology may be relevant to several commodities. Pre-technology science findings may spill-over across commodities because they enhance the invention potential of several commodity technology programmes. A technology to control insect pests (stalk borer control in maize and sorghum) or correct a soil problem (soil acidity) will spill-over across a number of commodities.

INTER-SECTORAL SPILL-OVER. At times a technology developed within the agricultural section may have an effect on the non-agricultural sector such as in the input-supply and manufacturing industries.

According to Davis *et al.* (1987), the spill-over effect is a combination of four effects:

- Price effects from increased production or reduced costs that are captured in the supply and demand framework.
- Spill-over technology from country 'X' that can be adapted without any research in country 'Y'.
- Spill-over of technology from country 'X' that requires adaptive research before it is applicable in country 'Y'.
- Spill-over of scientific knowledge that ultimately enhances future research in many areas. This is the spill-over effect of basic research that is difficult to quantify.

Based on the above discussions in empirical analysis one could identify three major types of spill-overs, namely economic spill-over, technological spill-over and knowledge spill-over.

ECONOMIC SPILL-OVER. Economic spill-overs refer to the price effects due to increased production and/or reduced costs captured in the supply and demand framework. Within the global and regional context, these spill-overs will affect the inter- and intra-regional production, consumption, trade and prices. These benefits can be captured and estimated using trade models.

TECHNOLOGICAL SPILL-OVER. Technological spill-overs refer to the spill-over of technology from one country to another or from one environment to another. Spill-over of technology can occur with or without adaptive research in the recipient region or country. Technological spill-overs increase the returns to research – this can happen either through 'spill-ins' or 'spill-outs'.

Spill-ins refers to a situation where a country is adopting a technology developed elsewhere. Spill-outs refer to a situation where the research findings are used by other countries.

KNOWLEDGE SPILL-OVER. Research programmes can produce both production technologies as well as R&D technologies. The benefits of the R&D technologies can spill-over from one commodity to another. Often the same individual trained by a project in a specific commodity area is also requested to handle other commodities. Another example may be the methods and technologies that are developed by the project for a given commodity could be applied to other commodities also. It is often difficult to estimate the monetary benefits of such spill-overs, but it is important to recognize them.

The generally accepted economic surplus approach using a multi-regional traded goods model can be used to estimate the economic spill-overs of R&D programmes (Edwards and Freebairn, 1981; Davis *et al.*, 1987).

As discussed in the previous sections, due to the wide-ranging implications of agricultural research to society, no single method is sufficient to adequately capture these impacts. Therefore, a multi-criteria analysis is often recommended for assessing the impact, which may also use a variety of methods. In this way one could use more than one measure to assess the impact. The various methods used to assess the impacts of R&D activities are summarized in Table 24.2.

Tracing the Impact of Agricultural R&D Investment

There is a long time lag between the time that R&D investments are initiated and the time it takes to realize the developmental impacts of the investment. The output, outcome and impact are generally sequentially produced over a period of time and become more difficult to articulate. Very often two approaches are used to trace the output and outcome effects of agricultural R&D: the impact chain and the outcome mapping.

Impact chain

The typical impact chain starts from the set of inputs and activities of a project/programme to the most highly aggregated development results, such as poverty reduction, food security, environmental protection, etc. The chain also specifies all the main intermediate steps: the activities of a project, the output, the use that others make of this output, the direct as well as possible indirect effects, and the implications of the use of these outputs on the ultimate beneficiaries – society (see Fig. 24.2).

Collaborative activities

These are the joint actions undertaken by the collaborators, for example a training workshop. Here one is expected to identify all collaborative activities undertaken by the individual organization/stakeholder group. List activities, key collaborators, as well as the contributions of each group. Clearly state the objectives of the collaborative activities.

Outputs

This refers to the results of the programme activities, i.e. goods and services produced by the set of collaborative activities. In the case of training activities the outcomes may be trained individuals with acquired skills (are able to apply the skills taught), a set of training materials and/or trained trainers.

Table 24.2. Applicability and characteristics of different impact assessment methods.

Methods	R&D time frame	R&D	Strengths	Weaknesses
Partial indicators	Past, ongoing and future	All	The information required to specify the indicators is relatively easy to collect Probably the best method for ongoing monitoring. Relative cost – low	The individual indicators can generally only be 'added up' on a judgemental basis, making overall impact assessment more difficult Potential for bias in assigning weights to different criteria
Integrated partial indicators	Future	Applied research and development	An easy but structured way to identify research priorities Forces the decision makers to explicitly consider the key determinants of impacts Relative cost – low	Totally relies on the judgement of (usually a few) individuals Potential for bias in assigning weights to different criteria
Mathematical programming	Past, ongoing and future	Applied research and development	More powerful and sophisticated Enables one to select 'optional' portfolio Can handle simultaneous change in many variables	Demanding in terms of data requirements Relative cost – high Not particularly useful for evaluating too diverse a set of R&D projects. If either the criteria or constraints are not well defined, there is a risk of arriving at a nonsensical 'optional' solution
Simulation method	Past and future	Applied research and development	Flexible Can be used to estimate optional level of research at national, commodity or programme level Can estimate the effect of research on prices, income, employment or other parameters Can handle simultaneous change in many variables	To be useful, they must accurately reflect the relationship between technological advancement and economic development Requires extensive amount of time to construct, validate and data to operate Relative cost – medium to high

(Continued)

Table 24.2. *Continued.*

Methods	R&D time frame	R&D	Strengths	Weakness
Production function approach	Past	Applied research and development	Offers a more rigorous analysis of the impact Estimates marginal rates of return Statistically isolates the effects of R&D from other complementary inputs and services	Uncertainty in projecting past rates of returns to future Demanding in terms of data Selection of suitable functional form Serious econometric problems Relative cost – high
Modified peer review	Past, ongoing and future	All	Relatively easy to organize Can provide valuable information on potential impacts Probably the best method for basic/strategic R&D Cost – low to medium	Relies on the opinions of a small number of people Qualitative information only
User surveys	Past and ongoing	Applied research and development	Overcomes the problem of a small number of respondents Possible to develop quantitative indices Relative cost – medium	Structuring the survey and analysing the results can be tricky Often requires considerable time to identify users, develop survey methodology and analyse results
Benefit–cost methods	Past (can be used for ongoing and future R&D in certain circumstances)	Applied research and development	Can provide reasonable defensible estimates of potential benefits Provides a structure and a framework for assessing R&D projects, which forces the right questions to be asked	Can be very time consuming and labour intensive Results are critically dependent on assumptions, which can be highly uncertain Because of cost and time requirements, can only be used for a limited number of projects Relative cost – high, data collection requirements are demanding

Cost-effectiveness analysis	Future, past (to a certain extent)	Applied research and development	Simplest Does not require benefit information Relative cost – medium	Nothing to prove that any of the alternatives compared can yield benefits over and above costs If one of the alternatives costs less, but produces a low quality product or has a different impact, then the assessment becomes more complicated
Case studies	Past	Applied research and development	Can provide good illustrations of the relationship between R&D and its impacts Probably the best method for basic/strategic R&D Relative cost – medium	Generally there is no way to 'add up' the results of a group of case studies to obtain a measure of the total impacts of the group The results cannot be extrapolated to other R&D projects that are not in the group

Source: based on Williams (1993).

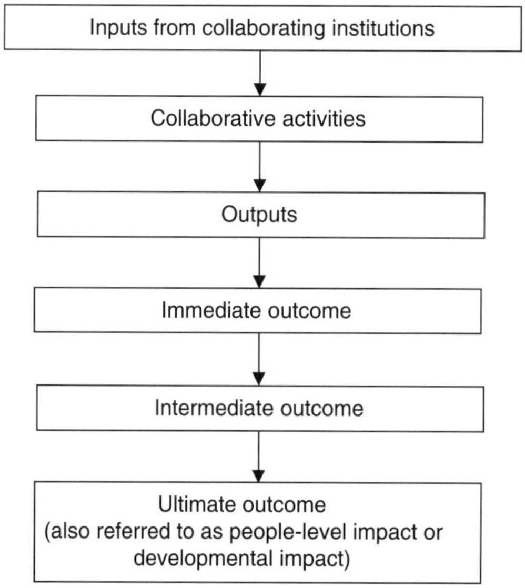

Fig. 24.2. Impact chain.

Immediate outcome

This refers to the first-level effect of the outputs: the observed or documented behavioural changes in those directly affected by the programme. In the case of a training programme, how did the training affect the behaviour of the trainee? Did they make any changes in the way of doing business as a result of the training? Did they apply the skills acquired?

In the case of research the first immediate outcome may be a change in the recommendations provided by the extension staff or even the behavioural change to use the direct product, i.e. adoption by the direct collaborator/contact farmers.

In the case of a technology the intermediate outcome may be the effects at the farm/household level, i.e. increased yield, reduction in cost.

It is worth noting that in order to bring about an outcome, the programme has to change people's behaviour. By trying to identify and then document the changes in attitudes, knowledge, perceptions and decisions taken by programme target groups, which logically link to the outcomes being observed, we can often acquire a good understanding of the actual impact that the programme has. Often, immediate and intermediate outcomes can be measured and documented directly. This requires clearly identifying the various clients of the programme and the way in which their behaviour is expected to change. If an expected outcome has been observed after the programme activity has started up, then this suggests that the programme is having an effect. If we can observe these short-term changes, then the logical case for the programme's attributions can be enhanced.

Outcomes are measures of the use that is made of the output by clients and partners. They reflect the value they place on them as intermediate product, which in turn are input in their management decision-making.

Ultimate outcome (impact)

Impact refers to measurable effects of the outputs and immediate outcomes on the well-being of the ultimate beneficiaries of the R&D efforts, namely the poor, the food and nutrition insecure, and the environment. Most socio-economic impacts and developmental impacts fall under this category. Very often the ultimate outcomes are closely linked to the sectoral/regional/national developmental goals.

Since there is considerable time-lag between the realization of outcome and impact, often one could use proxies or partial indicators in terms of assessing the people-level impact. In addition to programme output, a number of other factors may contribute to the realization of people-level impact. Thus attribution may be more difficult. This issue is discussed separately.

In assessing the outcome and impact, it is important to focus the analysis at three levels: the individuals who are directly involved in the project, the organizations participating in the various activities as well as the ultimate beneficiaries. The field-level observations can be complemented with expert opinion from people outside the project/programme who are knowledgeable about the programme area and the environment in which the programme operates, as well as the outcome and impact of the programme.

If there is documented evidence available (secondary sources such as evaluation reports) about the programme output, outcome and impact, it should be collected, analysed and documented. It is important to show evidence for any claims with respect to outcome and impact, as well as indicate where such evidences can be found.

To sum up, there are four products of concern of collaborative R&D activities: outputs, outcomes, changes in institutional performance and the final welfare impacts. They are sequentially produced and more difficult to document, articulate, measure and attribute as one moves from outputs to impacts. Attribution remains one of the methodological challenges in impact assessment studies. This is critical, especially where partnerships and collaborations are an increasing feature of collaborative activities. Therefore, as far as possible joint impact of various players should be measured rather than trying to separate out the contribution of individual institutions, which may not be feasible in most cases. However, it is important to make sure that the inputs and contribution of all partners are appropriately acknowledged.

Outcome mapping

Outcome mapping is a methodology for planning, monitoring and evaluating development initiatives, which aims to bring about social change. The process of outcome mapping helps a project team or programme to be specific about the actors, its targets, the changes it expects to see and the strategies it employs. Results are measured in terms of changes in behaviour; actions or relationships that can be influenced by the team or programme. It enhances the team and programme understanding of change process, improves the efficiency of achieving results and promotes realistic and accountable reporting.

The key terminologies/concepts used in outcome mapping are: boundary partners, intentional design, outcome challenges and progress markers.

- *Boundary partners:* individuals, groups or organizations with which the programme interacts directly and which the programme hopes to influence.
- *Intentional design:* the planning stage, where a programme reaches consensus on the macro-level changes it wants to influence and the strategies to be used.
- *Outcome challenge:* description of the ideal changes the programme intends to influence in the behaviour, relationships, activities and/or actions of a boundary partner.
- *Progress markers:* a set of graduated indicators of changed behaviour of a boundary partner, which focus on the depth or quality of change.

Outcome mapping is a tool that assists programme teams to learn from and to report realistically on their achievements by tracking the connections between what they do and what happens, focusing on change process and outcomes. It defines the limits of the programme's influence, promotes strategies that are appropriate to the context and recognizes the potential contributions of other actors. Development results (or outcomes) are measured as changes in behaviour and relationships of actors with which the programme interacts directly. Performance is assessed as the programme's contribution to influencing those changes with outcome mapping, it is possible to develop and use indicators that facilitate comparison and learning while retaining the relevant contextual details of the story at each site or in each case. Outcome mapping is especially useful in projects where success depends on behavioural change. It provides tools that help a development programme to think holistically and strategically about how it intends to achieve results.

Outcome mapping is usually initiated through a participatory process at a design workshop led by internal or external facilitators who are familiar with the methodology. It is useful to include boundary partners in the initial workshop for their input on the relevance, activities and direction of the programme. Ideally, the M&E system would have been outlined at the planning stage of the programme. It is a three-stage 12-step process as shown in Fig. 24.3. The three stages are: intentional design, outcome and performance monitoring, and evaluation planning.

Intentional design

The four basic questions to be asked at the intentional design stage are: why? (vision statement), how? (mission, strategy maps, organizational practices), who? (boundary partners) and what? (outcome challenges, progress markers). This stage helps the team to clarify and reach consensus on the macro-level changes they would like to support and to plan appropriate strategies. The long-term goals provide reference points to guide strategy formulation and action plans

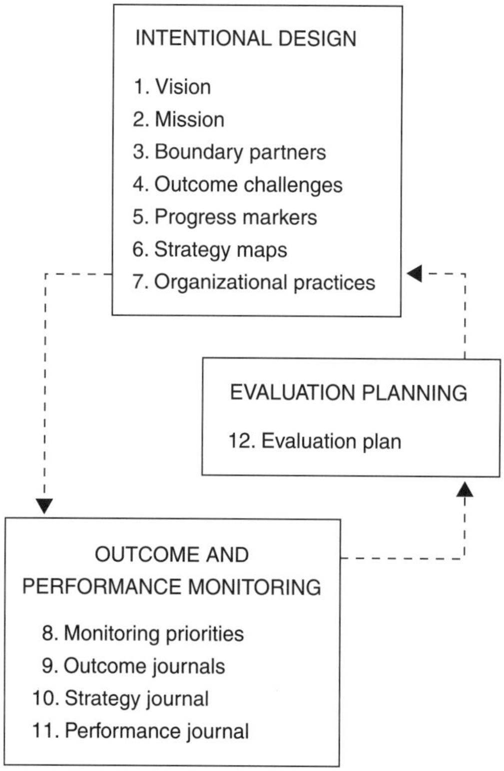

Fig. 24.3. The three stages and 12 steps of outcome mapping. Source: Smutylo (2005).

rather than acting as performance indicators. Progress markers that are used to track performance are developed for each boundary partner.

Outcome mapping does not help a team identify programme priorities. It is appropriate and useful only when a programme has already chosen its strategic direction and wants to chart its goals, partners, activities and progress towards anticipated results.

Outcome and performance monitoring

Provides a framework for monitoring actions and boundary partners' progress towards outcome/goals. The three data collection tools that can be used in this process are:

- An outcome journal monitors boundary partners' actions and relationship;
- A strategy journal monitors strategies and activities;
- A performance journal monitors the organizational practice that keeps the programme relevant and viable.

These tools provide a workplace and processes and help the team reflect on the data they have collected and how it can be used to improve performance. Select only the information that they can afford to collect.

Evaluation planning

Helps the team set priorities so they can target evaluation resources and activities where they will be most useful. This stage outlines the main elements of the evaluations to be conducted.

It is worth noting that 'outcome mapping' and 'result-based management' are compatible and outcome mapping can contribute important elements to results-based management; such as supporting stakeholder learning in relation to the management of the programme, fostering social communication as a basis for interactive participation, and strengthening local organizations and institutions.

Approaches to Estimating Rates of Returns

Efficiency analysis is a systematic way of comparing the cost and benefits of R&D investment. The analysis can be ex-ante or ex-post. The commonly used ex-post methods for ROR estimation can be divided into two broad groups, as shown in Fig. 24.4. The econometric method uses the production function in which research and transfer activities are considered inputs and give the marginal rate of return (MRR) to agricultural TDT. The MRR quantifies the returns to the last dollar spent in the research project. To determine the optimal allocation of funds, it is necessary to know the marginal benefit of the last research dollar invested. This is the only method that allows for the separation of the effects of research from those of extension and other support services. However, the data requirements have reduced the extensive use of this method.

The second groups of methods are the surplus approaches. These methods calculate the benefits of TDT as the net change in producer and consumer surplus, employing a partial equilibrium analysis. The different techniques are based on the difference in the assumed nature and elasticities of the supply and demand functions. The benefit–cost approach has various combinations of the nature of the supply shift and the functional form of the supply and demand curves. The cost-saving approach is in between these two approaches, but based on the same theoretical foundation.

These methods calculate the average rate of return (ARR). The ARR or internal rate of return (IRR) takes the research expenditure as given and calculates the ROR for the project or programme in its entirety. This provides information to assess the success of the project in terms

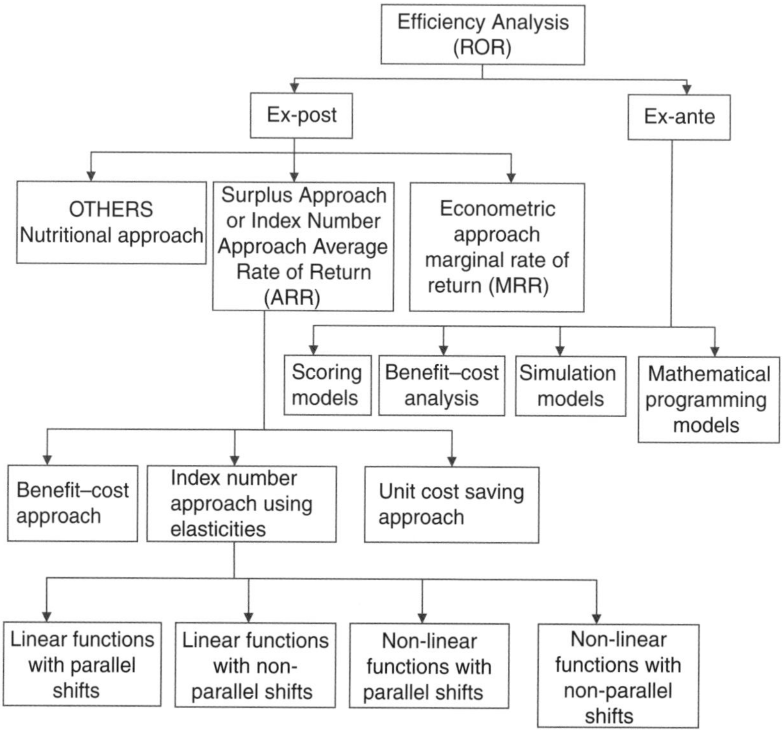

Fig. 24.4. Efficiency analysis. Source: Anandajayasekeram *et al.* (1996).

of generating adequate returns. However, the ARR measure is not always helpful in determining if the allocation of research funding to the project was appropriate. Because of the historic nature of ex-post evaluation, the results of these studies have mainly been used as political instruments to secure future funding. They demonstrate how efficient past investments were, but not necessarily where research resources should be allocated in the present, or the future. For a detailed description of the various techniques see Anandajayasekeram *et al.* (1996).

Economic surplus approach

A range of possible methods can be used to measure economic costs and benefits, but in practice the economic value of changes caused by research are most often evaluated using a 'partial equilibrium' approach in which all costs of production are summarized in a supply curve, and all its benefits are summarized in a demand curve. The area between the supply and demand curve is known as the economic surplus, which may be divided into 'consumer surplus' (area between the demand curve and the price level) and 'producer surplus' (area between the supply curve and price level) as shown in Fig. 24.5. Research or other public sector intervention aimed at producing new technologies can shift these curves, thus changing consumer and producer surplus, i.e. total welfare to the society.

Consumers' surplus measures the benefits to consumers of the purchase and consumption of agricultural products at a specified price – usually the market price. These benefits occur because the intrinsic value to the consumers for the products exceeds what they pay for the products (except at the margin). Successful technology development and dissemination (TDD)

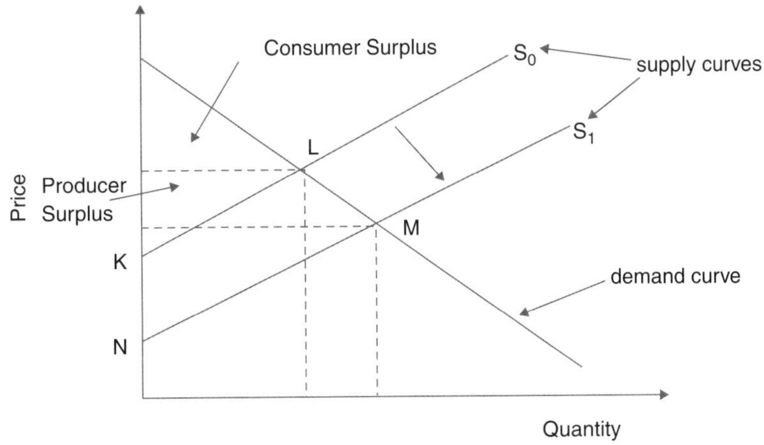

Fig. 24.5. Welfare gains due to technological change.

affects consumers' surplus by lowering the market price and increasing the quantity supplied to market. For staple food crops consumers' surplus captures the benefits to consumers of lower food prices. These benefits include the ability to reallocate income previously spent on food to other needed items such as healthcare, education, etc. Thus, consumers' surplus is a complex measure that captures many of the improvements in humanitarian objectives arising from successful agricultural TDD.

Producers' surplus measures the benefits to producers of increases in productivity and production. The change in producers' surplus is equivalent to the change in net farm incomes arising from the productivity increase.

Total economic surplus to the society = producer surplus + consumer surplus

With the new productivity increasing technology, the supply curve will shift to the right to S_1. The nature of the shift may vary depending on the assumptions made. The area KLMN is the welfare gains to the society as a result of the improved technology, i.e. R&D efforts. This area can be estimated. There is also the cost associated with the development and dissemination of the technology. Therefore:

The net gains to the society = the welfare gains due to the new technology (KLMN)
− cost of technology development and dissemination

As shown in Box 24.2 the TDT costs in general include the technology development cost (research cost), the technology transfer cost (extension cost) and technology adoption cost (costs incurred by the farmers who adopt the new technology).

The change in economic surplus is a measure of the social benefit derived from research. The empirical information needed to estimate the changes in economic surplus includes:

- How much the technical change shifted the supply curve;
- Knowledge of the parameters that describe the conditions of demand and supply for products;
- The information on the costs of the R&D programme required to induce a shift in supply curve.

As the cost and benefit flow occur over a period of time the incremental net benefit flow is calculated by subtracting the research, transfer and adoption costs from the gross benefits. The incremental net benefit flows over the years are converted into values that can be compared by discounting/compounding. This allows taking into account the time value of

Box 24.2. Research development and transfer costs.

Research (technology development) costs

Research and development costs include:

- Staff salaries and benefits;
- Recurrent expenditures;
- Overhead and administration expenditures;
- Depreciation of capital assets.

Technology transfer (extension costs)

Technology transfer costs include:

- On-farm research and demonstration trials;
- Costs of running the commodity training centre;
- Expenditures by the public extension institution on extension activities on a particular commodity (estimates);
- Expenditures by chemical and other input companies on extension and promotion activities;
- Expenditures by public and private product marketing firms on extension;
- Expenditures by farmer organizations (commodity associations and farmers unions) on extension.

Adoption costs

Adoption costs depending on the technology may include:

- Difference in the cost of seed between new and old varieties;
- Difference in the costs of chemicals between new and old pest and disease control methods;
- Difference in the use of labour and equipment between the new and old production practice;
- Differences in fertilizer use between the new and old varieties.

money. The discount rate used is the approximation of the social preference rate. This is normally taken as the rate of interest on long-term government bonds.

The commonly used measures to express the net benefit over time are the net present value (NPV), IRRs (IRR), and benefit–cost ratio.

- $\text{NPV} = \sum_{t=1}^{n} DB - \sum_{t=1}^{n} DC$

  Where DB: discounted benefit
  DC: discounted cost

- IRR is the discount rate at which $\sum_{t=1}^{n} DB = \sum_{t=1}^{n} DC$

- Benefit–cost ratio $= \dfrac{\sum_{t=1}^{n} DB}{\sum_{t=1}^{n} DC}$

The stream of net benefit flow for a typical R&D programme is shown in Fig. 24.6.

The summations of the discounted incremental net benefits yield the NPV, which indicates a positive benefit when the value is greater than zero. Of much interest is the interest rate

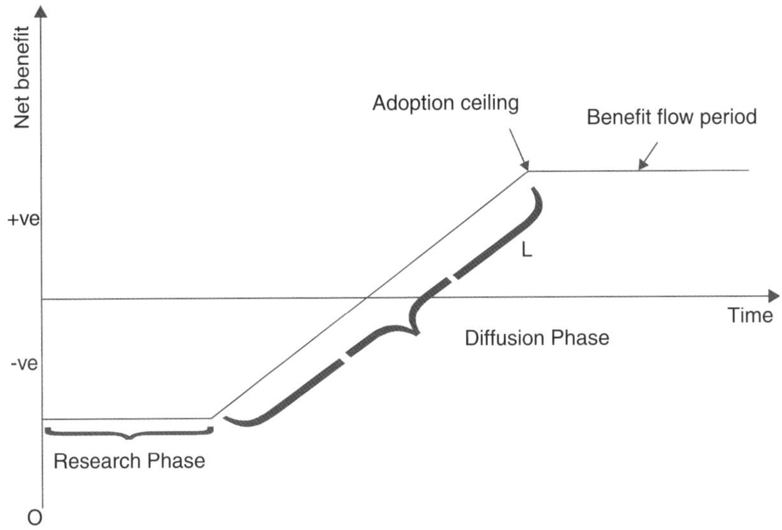

Fig. 24.6 Costs and benefits of R&D over time.

that reduces the sum of the discounted incremental net benefits to zero. This discount rate is the IRR (also often referred to as ROR). When this rate is greater than the social time preference rate then past investments on that particular R&D activity have been socially profitable. Similarly, when the benefit–cost ratio is greater than 1, it represents that R&D investment is profitable. The selection and ranking principles used in investment analysis are summarized in Table 24.3.

There are several modifications of the surplus approach (often termed as index number methods) used in empirical analysis. These include:

1. Benefit–cost approach using 'with' and 'without'. This is the most simplified version of the index number approach.
2. Index number approach assuming linear demand and supply functions with parallel shifts in the supply function.
3. Index number approach with linear functions and non-parallel shift.
4. Index number methods with non-linear functions with parallel shifts.
5. Index number method with non-linear functions and pivotal shifts.

The most commonly used approaches are the benefit–cost approach using 'with' and 'without' scenario; linear function with parallel shifts; and non-linear functions with pivotal shifts (Akino and Hayami approach). These approaches are discussed in a little more detail in the following sections.

Benefit–cost analysis approach

This is the simplest form of index number approach used to estimate the ROR for R&D investment. The theoretical framework assumes a perfectly inelastic supply curve and a perfectly elastic demand curve as shown in Fig. 24.7.

The original supply curve is S_0 and it shifts to the rights as a result of the new technology. The new supply curve is denoted by S_1. The entire shaded area Q_0Q_1LM is the increase in surplus due to the improved technology. The model assumes that the price will remain constant at OP^*.

Table 24.3. Selection and ranking principles for investment analysis.

Approach	Selection principles	Ranking principles
Internal rate of return	Select the investment if and only if the internal rate of return exceeds the cost of raising investment funds	Rank all investments in order of decreasing internal rate of return
Net present value	Select the investment if and only if the net present value is positive	If investments are substantially the same size, rank in order of decreasing net present value
Benefit–cost ratio	Select if and only if the ratio is greater than 1	Rank all investments in order of decreasing benefit–cost ratio

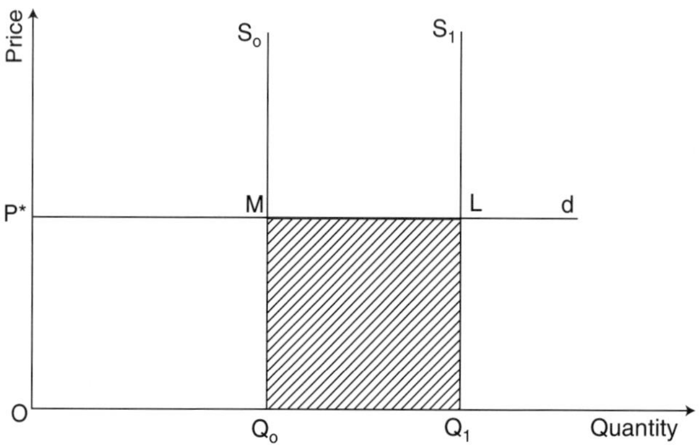

Fig. 24.7. Benefit–cost analysis approach.

$$Area Q_0 Q_1 LM = OP^* \times (Q_1 - Q_0)$$

$(Q_1 - Q_0)$ is the increase in total production as a result of improved technology. This is obtained by multiplying the farm-level yield gains from the new technology by the area cultivated using the new technology.

$$\text{Net benefit} = OP^* \times (Q_1 - Q_0) - \text{Cost Associated with the TDT}$$

$$\text{Net benefit} = OP^* \times (Q_1 - Q_0) - (\text{cost of technology development} + \text{cost of technology transfer} + \text{cost of technology adoption})$$

Note that the entire benefit in this case will go to the producers. The total welfare gains could be easily estimated by multiplying the increased production due to the improved technology by the constant price OP^*. The net benefit can be estimated by simply subtracting the cost of technology development and dissemination. Information on elasticities is not required to estimate the net benefit. The main feature of a benefit–cost analysis framework is the clear specification of 'with' and 'without' scenario and grouping relevant cost and benefits accordingly over the period.

Akino–Hayami method (A–H method)

The A–H method assumes a constant elasticity of demand and supply curves and supply shift is considered to be pivotal as shown in Fig. 24.8. The original work on this model was done by Akino–Hayami, where they attempted to measure the social returns to rice breeding in terms

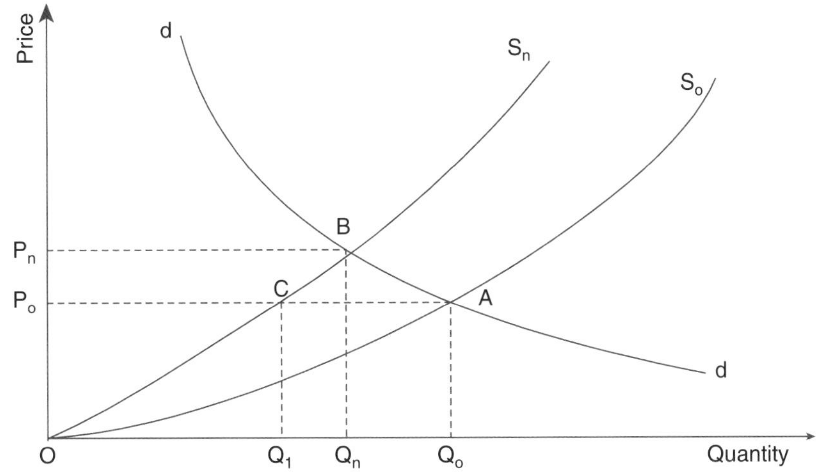

Fig. 24.8. Akino–Hayami method.

of changes in consumers' and producers' surplus, resulting from the shift in supply curve corresponding to a shift in the production function. Lines dd and S_o represent the actual market demand and supply curves, whereas S_n represents the original supply curve without the improved technology.

The increase in social benefit due to improved technology = Area ABC + Area AOC

Area AOC is calculated as:

$$AOC = (K\text{-}factor) \times (P_0 \times Q_0)$$

Where *K-factor* is calculated by dividing the product of proportion of area planted in improved varieties and yield gains from improved varieties by improved variety yield.

$P_0 \times Q_0$ is the total production value.

Area ABC is calculated as:

$$ABC = 0.5 \times \text{Area AOC} + \text{K-factor} \times \frac{(1 + \text{price elasticity of supply})^2}{(\text{price elasticity of supply} + \text{price elasticity of demand})}$$

Net benefit = (Area ABC + Area AOC) − cost of technology development and transfer

Therefore in order to estimate the social RORs using the A-H method, one needs the following information:

- Price elasticity of demand and supply;
- Rate of shift in the production function;
- Value of the commodity output ($p_o q_o$);
- Technology dissemination and development cost.

Although the relative magnitude of changes in consumers' and producers' surpluses are critically dependent on the choice of specific values of demand and supply elasticities, the overall social benefit is not so sensitive to the absolute value of the elasticity estimate as 'K' is a small fraction of the output.[2]

Cost-saving method

A new technology can affect the cost of production in a number of ways. For example, it could provide increased output for the same level of input (productivity increasing technologies), it could simply reduce the input required or a change in the proportions of input used (cost reducing technologies) for the same level of output; or the technology can affect both inputs and outputs simultaneously. The implication of this entire scenario is the overall reduction in per unit cost, i.e. the average total cost per unit of production will decline.

Let us assume that the average per unit cost with the old technology is ATC_0 and the average cost per unit with the new technology is ATC_1. Note with the new technology ATC_0 is greater than ATC_1. Therefore the unit cost saving associated with the improved technology (often called 'R' factor) is estimated as follows:

Unit cost saving = $ATC_0 - ATC_1$

This reduction in unit cost or the 'R' factor is used to estimate the RORs. Remember that the total fixed cost may not vary but the average fixed cost will vary depending on the level of production.

The theoretical model assumes a linear demand and linear supply curve with parallel shift as shown in Fig. 24.9. The model also assumes that the area BCD is small and negligible. The welfare gain to the society is approximated to the area ABDE. The area ABDE is equal to the area P_oBDF. Thus, the net gain could be estimated by the area P_oBDF, which is the cost savings accrued to the industry at the original level of output OQ_o.

Once again the estimation of the area P_0BDF does not require any elasticity estimates. All that is needed is the cost saving associated with the new technology; the 'R' factor and the quantity produced using the new technology. Remember the approximation made (i.e. the area BCD is small and negligible) underestimates the benefits. By using enterprise budgets one could estimate the unit cost-savings associated with the technology. Adoption studies will provide information on the quantity produced using the new technology. This simple method of estimating the benefit is called the 'cost-saving' method.[3]

In this case:

Net benefit = gross benefit (area P_oBDF) – (cost of technology development
+ cost of technology transfer)

Note that technology adoption cost is included in the cost of production. To be consistent with the model, the appropriate level of output that should be used in the calculation is the level of output prevailing in the industry prior to the innovation becoming available. If experimental

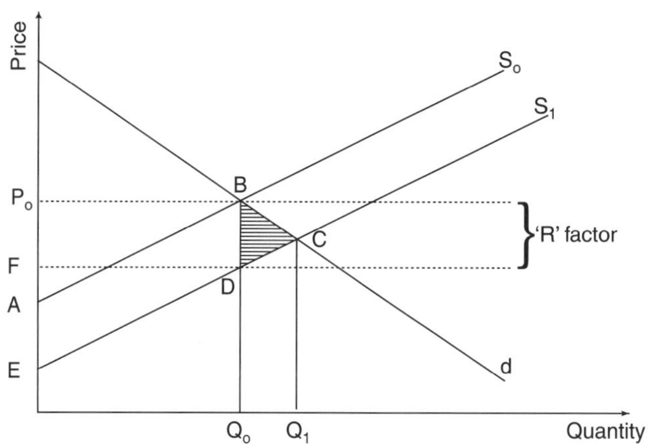

Fig. 24.9. Unit cost saving method.

data are used to construct budgets, then there is a need to adjust yield to reflect farm-level conditions, and the appropriate price to use is the 'field price' (CIMMYT, 1988).

Field price can be defined for both inputs as well as outputs. The field price for a variable input is the value, which must be given up to bring an extra unit of input into the field. The field cost is the field price multiplied by the quantity of input needed for a given area, usually an acre or hectare. Field price of the crop is defined as the value to the farmer of an additional unit of production in the field, prior to harvest. It is calculated by taking the price that farmers receive (or can receive) for the crop when they sell it, and subtracting all costs associated with harvest and shelling that is proportional to the crop yield and the marketing cost. (For more details on the estimation of field price see CIMMYT, 1988.)

Advantages and disadvantages of surplus approach

There are a number of advantages of the surplus approach. These include:

- The surplus approach is flexible, there are a number of derivatives of this approach – it provides a mechanism to analyse how the benefits of research are distributed between producers and consumers.
- The model can be applied to both 'closed' as well as 'open' economies.
- Approach is useful for both ex-ante and ex-post impact studies.

The approach estimates the average ROR, thus it is a less powerful tool than the econometric approach, which estimates the marginal RORs. The approach estimates the ROR for the combined investment, thus, it is difficult to separate the contribution of research from other services.

Using economic surplus approach to measure the value of R&D or other investments is popular because it exploits readily available data on prices and quantities to permit reliable comparisons of overall costs and benefits, without having to investigate each specific effect in detail. However, these ROR estimates should be supplemented with detailed evidence on specific effects, whenever policy makers are interested and the data can be made available. In most empirical studies, adoption information is collected often using adoption surveys. This will also provide useful feedback to the R&D community.

The IRRs (ROR is also known as IRR) concept is used to summarize the year-to-year stream of costs and benefits using a single number, which is the percentage 'interest rate' earned on the R&D investment over a period of time. Thus 'ROR' results can be compared directly with other kinds of interest rates, particularly the rates paid for loans or received from other investments. The ROR is perhaps the most useful single number summarizing the efficiency of investment in agricultural TDD and is a useful tool for policy decisions regarding the levels of investment in agricultural TDD.

Econometric approach

The majority of the reported empirical ROR models are based on either economic surplus calculations (following Griliches, 1958), or on econometric estimation of output elasticity of R&D derived from the production relationship (also pioneered by Griliches, in the early 1960s). The production function approach is the most commonly used econometric method to estimate the economic benefit of R&D. The production function is based on the idea that the amount of output of a production process depends upon the amounts of input in such a way that there is one unique amount of output resulting from each possible amount of input. For any production process this unique relationship between output and input is referred to as the production function for the process.

In the production function approach, the R&D elasticity is derived directly from estimation of a production, cost or profit function, with R&D included as a variable, or by first constructing a total factor productivity index, changes in which are then explained by the technology-related variables. In either case, it is the effect of the technology input on output or on the productivity of the sector adopting the technology that is measured.

A conventional general production function as a mathematical function could be expressed as follows:

$$Y = f\left(X_1, X_2, \ldots, X_g, X_{g+1}, \ldots, X_k, X_{k+1}, \ldots, X_n\right) + E$$

Where Y is the output, a dependent variable and depends on a number of factors (the independent variables)

$\underline{X_1 \ldots X_g} \rightarrow$ The variable inputs

$X_{g+1} \ldots X_k \rightarrow$ Fixed inputs (does not vary during the production process)

$X_{k+1} \ldots X_n \rightarrow$ The random variables

$E \rightarrow$ Error terms

Random variables are those variables that cannot be controlled in an actual production situation. Variable and fixed resources are of primary concern in production economics. In empirical studies, several forms of mathematical functions (such as linear, power, quadratic and cubic) are being used.

The advantage of a production function model is that it can be extended to include these technology variables, where Y_i can be expressed as a function of traditional agricultural inputs X_j, and the variables which shift the production function over time T_k. The simplest form is the single output production function.

$$Y_i = f(X_j, T_k) + C$$

Where X_j is the traditional input variables and T_k vector normally accounts for technology. It includes lagged R&D expenditures that generate new technology; extension expenditures that disseminate the results to the farmers (including cost of on-farm research) diffusing the technology and the education level of the farmers, which affects both their own creative and managerial abilities and their skill in appraising and adopting exogenous technologies. The weather is also normally included as it explains some of the residual errors.

Simplest econometric techniques, using linear regression analysis, have been in use at least since 1928, when Cobb and Douglas fitted a production function to data for the US manufacturing industry. By assuming that X and Y represent input and output respectively in a production function, the inclusion of the logarithm of R&D and extension expenditures (T) in the Cobb–Douglas gives:

$$L_n Y = L_n \alpha_0 + \alpha_1 L_n X_1 + \alpha_2 L_n X_2 \ldots \alpha_n L_n X_n + \alpha_t L_n T$$

Where X_1 to X_n are the respective input variables and α_1 to α_n are the coefficients (elasticities) to be estimated. The above equation is linear in logarithms, so that the coefficient of T (α_t) is the output elasticity of R&D and extension expenditures.

In order to derive an ROR, the elasticities were converted to value of marginal product. In the case of time-series analysis, each lag coefficient, β_i, is the output elasticity of R&D for that year:

$$\beta_i = \frac{\delta \ln Y_t}{\delta \ln RD_{t-1}} = \frac{\delta Y}{\delta RD_{t-1}} \times \frac{RD_{t-1}}{Y}$$

Where Y is yield.

Thus, the marginal physical product of R&D was the elasticity multiplied by the average physical product:

$$MPP_{t-1} = \frac{\delta Y}{\delta RD_t} = \beta_i \frac{Y}{RD_{t-1}}$$

Changing from continuous to discrete approximation gives:

$$\frac{\Delta Output}{\Delta RD_{t-1}} = \beta_i \frac{\overline{Y}}{\overline{RD}_{t-1}}$$

$$VMP_{t-1} = \frac{\Delta Value_t}{\Delta RD_{t-1}} = \beta_i \frac{\overline{Y}}{\overline{RD}_{t-1}} \times \frac{\Delta Value_t}{\Delta Y}$$

The marginal internal rate of return (MIRR) can be calculated from:

$$\sum_{i=1}^{n} \frac{VMP_{t-1}}{(1+r)^i} - 1 = 0$$

Where n is the lag length, by solving for r (the MIRR), one could estimate the ROR.

Note: the value of the marginal product of research is found by multiplying the R&D coefficient by the average product of research.

$MVP_R = \beta_{RD} \times$ average product of research

IRR (r) is obtained by solving the equation:

$$\sum_{i=1}^{n} \frac{MVP_{t-1}}{(1+r)^i} - 1 = 0$$

Evenson (1967) estimated that research investment requires 6–7½ years before affecting production. In the case of agricultural technology, the lag period can be estimated empirically and is equal to the duration between commencement of research activity and the realization of research benefit.

A number of issues need to be considered in using the production function approach. These include:

- Choice of explanatory variables;
- Choice of algebraic form;
- Econometric problems associated with the estimation;
- Availability of data.

Advantages and disadvantages of production function approach

The major advantage of the production function approach is that it can statistically isolate the effect of the various R&D components on productivity growth. One can study the effects of research, extension as well as their first order interaction effects. The production function methodology has the capacity to assess the impact of changes in funding levels, policy and institutional structures as well as the effects of complementary programmes such as seed multiplication and distribution, extension and credit programmes on R&D productivity.

There are a number of limitations associated with the production function approach. It is useful only for ex-post analysis. The method is data intensive and has its utility where a sufficient amount of good quality data exists. The estimation procedure also encounters a number of econometric problems. The major ones are multi-collinearity and serial correlation (auto correlation). The comparison of the approaches described above is presented in Table 24.4.

Table 24.4. Comparison of approaches.

	Surplus approach			Production function approach
	Benefit–cost approach	Cost-saving method	Akino–Hayami method	
Accommodate price change	Yes	Yes	Yes	No
Time value	Yes	Yes	Yes	Yes
Time perspective (ex-ante, ex-post)	Both	Both	Both	Ex-post only
Average/marginal rate of return	Average	Average	Average	Marginal
Data needed				
• Elasticities	No	No	Yes	No
• Research cost	Yes	Yes	Yes	Yes
• Yield response	Yes	Yes	Yes	Yes (aggregate)
• Adoption cost (data)	Yes	Yes	Yes	Yes
• Extension cost	Yes	Yes	Yes	Yes
Distribution of benefits	No*	Yes	Yes	No
Separation of effects of research vs. complementary services	No	No	No	Yes
Econometric problems in consideration	No	No	No	Yes
Inclusion/consideration of externalities	Yes	Yes	Yes	No
Shift in supply curve	Parallel	Parallel	Pivotal	NA
Can quantify research, extension, interaction	No	No	No	Yes
Can handle lags involved in research and adoption	Yes	Yes	Yes	Yes
Can quantify spill-over effects	Yes	Yes	Yes	No

*All benefits accrue to the producers.

Environmental Impact Assessment

An EIA is an activity designed to identify and predict the impact of an R&D intervention on the geophysical environment, on human health and well-being and to interpret and communicate information about the impact (Munn, 1979).

To be effective EIA should be an integral part of planning research and development projects and should be carried out at the same time as technical, economic and socio-political assessment. EIA is becoming increasingly important due to concerns for ecologically sustainable development.

The environmental impacts of technologies can be on-site market impacts, on-site non-market effects, off-site market impacts and off-site non-market effects.

On-site market impacts

The on-site market impacts are those that affect only one site, do not have downstream effects and can be evaluated using conventional markets. An example of that is soil mining (soil degradation) – loss of nutrients when farming systems do not adequately replenish the nutrients used. These effects are specific to the site that is affected, but they have intra-temporal as well as inter-temporal effects on the productivity of the soil. These impacts are reflected in yield losses and can be valued using the market prices for the relevant crops.

On-site non-market impacts

On-site non-market effects are those that affect only one site but are not reflected in the marketplace, e.g. slash and burn systems. These affect the biological diversity on-site, but the loss of biodiversity cannot be valued using conventional markets. Contingency valuation techniques are needed to evaluate such an impact.

Off-site market impacts

Off-site effects concern individuals and communities downstream from where the activity generating the impact is undertaken. Examples are:

- Downstream effects including silting of reservoirs, rivers and irrigation canals;
- Reduction in water storage capacity of reservoirs and irrigation capacity.

These render water more costly and/or increase dredging costs for rivers and harbours.

Off-site non-market impacts

Off-site non-market impacts are effects that affect downstream communities on sites different from where the impact originated and they will affect individuals of generations that succeed the one that undertook the pollution activity. A good example of this is the atmospheric pollution resulting from an agricultural activity, such as the use of methyl bromide for soil fumigation leading to the depletion of the ozone layer. Another example is deforestation and its effects on the catchment area.

In order to quantify and value the environmental impact of an agricultural research initiative, it is important to understand the source of the impact, the nature of an impact and the relationship between the impact and those variables that can affect current, potential or future producers and consumers. The best-known attempts at quantifying environmental impacts are concerned with soil erosion. In terms of valuation, the most difficult ones are those dealing with biodiversity and their benefits and costs requiring the use of contingency valuation. This may be due to the fact that the costs of establishing these values for inclusion in project evaluation may be prohibitive and/or may require longer time to do a proper contingency evaluation study.

Components of environmental impact assessment

A typical EIA should contain the following:

- A description of the proposed activities and of alternatives;
- Prediction of the nature and magnitude of environmental effects (both positive and negative);
- An identification of human concerns;

- A list of impact indicators as well as the methods used to determine their scales of magnitude and relative weights;
- A prediction of the magnitude of the impact indicators, of the total impact for the projects, and for alternatives;
- Valuation of the physical effects identified.

Environmental impacts should be assessed as the difference between the future state of environment if the action took place and the state if no action occurred. A reference condition is the 'without action' condition and, because of naturally occurring changes, it is not necessarily the present condition. An EIA should contain three subsections related to the environment, namely: determination of the initial reference state; an estimate of the future state 'without action'; and an estimate of future state 'with action'.

EIA is normally associated with four kinds of uncertainty. These include: natural variability of the environment; inadequate understanding of the behaviour of the environment; inadequate data for the region or country being assessed, and socio-economic uncertainties which are most difficult to estimate.

A number of conceptual and methodological issues need to be resolved in assessing the environmental implication of agricultural R&D and the impact assessment of NRM research (Izac, 1998; Pachico, 1998). Assessing the full range of impacts requires that appropriate information sources be tapped at various levels, such as field, watersheds and regions. It also raises technical difficulties of measuring and valuing the resource degradation (e.g. soil loss) or environmental pollution (chemical contaminations and residual effects).

At present there appears to be consensus that estimates of the quantities of resource depletion or environmental damage are easier to get (although often costly) than estimates of the correct prices to attach to the quantities (Crosson and Anderson, 1993). Even in valuing private shadow prices for on-site effects of resource quality, there are valuation issues such as whether to use foregone output, replacement costs, user costs, or option and existence value to measure the costs of resource degradation (Harrington *et al.*, 1994). Problems are compounded with externalities where costs of agricultural chemical pollution may include regulatory costs, health-related costs and unpriced environmental costs (Steiner *et al.*, 1995). A number of techniques now exist for placing economic values on environmental changes (Winpenny, 1991; Malik and Faeth, 1993; Rola and Pingali, 1993; Brandon, 1995). A final complication with valuing changes in natural resources and environmental quality is the effect of policy distortions, such as subsidized inputs that foster overuse of chemicals, or institutional policies such as land tenure that affect adoption of improved management practices (Maredia *et al.*, 2000).

A comprehensive impact assessment of any R&D programme should include the environmental implications. The main problem is that they are data intensive and require highly developed skills in collecting and statistically manipulating large quantities of data. Therefore, very often qualitative assessments are made. Methodologies for EIA should be selected as appropriate to the nature of the action, available database and the geographic setting.

Guidelines for environmental impact assessment

The basic question that needs to be asked is: will action significantly affect environment? If the answer to the question is yes, then there is a need to conduct an EIA. For every proposed R&D project, the environmental assessment should include:

- A prediction of the nature and magnitude of environmental effects (positive and negative);
- A listing of impact indicators, whereby effects can be monitored;

- The identification of human concerns regarding environmental changes;
- A statement whether these could be incorporated within a market framework and their values elicited.

The level of detail depends on:

- Sensitivity of the affected environment and their social values;
- Scale of the proposed technology;
- Types of effects it could have;
- The resources, scientific expertise and time available;
- The cost of impact assessment relative to likely environmental value.

Full EIA requires complex analyses of physical, biological, social and economic procedures. This may also require long-term commitment to the relevant areas of cross-disciplinary research.

In the past, the multidisciplinary nature of environmental issues has caused problems with the quality and general availability of data. For example, R&D efforts might lead to the development of fertilizers and have long-term negative effects on the soil. To incorporate such externalities in the R&D framework, the physical effects on soil would need to be monitored closely by scientists before their economic impact could be estimated.

Another problem is obtaining statistically reliable field-specific data. Reconciling different levels of aggregation to obtain reliable estimates is another issue confronting the analysts. Many physical and biological models provide information at the low level of aggregation. For example, movement of pesticides through soils is determined by several factors, such as specific soil characteristics (physical and chemical), properties of the soil, the climate, crop management practices, etc. The problem is how to generate information that reflects the physical, biological and economic diversity of the region/nation under study, and how to combine this information to yield reliable information about the region/nation.

It is important to remember that in the absence of data required for thorough analysis, it may still be possible to identify the nature of the social costs and benefits, together with the gainers and losers. The prediction of negative environmental side-effects does not necessarily mean that the new technology should not be used. The net benefit may be sufficiently large to provide compensation to those who are harmed and still leave a net surplus to the society. This is a policy question that needs to be addressed.

Assessing the Impact of Policy Research

There are very few impact studies (quantitative studies) related to policy and socio-economic research. This is partly because it is much more difficult to do them than it is for the biophysical research. There are virtually no best practices available for assessing the impact of socio-economic research (Maredia *et al.*, 2000). The problems of measurement, sampling and attribution still remain but are more complicated.

Babu and Mcthindi (1995) separate the benefit of policy research into pre- and post-decision-making benefits. The former involves improved processes related to capacity building and institutional strengthening. The latter are evaluations of the primary and secondary impacts of the policies that emerge.

Similarly, Ryan (2001) also identified what is called 'process benefit' and the 'socio-economic impact'. In order to assess the overall impact one need to examine how well the activity/project/programme is able to influence the process and capacity of the institutes, and derive measures of the impact of these changes, especially on the poor. It has been argued that the process benefits are necessary but not sufficient outcomes to signify that the programme has been effective, while socio-economic impacts are both necessary and sufficient.

Garret and Islam (1998) suggest that social science research evaluation should only look at outputs, processes and potential outcomes, rather than focusing on actual policy outcomes. They maintain that it is difficult to establish a link to the policy impact of social science research and that often the research contributes to a body of knowledge that policy makers access when and if they see fit. Weiss and Bucavalas (1980) claims that the following attributes are what policy makers find useful: research quality, conformity to expectations, action orientation and challenge to the status quo. According to Ryan (2001) there are four products of economic policy research and related activities, namely outputs, outcomes, economic policy responses and welfare impacts. They are generally sequentially produced and more difficult to document, articulate, measure and attribute as one moves from outputs to impacts. In addition, Smith and Parday (1997) allude to the 'Cassandra problems' (this is the value of 'good research advice' not taken) in assessing the impact of policy research. As is often the case, there are long lead and lag times between the completion of research and the adoption of results. Impact studies that are conducted too soon after the results are available and disseminated may simply be premature.

Ryan (2001) identified a number of issues that need to be considered in the conduct of impact evaluation of economic policy research. These include: scale and scope, time horizon, supply versus demand approaches, importance of surprises, attribution, choice of indicators and time lags.

The products of socio-economic research may be new economic information (knowledge related to policies and institutional reforms); some output that contributes to technological change (innovation in management and management tools for R&D); output that contributes to public policies (institutional innovation); each of which has its own users, impacts and benefits. Some of the disembodied innovations produced by economic research are institutional innovations that provide new organizational structure to address social, economic and environmental problems. Some are managerial and decision-making innovations that improve the choices made by firms, consumers and the public sector.

Research on the productivity of social science research is in its infancy. The productivity of research in this field depends strongly on how the research results are distributed and utilized. Thus the value of economic research depends on the transmission and interpretation of research results to final users and the capacity of the users to take advantage of them.

In assessing the benefits of economic research non-economists will pay less attention to the analysis and more attention to the credibility of the evidence. Therefore, careful documentation is important.

The empirical methodology proposed here follows from the chain that leads from policy-oriented social science research to economic growth. Research produces policy-relevant knowledge for policy makers, who can then be expected to design and implement policies that in turn will promote greater efficiency and growth. Empirically, the research–knowledge nexus is omitted for obvious reasons. The focus is on the nexus between knowledge and policy quality and economic growth. More precisely, a three-stage empirical approach is proposed that: (i) estimates the effects of research-induced knowledge on policy quality; (ii) estimates the effect of policy quality on growth; and (iii) uses the two coefficients to deduce the effects of knowledge on growth.

Finally, no matter how social welfare is defined, the production of useful policy-relevant knowledge will not yield tangible social benefits if policy makers are unable or unwilling to make use of knowledge, for the same reason that knowledge that can cure cancer is socially unproductive until it is translated into an actual drug or treatment. Thus, in the final analysis, the value of policy-oriented social science research depends critically on the policy makers and the constraints and incentives they face. For it is they, not social scientists, who must build and run the tools called policies.

A framework for measuring the benefits of policy research

The evaluation of policy research requires an assessment of the value of policy changes. It also requires, for ex-post assessment, an estimation of the likelihood that a proposed policy change will be adopted. Several steps must be undertaken in order to assess the impact, whether it is done ex-post or ex-ante.

1. The problem must be defined in terms of the objectives of the evaluation, the policy research programme to be evaluated, objectives for the policies and path followed from the research to the change in policy (actual or projected). Formulae must be defined for measuring economic benefits.
2. Market data (such as prices, quantities and elasticities) and other data needed for research must be compiled. Estimates must be made of potential changes in costs, likelihood that a policy will be implemented, lags in research and implementation and so on. For ex-post evaluation, the policy research itself may have already supplied some of this information. If not it may be necessary to elicit peoples' subjective estimates. Such subjective estimates are needed for all ex-ante evaluations. For ex-post analysis, an attempt can be made to talk to people involved in the policy decision to determine the influence of the research. For ex-ante analysis it may be possible to assess historical probabilities for factors such as adoption of a policy, in order to place rough bounds on future probabilities. Uncertainty of parameter estimates requires that information should be gathered in a way that will allow for distribution around those estimates.
3. Analyse the data by combining the data and the formulae for economic surplus derived to assess specific policy research applying capital budgeting methods to streams of benefits and costs; and then using the results to help justify programmes or choose among alternatives.
4. Interpret the results and establish effective mechanism for communicating the results.

Measuring the benefits of policy-oriented social science research is difficult for a number of reasons. Causality between completed policy research and changes in policy is nearly always uncertain. Predicting the adoption of policy recommendations is highly uncertain in the ex-ante evaluation of policy research. The complexity of the effects of policy usually implies that evaluation is forced to use approximate measures lest the evaluation of each policy research programme or project becomes a major policy research project itself. Lastly, certain types of policy research generate benefits that are not priced in the market. These difficulties render any aggregate econometric analysis of the benefit of policy research highly suspect, but do not preclude the use of economic surplus analysis. Economic surplus analysis has the advantage that it can facilitate evaluation of diverse types of policies, assess the distributional effects of policy research, generate results that are directly comparable with evaluations of technology-oriented research, calculate ex-ante, or ex-post research benefits and provide an assessment that is consistent with economic theory.

Owing to the uncertainty of many parameters used in the economic analysis, there is a need for carefully structured questions posed face-to-face with those most knowledgeable about the policy process and the proposed or completed research.

Understanding why a society adopts its policies is crucial for predicting whether a proposed piece of policy change occurs in agriculture. However, there should be no illusions about the difficulty of making a quantitative evaluation policy research ex-ante. The uncertainty surrounding the estimated benefits of such an evaluation will be inevitably high. Modelling the market effect of policies graphically and in formulae for economic surplus is important.

Assessing the Impact of Training

Many national and international agricultural research institutes devote sizeable resources to activities such as training, networking and advisory services. Such activities are aimed at improving organizational performance in agricultural and natural resources research and, over the longer term, enhancing the impact of these institutes on agricultural productivity. These are difficult to fit into the economic surplus model. Consequently, there has been little methodological and practical work in the area of economic impact assessment of these activities.

Related to training and capacity building initiatives, separation of skill building from its application is not appropriate when trying to assess impact, as without the application of the skills there is unlikely to be impact.

Ultimately, the impact of a training programme occurs through changes in the efficiency of operation of agricultural and related sectors brought about by changes in the way that the trainees work. There may also be changes brought about in the host institutes as a result of the implementation of the programme, in cases when the training is conducted in collaboration with learning institutes. However, these are likely to be of secondary importance. The sequence and the flow of benefits are discussed in the following sections. These sections provide a fairly complete coverage of possible impacts, but it is only illustrative. For any element of the programme, only parts of the sequence may be relevant and there may be overlap between some parts. The practicality of estimating elements of the impacts should be addressed in each individual situation.

Logical flow of impacts

The flow of impacts from the programme can be conceived as starting from the trainees and ultimately having an international dimension in the following progression:

- Trainees;
- Host training institution (if the training is conducted in collaboration with local learning institutions);
- Employer institution;
- National agricultural sector;
- National economy;
- Regional agricultural economy;
- International agricultural economy.

Identifying outputs and impacts

Potential outputs and impacts of the training programme are identified for each level of impact assessment as outlined in the previous section. The elements are clearly not always mutually exclusive and many are intermediate impacts or are only elements of final impacts. For example, it is through the application of new skills and tools by trainees that the impact on employer institutes and on the national and regional economy is likely to occur.

The potential impacts at each of the levels identified are listed below.

1. Trainees:
   - academic qualifications
   - new skills and tools
   - career prospects
   - better remuneration
   - mind-broadening international, professional and cultural experience
   - new professional contacts.
2. Host academic training institution:
   - sustainable strengthening of capacity
   - image and recognition
   - contribution to national development
   - contribution to regional development
   - quality of academic programmes
   - quality and quantity of research output.
3. Employer institution:
   - strengthening of organizational capacity
   - quality and quantity of research output
   - contribution to national and regional development
   - recognition of the institute.
4. National agricultural sector and economy:
   - increased productivity
   - improved sectoral, intersectoral linkages.
5. Regional and international agricultural economy:
   - increased productivity
   - improved collaboration
   - improved regional capacity.

Measuring impacts of training programmes

The above discussion of benefits presented in terms of outcomes of training in general, would be applicable to any training programmes. However, given the limited time frame, meaningful quantitative measure of the overall programme outcome are not attainable. Estimating what would be the final impact of the improvement in the skills of a trainee would be an extremely difficult task. Even at a conceptual level, it is difficult to define the limits of the impacts, because those impacts are likely to be dispersed over many actions and over time. The skills and work habits learned by a trainee during participation in the programme may affect their direct productivity and influence other people in a sequence of jobs over many years. Given many trainees and several fields, measurement of impact of this nature is obviously impractical.

The nature of the ultimate potential impact of the programme needs to be kept in mind. However, many empirical studies seek qualitative indicators of the intermediate impacts of the programme from employers, graduates, current students and host institutions, and a measure of the cost-effectiveness of the programmes. These are used as proxy measures of the ultimate impact.

Organizational Impact Assessment

'Organizational impacts' are defined as the effects of R&D activities and outputs on the operational environment, organizational motivation and capacity of client organizations. Organizational impact refers to the effects of R&D programmes on capacity building, human resources development and performance of the R&D institutes. This model as shown in Fig. 24.10 views organizational performance as a function of the organization's operational environment, its organizational motivation and its organizational capacity (Lusthaus *et al.*, 1995; Horton *et al.*, 2000). The four dimensions for organizational management are summarized in Box 24.3.

The framework for assessing organizational impact consists of a variety of methods such as surveys, case studies and document analysis to understand the four dimensions of organizational assessment: the operational environment, motivation, capacity and performance. The term operational environment refers to the external environment in which the organization carries out its activities. Motivation refers to internal factors that influence the direction of the organization's activities and the level of energy it displays in its activities. The term capacity refers to the resources, knowledge and skills of the organization. The term performance refers to the achievements of the organization in relation to its objectives in terms of effectiveness, efficiency, relevance and sustainability.

The term organizational capacity development refers to the systematic process of planned organizational change that is intended to enhance the efficiency, effectiveness and sustainability with which the organization pursues its strategy, accomplishes its mission, achieves its goals and delivers value to its stakeholders. It is important to distinguish between resource provision and resource acquisition from capacity development. In order to achieve superior performance an organization must learn to do more than merely identify its needs and acquire the missing resources; its managers and staff must learn how to nurture, integrate and deploy their resource to create the capabilities needed to accomplish strategic goals.

A theory of action also called programme theory is very useful as framework for assessing the organizational impact. Programme theory refers to the set of beliefs that underlie action. Programme theory is a 'plausible and sensible model of how a programme is supposed to work'

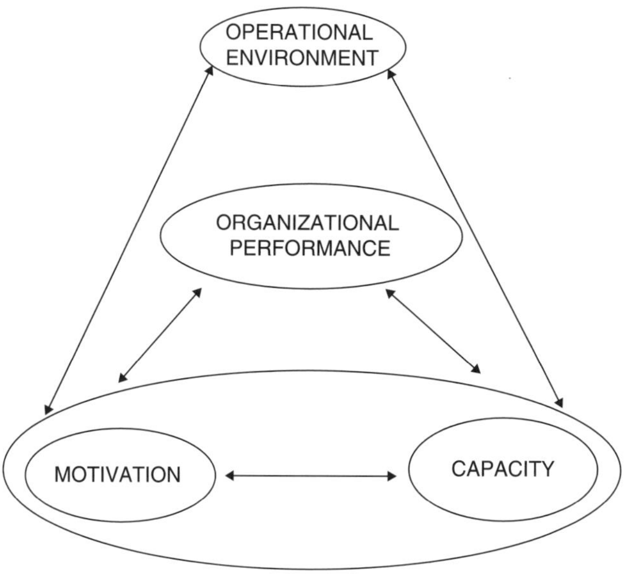

Fig. 24.10. Framework for organizational assessment. Source: adopted from Lusthaus *et al.* (1995).

Box 24.3. Four dimensions for organizational management.

The external environment

- Administrative and legal environment
- External political environment
- Socio-cultural environment
- Technological environment
- Economic environment
- Stakeholders
- Infrastructure
- Policy/natural resources management

Organizational motivation

- The organization's history
- Mission
- Culture
- Incentives/reward (systems) schemes

Organizational capacity

- Strategic leadership
- Programme planning
- Management and execution
- Resource allocation and management
- Linkages and co-ordination of clients, partners, government policy makers and external donors

Organizational performance

- The effectiveness with which the organization achieves its mission and goals
- The efficiency of resource use
- The organization's sustainability in terms of its continued relevance to its stakeholders
- Relevancy
- Quality

Source: Lusthaus *et al.* (1995)

(Bickman, 1987). It identifies programme resources, programme activities and intended programme outcomes, and specifies a chain of causal assumptions, linking programme resources, activities, intermediate outcomes and ultimate goals (Bickman, 1987). The theory contained in a programme's official proposal, or other documents, often differs from the actual 'theories in-use', i.e. the bases on which people actually act. Hence, before conducting an evaluation it is important to review documents, discuss goals and assumptions with key stakeholders, and finally to agree on an up-to-date theory of action for the current capacity development effort or the organizational impact.

Methods Assessment

Over the last two to three decades a number of methods and tools have been introduced in order to improve the relevance, effectiveness and efficiency of the Agriculture and Natural Resources technology development and dissemination process. Thus, the issue of methods assessment is becoming increasingly important.

The following pre-conditions should exist in order to assess 'methods' being employed:

- A clear understanding of the problem that is being addressed, as well as the objectives of the project/programme.
- Full knowledge or description of the method being tested as well as the alternatives available to achieve the same objective. This information is essential to establish the 'with' and 'without' scenario.
- Complete inventory of resources used in the method(s) being tested as well as the alternatives.
- A functioning M&E system to collect the relevant information during the implementation of the programme. This requires an operationalized logical framework.

Proposed criteria

Generally, methods and technologies are assessed in terms of relevance, effectiveness, efficiency and impact. The following criteria can be used in assessing the field methods.

1. Relevance and adequacy. This criterion is closely linked to the problem being addressed and the desired objectives. Relevancy deals with the appropriateness of the method to address the problem that is being dealt with. Adequacy refers to the ability of the methods to achieve the objectives. The methods should be simple enough to be understood by all stakeholders especially the beneficiaries (as it deals with field methods) and at the same time adequate enough to address the objectives. Beneficiary assessment is a useful tool to measure this.

2. Effectiveness measures the degree to which the method attained its objectives. The basic set of questions to be answered here is:

- How effective the method being tested is in achieving the set of objectives being pursued.
- In terms of effectiveness, how does this method compare with the alternative(s) being practised.
- Number of farmers reached, as well as adoption and utilization by various stakeholders can be used as measures of effectiveness.

3. Efficiency measures the degree to which an activity/method produces the desired output at least cost. This criterion deals with cost-effectiveness – how does this method perform compared with the alternatives in terms of:

- Costs and benefits;
- If the output of the alternatives is the same in terms of quantity and quality, then least cost criteria is adequate to assess the relative efficiency, i.e. select the method that is least expensive;
- Cost per farmer can be another criterion that could be used.

In order to assess the efficiency we need information on:

- Human resources used/needed in terms of quality and quantity;
- Financial resources involved in implementing the method and the alternative;
- Physical inputs required – equipment;
- All other relevant costs;
- Some estimate of the quantity and quality of the outcome.

4. Potential impact and project or activity can be measured in terms of institutional (intermediate), economic, social and environment. Methods, being R&D technologies, may generate significant institutional impact and hence one needs to carefully identify the relevant and critical indicators. It may be difficult to assess the people-level impact, i.e. impact of the outcome of the methods on the ultimate beneficiaries, the farmers. It is also important to identify both quantitative and qualitative indicators. Assessment by the direct beneficiaries may be very useful in determining the impact. Often, a formal or informal survey of the stakeholders can be used as an instrument to measure the impact. Perception is an equally acceptable measure.

5. Ability to address gender and equity related issues. One of the greatest concerns of development practitioners and policy makers is the issue of poverty alleviation, income distribution, environmental consequences and gender. Thus the method being tested should be able to handle these and should have negative consequences with respect to these issues. It may be difficult to get quantitative assessment in all the aspects. In addition to quantitative assessment, where necessary qualitative assessments and perceptions of the stakeholders could be used to assess methods against this criteria.

6. Suitability to local socio-institutional environment. This aspect is critical for institutionalization. The method should be simple enough and should be implemented within the existing institutional framework and human capacity. If the human capacity is not there, then it should be developed. The question here is how easy or difficult it is to create this necessary additional capacity.

7. Sustainability beyond the project period. Continuity and sustainability are important criteria in assessing externally supported programmes. The basic question here is: what is the likelihood that the institutes and/or the communities will continue to employ or utilize the method being tested beyond the project period. Here the assessment is related to the direct target community of the project/programme.

8. Potential for institutionalization. The term institutionalization is defined as a process that impersonally establishes a mechanism to integrate the method into the day-to-day activities of the institutes and/or community. It refers to the permanent integration of the procedure and method into the regular activities of the organization at the national level. Some comment on the prospects and problems for institutionalization at the national level is also important in assessing the method.

9. Regional relevance refers to the wider adaptability of the method outside the national boundaries within the Eastern and Southern Africa region or the country and the implementation and consequences of this.

This list may not be exhaustive. Depending on the specific circumstances one may add additional criteria for assessment. It is suggested that at least one should be able to use these as a minimum set of criteria in assessing methods.

Major Issues in Assessing Impact of R&D Activities

Impact assessment is directed at establishing, with as much certainty as possible, whether or not an intervention is producing its intended effects. The three basic issues that need to be taken care of in any empirical impact study are causality, attribution and incrementality. All these three aspects are somewhat interrelated.

Causality

In measuring the impact of R&D investments it is important to ensure that the impacts measured are all a result of the technologies and activities undertaken within the programme/project. However, as one moves from the direct product to broader economic, social and environmental effects, the chain of causal events is too long and complex and the variables affecting ultimate outcomes are too numerous to permit the identification and measurement of impacts of specific interventions (Biggs, 1990; Rossi and Freeman, 1993).

Incrementality

R&D programmes operate in environment in which ordinary or 'natural' sequences of events influence outcomes, e.g. a programme to reduce poverty has to consider that some families and individuals will become economically better off without the help of the project, or careful seed selection by farmers will gradually increase the yield levels even without improved varieties. Thus it is important to make a distinction between the 'gross outcome' and 'net outcome' of interventions. Gross outcomes are changes in an outcome measure that are observed after a programme has been operating. Net outcomes are those results attributable to the intervention, free and clear of the effect of other causes in the programme's context.

Gross outcome = Effect of intervention or Net outcome + Effects of confounding factors

Impact assessment must arrive at estimates of net intervention effects, i.e. should measure the incremental changes attributable to the intervention. The compounding factors are the extraneous 'causes' that compete with intervention efforts to explain changes in the target problem or population after the programme has been put into operation. Thus in measuring the impact an analyst must pay attention to endogenous changes, secular drifts (long-term trends), interfering events (short-term events), self-selection and stochastic effects, as well as unreliability effects.

In the literature one could identify three different types of comparisons: 'before' and 'after comparisons'; 'with' and 'without' comparisons; and 'target' versus 'actual' comparison.

The 'before' and 'after' (also referred to as historical comparison) comparison fails to account for changes in production and other outcomes that would occur without the project, thus leading to an erroneous statement of the benefit attributable to the project investment.

Consider a situation where the production in the area is already growing, but only slowly and probably will continue to grow during the life of the project. The objective of the project is to increase growth by intensifying production. In this situation, if the analyst had simply compared the output before and after the project, (s)he would have erroneously attributed the total increase in production to the investment. Actually what can be attributed to the project investment is only the incremental increase in production that would have occurred as a result of intervention.

'With' and 'without' comparison is the most appropriate measure as it accounts for the other endogenous changes. This is almost like the controlled experiment but difficult to achieve in the real world. However, it is important to specify the 'with' and 'without' scenario very clearly to avoid wrong estimation of benefits.

As it is difficult to actually simulate the 'with' and 'without' scenario, very often comparison is made between the 'target' set at the time of planning and the 'actual' realized at the end of the project. A logical framework approach once again can assist in the process. Here the challenge is to set a realistic, measurable, quantified target at the time of planning.

Attribution

Attribution problems arise when one believes or tries to claim that a programme has resulted in certain outcomes and there are alternative plausible explanations. The more obvious these factors are, the less credible is the performance or impact information. Mayne (1999) suggested a number of strategies that can be used to address attribution through performance

Box 24.4. Steps in contribution analysis.

Contribution analysis – addressing attribution with performance measures:

- Acknowledge the problem
- Present the logic of the programme
- Identify and document behavioural changes
- Use discriminating indicators
- Track performance over time
- Discuss and test alternative explanations
- Gather multiple lines of evidence
- When required refer to the need for an evaluation

measurement (see Box 24.4). Collectively these are elements of a contribution analysis. The various elements of the contribution analysis are in Box 24.4 and discussed in the following sections.

Acknowledge the problem

The first step in contribution analysis is simply acknowledging that there are other factors at play in addition to the programme and that it is therefore usually not immediately clear what effect the programme has had or is having in producing the outcome in question. Acknowledging the other factors at play is more honest and hence more credible than pretending they do not exist.

Analyse and present the logic of the programme

There is some logical reasoning behind the programme that explains what it is supposed to be accomplishing and how. A logic chart for the programme tries to display on a page how the programme is supposed to work, i.e. how the various outputs of the programme are believed to produce a number of results that will lead to the intended final outcome of the programme. Logic charts can also discuss the key external factors influencing outcome.

A logical framework used in the planning process can be effectively used for this purpose. Fig. 24.11 illustrates in a generic fashion what a logic chart can look like and Fig. 24.12 is a specific logic chart for a training-related project. Note that the logic chart:

- Illustrates the linkages between specific outputs, specific intermediate outcomes and specific end outcomes;
- Explicitly includes the idea of reach (who the programme is expected to reach) and immediate outcomes. It is often at these levels in the results chain the performance indicators can do a good job of measuring, and it is at this level the programme typically has most control. The evidence that the intended immediate outcomes have in fact occurred is a critical step in demonstrating the larger performance and impact story.

Developing and using a logic chart has a number of benefits for programme managers, such as developing consensus on what the programme is trying to accomplish, developing an understanding on how it is believed to be working, clearly identifying the clients of the programme, seeking and getting agreement on precisely what results are intended and identifying key measures of performance. One is also interested in the additional benefits of identifying:

- The cause–effect relationships implicit in the programme's theory;
- The outside factors at play;
- Areas where understanding about the impact of the programme is weak.

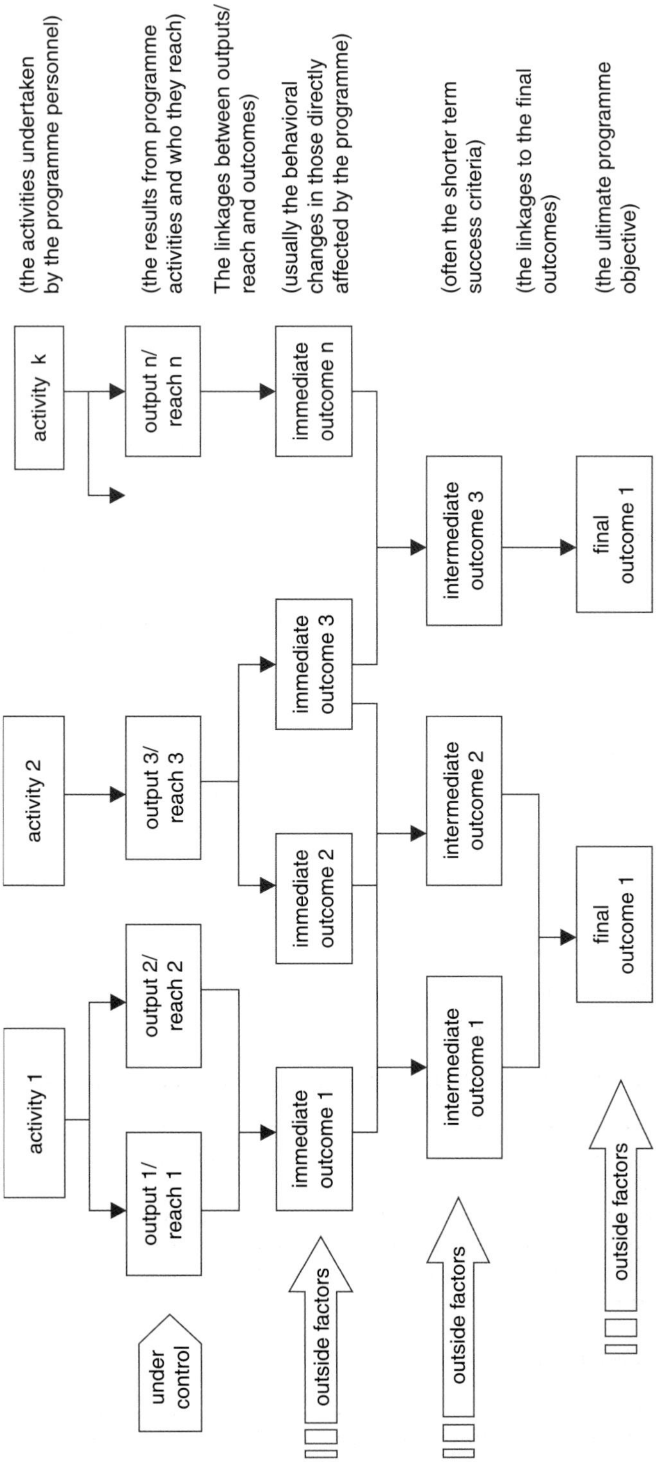

Fig. 24.11. A programme logic chart. Source Mayne, 1995.

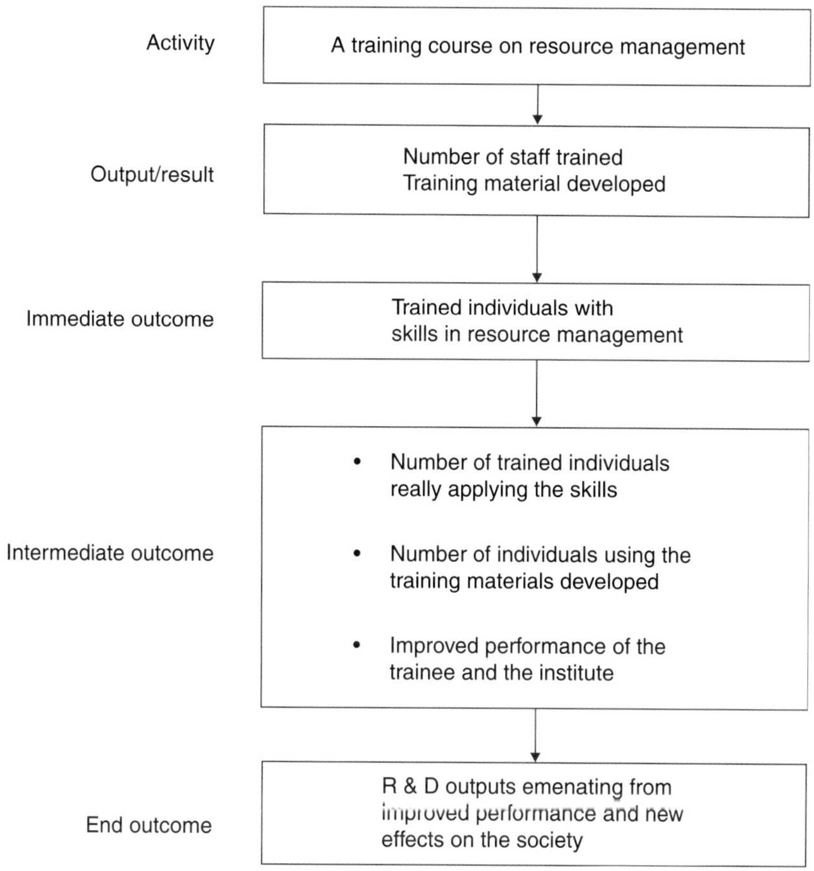

Fig. 24.12. Logical chart for a training project.

If significant outside factors are identified as possibly having an effect on the intended outcome, then evidence to refute or determine the extent of influence of those claims will be useful in addressing the attribution question. As mentioned earlier, a logical framework included in most of the project proposals can be extended to accommodate some of these aspects identified here.

Identify, measure and document expected behavioural changes

In order to bring about an outcome, programmes have to change people's behaviour. By trying to identify and then document the changes in attitudes, knowledge, perceptions and decisions taken by programme target groups, which logically link to the outcomes being observed, a good understanding of the actual impact the programme is having can often be acquired. This of course requires clearly identifying who the various clients of the programme are and how their behaviour is expected to change. If one could observe these short-term changes occurring, the logical case for the programme's attribution can be enhanced.

Use discriminating indicators

In assessing programmes, considerable care is needed in selecting indicators of performance. Here, we are considering the attribution issue where it is important to use performance indicators that best discriminate or focus on the outcomes in question.

Many indicators are ratios, where the denominator qualifies the numerator. Consider a programme designed to reduce air accidents by inspection of the airworthiness of aircraft. An indicator might be the number of air accidents per air-miles flown. A better indicator would be the number of air accidents due to structural failure per air-mile flown. But structural failures may occur regardless of inspections. Therefore, it may be better still to use two indicators: the number of accidents per air-miles flown due to structural failure in aircraft inspected and the number of air accidents per air-mile flown due to structural failure in aircraft not inspected. By comparing structural failures in inspected and uninspected aircraft, one can estimate what inspection does to reduce the problems that inspection is designed to address. Questions of attribution still exist, but the more refined indicators reduce the problem and improve the chance of providing useful information on the contribution of the programme.

Tracking performance over time or location

In cases where the programme activities have varied over time, showing that outcomes have varied in a consistent manner with the variation in activities can strengthen the argument that the activities have indeed made a difference. Hendricks (1996) identifies a number of such cases where by tracking performance measures we might show that:

- Outcomes appeared at an appropriate time after our efforts began;
- Outcomes appeared in different locations or with different people;
- Outcomes faded when our efforts stopped;
- Only those outcomes appeared that we should have affected;
- Outcomes appeared only where or when we were active;
- The biggest outcomes appeared where we did the most.

In some areas of programming such as the impacts from research activities, there is likely to be a significant delay before the intended outcomes occur, and the attribution picture portrayed through tracking performance over time will not be as evident. In these cases, one still needs to track outcomes over time to see if the intended outcomes have occurred, but demonstrating or understanding attribution is even more of a challenge.

Explore and discuss plausible alternative explanations

The attribution problem arises when one believes or is trying to claim that a programme has resulted in certain outcomes and there is an alternative plausible explanation, i.e. there may be other reasons for the observed outcome. Dealing with these alternative explanations in a systematic way is one way of handling the attribution issue. This entails:

- Identifying the most likely alternative explanations;
- Presenting whatever evidence or argument you have to discuss and, where appropriate, discounting these alternative explanations;
- Presenting whatever evidence there is that the programme is a more likely explanation.

Two generic types of evidence could be used to counter arguments for alternatives to the programme and its situation:

1. The logic argument – one might refer to the theory behind the programme and the kind of theory that would be needed to support claims for rival hypotheses.
2. One can also bring actual evidence to bear concerning the alternative explanations.

Addressing attribution problems in this way demonstrates that:

- You are aware of the complexity of the situation;
- You acknowledge and understand the other factors at play;

- You are nevertheless concluding (assuming you are) that the most likely explanation for the observed outcome is that the programme has made a significant contribution.

Unless one discusses alternative explanations, the claim about simply pointing out the existence of alternative explanations can effectively challenge the programme's efficacy.

Gather additional relevant evidence

One might also gather evidence concerning alternative explanations of the observed outcome. The data might be part of the routine performance measurement system, but more likely would be collected from time to time when the analysis of the programme's contribution is undertaken. Data collection might entail a review of the relevant literature, surveys, tracking of relevant external factors, field visits or focus groups.

Additional information may be collected in a number of ways: expert opinion, structured survey, reviewing programme files and secondary data analysis including meta-analysis.

EXPERT OPINION. In many situations, there are persons outside the programme who are seen as knowledgeable about the programme area, the programme's impacts and the environment in which the programme operates. A focus group of experts may be another approach that would allow some probing as to why such views are held.

STRUCTURED SURVEY. A structured survey may be able to provide some evidence, albeit subjective in nature, of the extent to which the programme is influencing an outcome. Surveying such individuals is often done to find out other information about the programmes, in which case adding questions on attribution is not very expensive.

PROGRAMME FILES. There is frequently considerable existing data available from programme files, some of which might be useful to provide information on the contribution of the programme. This type of existing data, which probably has been collected for other purposes, can often contain valuable information, particularly if used in conjunction with new data collected.

AVAILABLE SECONDARY ANALYSIS. In other cases, there may be useful secondary analysis that the others have done in the programme area that might clarify measurement and attribution issues. In still other cases, there may be meta-analysis that has been done – analysis that synthesizes a number of studies in an area.

CASE STUDIES. One could also make use of specific case studies. They can reveal the real nature of the programme and also demonstrate, at least in these cases, that one can be fairly confident about the impacts of the programme's activities. This type of evidence can be quite persuasive but appropriate cautions are a must, especially when it is quite anecdotal. Case study and anecdotal evidence is best when illustrating a concrete case to complement other evidence that has been collected. On its own, however, it can be quite misleading as it may merely be one of the few cases that appear to have worked while the majority has not. Nevertheless, if the context and limitations are made clear, there is often a useful role for individual case studies.

Use multiple lines of evidence

The 'multiple lines of evidence' argues that while no one piece of evidence may be very convincing, a larger set of different and complementary evidence can become quite convincing.

Thus, in trying to reduce uncertainly surrounding attribution, using as many lines of evidence as possible is a sensible, practical and credible strategy.

Refer to the need for an evaluation

If it is critical to have good information on attribution, and the available evidence points in different directions, then the best strategy is to simply acknowledge that one does not know and suggest that a thorough study be carried out to address the attribution question.

In most cases, however, the programme has indeed made a significant contribution; the various lines of evidence will confirm this. Remember what 'contribution analysis' helps us to do is to reduce the uncertainty about the contribution made and not proving the contribution made by the programme. Mayne (1999) argues that undertaking a contribution analysis would examine and present the best case possible – a credible convincing performance story – for attribution with the available evidence.

Credible convincing stories have the following characteristics:

- Well-articulated presentation of the context of the programme and its general aims (presenting a plausible theory).
- Highlighting the contribution analysis indicating there is association between what the programme has done and the outcomes observed.
- Pointing out that the main alternative explanations for the outcomes accruing, such as other related programmes or external factors, have been ruled out or clearly have only had a limited influence.

If there are too many gaps in the story, an in-depth analysis is needed to better understand the contribution of the programme.

Best Bet Practices

R&D institutes may execute a large number of projects and programmes at any given point in time. As pointed out earlier, ex-post impact studies have very limited use. Thus, trying to trace the impacts of these large number of activities may be both time consuming and prohibitively expensive. Studies on the distribution of benefits and the effects of research programmes suggest that these distributions are skewed. A small number of projects may account for most of the effects of a research programme (Parker *et al.*, 1998). Out of the several royalty generating research projects at the University of California, the top two generated 70% of the technology transferred in 1994. This suggests that an assessment of the economic impact of a research programme should concentrate on identifying the most effective research projects and assessing their benefits. The aggregate benefits of these projects provide a lower bound for the benefit of the programme.

A 'cherry picking' approach to estimating research benefits is to identify a small number of projects and policies that a priori seem to have the highest discounted net benefit and estimate only their benefits. Three possible approaches for identifying productive research projects are:

1. *Method 1: screen projects sequentially on their assessed productivity*. This procedure was applied by Just *et al.* (1988) in their study on the economic benefits of the US–Israel Notational Agricultural Research and Development Fund (BARD) for the USA. In this study they used a two-step procedure to obtain a lower bound estimate of the economic benefits of the fund.

- Ask fund directors and senior staff to identify projects with commercial potential.
- Research proposal and final reports from each proposal was reviewed.
- The principal investigators were asked to provide basic information about the economic impacts of innovations resulting from their research and the names of those who adopted or likely to adopt the innovations.
- This screening identified projects with the highest potential for generating benefit.

This study revealed that the distribution of benefits was highly skewed and the top two projects provided 60% of the expected benefit.

2. *Method 2: individual researchers select a subset of successful projects.* A research team will then screen these projects to identify the most productive ones. Their benefits are then quantified. This procedure was used by Goldman *et al.* (1990) and McWilliams and Zilberman (1996) to evaluate the productivity of extension in two California counties. They asked the project leaders (farm advisors) to provide basic information on regions and populations that would benefit from the discovery, quantitative estimates of per unit benefits (per acre, per animal), and the names of individuals who could verify these claims. After initial screening, the research team for each study identified less than ten projects with significant economic effects, and they quantified the benefits from these.

3. *Method 3: a third approach would be to make the initial selection based on the volume of services used.* Where the amount of time allotted to each project or number of contacts with each client is documented, projects that require the most effort or service can be selected to have their benefits quantified further. This approach was used by Parker *et al.* (1996) to assess the California Irrigation Management Information system (CIMIS).

The key feature that these three approaches share is the collection of evidence from users (or potential users) of the system. Estimates of benefits provided by project leaders should be used only for the initial screening of projects and – when corroborated by testimony from users or objective experts – may be able to identify unexpected benefits. The cost, speed and accuracy of research evaluation are significantly improved if documentation procedures are established and followed as part of the research effort.

The 'best practices' for estimating the returns to R&D investment are summarized below (Maredia *et al.*, 2000).

1. *Good practices in measuring programme/project costs:*

- Include all research, development and technology transfer cost relevant to the development and dissemination of the technology being assessed.
- Include cost of all partners involved International Agricultural Research Centres, National Agricultural Research Systems, NGOs, etc.
- Include pro-rated overhead and administrative costs of the institutes involved.

2. *Good practices for calculating the K-factor:*

- Combine technical, scientific and economic information from the various sources. If available, use the results of previous experimental trials to assess changes in cost and yield.
- Reflect research-induced changes in the use of inputs and their opportunity cost in the measure of cost savings.
- Disaggregate supply-shift estimates by components of a research programme.
- Carefully construct the 'with' and 'without' scenarios to estimate the incremental net benefit truly attributable to the research investment under consideration.
- Make the underlying assumptions explicit, so that others can transparently assess the analysis.

3. *Good practices for estimating benefits:*

- Study the market situation of the commodities affected by R&D and determine tradable status appropriate for estimating total surplus.
- Use economic prices that are appropriately adjusted to reflect policy distortions in the output market. All monetary values should be expressed in real terms.
- Estimate of research benefits should be disaggregated by commodities, production environment or geographical basis if the parameter estimates are different for different components of the research programme.
- Ensure that spill-over effects are included.

4. *Good practices in data analysis and calculating returns to investment:*

- Create a different spreadsheet template for each of the commodity/products/zones affected by the R&D programme.
- Document the logical structure of the model/method used for calculating the benefit and cost stream, including clear specification of assumptions and sources of data used to derive the various model parameters.
- Calculate both summary measures, i.e. NPV and IRR to indicate the profitability of R&D investment.
- Carry out appropriate sensitivity analysis by varying some of the critical assumptions made in the model.
- When evidence of research spill-outs and spill-ins exists, good practice for ex-post impact assessment is to conduct spatially disaggregated analysis that explicitly treats spill-outs and spill-ins.

5. *General procedure:*

- Define the minimum data set and collect them on a regular basis. Collecting panel data on a regular basis is cost-effective.
- Combine both qualitative and quantitative assessment.
- Build in-house capacity for ongoing evaluation.
- Develop mechanisms to integrate information with decision-making.
- If the institute is genuinely interested in improving its quality, relevance and effectiveness, sampling of 'failures' as well as 'successes' may offer more insight.
- Establishing baseline conditions is a prerequisite to providing credible evidence of impact in terms of measuring the people-level impacts of R&D investment.
- Instil impact culture at all levels, i.e. institute, programme and project. This simply means having an impact orientation so that research is continuously conducted in relation to specific goals. As the payoffs from research are long term and inherently uncertain, in using impact assessment as a planning tool one must allow a certain margin of failure if innovation is not to be interrupted.
- Impact assessment needs to be driven by internal consideration of research efficiency, i.e. to generate feedback on research impact for purposes of project guidance, priority setting and resource allocation. The external (accountability) and internal (research efficiency) dimensions of impact assessment should be seen as complementary. Whether it is done by internal staff or outside experts, it is important to ensure that project assessment follows an accepted standard of good practices and produces credible results.

It is a well-known fact that the investors in public R&D are no longer satisfied with activity-based progress reports. They in fact expect outcome and impact evaluation, i.e. an objective assessment of the actual effects of the funded programme on the target populations. Now there is also general consensus that no single method or data set is capable of capturing what is necessary to understand complex systems involving policies, organizations, institutions, programmes and people through which goals are achieved. In order to ensure an adequate coverage of the information universe, both qualitative and quantitative data are required and multiple methods and approaches are advocated (Greene and Caracelli, 1997). Retrospective narratives are an essential component of the former and indeed provide the basis for quantitative estimates and the related issue of attribution. Case studies are essential for further refining approaches and methodologies and learning lessons, which can enhance the future impacts and help define 'best practices'. Use of independent evaluators in the process offers the advantage of objectivity and lend credibility to the impact evaluation, although the cost often means that only a small number of projects and programmes can be subjected to this type of evaluation.

Notes

[1] Accountability can cover: impact accountability, coverage accountability, service delivery accountability, efficiency accountability, fiscal accountability and legal accountability.

[2] For details of the functional form and its derivatives please refer to: Alston *et al.* (1995); Anandajayasekeram *et al.* (1996); Masters *et al.* (1996); Akino and Hayami (1975).

[3] A number of impact studies conducted by CSIRO in Australia used this approach. For more details see Johnson *et al.* (1992).

References

Akino, M. and Hayami, Y. (1975) Efficiency and equity in public research: rice breeding in Japan's economic development. *American Journal of Agricultural Economics* 57, 1–10.

Alston, J.M., Norton, G.W. and Pardey, P.G. (1995) Science under scarcity. *Principle and Practice for Agriculture.* Cornell University Press, Ithaca, New York.

Anandajayasekeram, P. and Martella, D. (1996) Conceptual framework to assess the impacts of agricultural research. *Journal of Agricultural Economics and Development* 1, 1–15.

Anandajayasekeram, P., Martella, D.R. and Rukuni, M. (1996) *A Training Manual on Research and Development Evaluation and Impact Assessment of Investments in Agricultural and Natural Resources Research.* SADC-SACCAR, Gabarone, Botswana.

Anderson, J.R. and Herdt, R.W. (1990) Reflections on impact assessment. In: Echeverria, R.G. (ed.) *Methods for Diagnosing Research System Constraints and Assessing the Impact of Agricultural Research*, Vol. II. International Service for National Agricultural Research, The Hague, The Netherlands, pp. 35–42.

Babu, S. and Mcthindi, G. (1995) Costs and benefits of informed food policy decisions: a case study of food security and nutrition monitoring in Malawi. *Quarterly Journal of International Agriculture* 34(3), 292–308.

Bickman, L. (1987) Using Program theory in evaluation. In: Bickman, L. (ed.) *New Directions for Program Evaluation*, No. 33. Jossey-Bass, San Francisco, California.

Biggs, S.D. (1990) A multiple source of innovation model of agricultural research and technology promotion. *World Development* 18(11), 1481–1499.

Brandon, C. (1995) *Valuing environmental costs in Pakistan: the economy-wide impact of environmental degradation.* Background paper for 'Pakistan 2010.' Asia Environmental Division, World Bank, Washington, DC.

Byerlee, D. and Moya, P. (1993) *Impacts of International Wheat Breeding Research in the Developing World, 1966–90.* CIMMYT, Mexico, DF.

CIMMYT, International Maize and Wheat Improvement Center (CIMMYT) (1988) *From Agronomic Data to Farmer Recommendations: an Economics Training Manual.* CIMMYT, Mexico, Revised Edition.

Cracknell, B. (1996) Evaluating development aid: strengths and weaknesses. *Evaluation* 2, 23–34.

Crosson, P. and Anderson, J.R. (1993) *Concerns for Sustainability: Integration of Natural Resource and Environmental Issues in the Research Agendas of NARS.* ISNAR Research Report No. 4. International Service for National Agricultural Research, The Hague, The Netherlands.

DANIDA (1994) Evaluation report: Agricultural sector evaluation. In: *Impact Studies*, Vol. 2: *Methods and Findings*. Danish International Development Agency, Copenhagen, Denmark, pp. 1–13.

Davis, J.S., Oram, P.A. and Ryan, J.G. (1987) *Assessment of Agricultural Research Priorities: an International Perspective*. ACIAR Monograph No. 4. ACIAR, Canberra, Australia.

Edwards, G.W. and Freebairn, J.W. (1981) *Measuring a Country's Gains from Research: Theory and Application to Rural Research in Australia*. AGPS, Canberra, Australia.

Evenson, R.E. (1967) The contribution of agricultural research to production. *Journal of Farm Economics* 49, 1415–1425.

Evenson, R.E. (1987) *The International Agricultural Centers: Their impact on spending for national agricultural research and extension*. CGIAR Study Paper No. 22. World Bank, Washington, DC.

Garret, J.L. and Islam, Y. (1998) Policy research and the policy process: Do the twain ever meet? *IIED Gatekeeper Series* No. 74. International Institute for Environment and Development, London.

Goldman, G., Shah, F. and Zilberman, D. (1990) *A preliminary report on the economic analysis of University of California cooperative extension activities: The case of Stanislaus country*. Department of Agricultural and Resource Economics, University of California at Berkeley, Berkeley, California.

Greene, J. and Caracelli, V. (eds) (1997) *Advances in Mixed-Method Evaluation. New Directions for Evaluation*. No. 74.

Griliches, Z. (1958) Research costs and social returns: hybrid corn and related innovations. *Journal of Political Economy* 66, 419–431.

Harrington, L., Jones, P. and Winograd, M. (1994) *Operationalizing Sustainability: A total productivity approach*. Paper presented at the Land Quality Indicators Conference, CIAT, Cali, June 1994.

Hendricks, M. (1996) *Performance monitoring: how to measure effectively the result of our efforts*. Presented at the American Evaluation Association Annual Conference, Atlanta, November 6, 1996.

Horton, D.E. (1990) Assessing the impact of international research: concepts and challenges. In: Echeverria, R.G. (ed.) *Methods for Diagnosing Research System Constraints and Assessing the Impact of Agricultural Research*, Vol. II. *Assessing the Impact of Agricultural Research*. ISNAR, The Hague, The Netherlands.

Horton, D., Mackay, R., Anderson, A. and Dupleich, L. (2000) *Evaluating capacity development in planning, monitoring, and evaluation: A case from Agricultural research*. Research Report No. 17. International Service for National Agricultural Research, The Hague, The Netherlands.

Izac, A.-M.N. (1998) *Assessing the Impact of Research in Natural Resources Management*. Synthesis of an International Workshop 27–29 April 1998. ICRAF Nairobi, Kenya.

Johnston, B., Healy, T., I'ons, J. and McGregor, M. (1992) *Rural Research – the Pay off: the Returns from Research Undertaken by the CSIRO Institute of Plant Production and Processing*. CSIRO Occasional Paper No. 7, Canberra, Australia.

Just, R.E., Zilberman, D., Parker, D. and Phillips, M. (1988) *The Economic Impacts of BARD research on the US*. Report Prepared for the Commission to Evaluate BARD. University of California, Berkeley, California.

Kumar, K. (1995) Measuring performance of agricultural and rural development programs. In: Picciotto, R. and Rist, R.C. (eds) *Evaluating Country Development Policies and Programs: New approaches for a New Agenda*. Jossey-Bass, San Francisco, California.

Lusthaus, C., Anderson, G. and Murphy, E. (1995) *Institutional Assessment: a Framework for Strengthening Organizational Capacity for IDRC's Research Partners*. IDRC, Ottawa.

Maredia, M., Byerlee, D. and Anderson, J. (2000) *Ex-Post Evaluation of Economic Impacts of Agricultural Research Programme: A Tour of Good Practice*. A Paper Presented to the Workshop on 'The Future of Impact Assessment in CGIAR: Needs, Constraints, and Options'. Standing Panel on Impact Assessment (SPIA) of the Technical Advisory Committee, Rome, 3–5 May, Rome.

Masters, W.A., Bedingar, T. and Oehmke, J.F. (1996) *The Impact of Agricultural Research in Africa: Aggregate and Case Study Evidence*. Mimeo (Department of Agricultural Economics), Purdue University, West Lafayette, Indiana.

Mayne, J. (1999) *Addressing Attribution Through Contribution Analysis: Using Performance Measures Sensibly*. Discussion Paper. Office of the Auditors General of Canada, Ottawa, Canada.

McWilliams, B. and Zilberman, D. (1996) Time of technology adoption and learning by using. *Economics of Innovation and New Technology* 4(2), 139–154.

Munn, R.E. (1979) *Environmental Impact Assessment: Principles and Procedures*. Scientific Committee on Problems of the Environment, SCOPE Report No. 5. Institute for Environmental Studies, University of Toronto.

Pachico, D. (1998) *Conceptual framework for natural resource management research and basic methodological issues in impact assessment.* Paper presented at the international workshop, Assessing the impact of research in natural resource management, 27–29 April 1998, Nairobi, Kenya.

Parker, D., Zilberman, D., Cohen, D. and Osgood, D. (1996) *The economic costs and benefits associated with the California Irrigation Management Information System (CIMIS): Final Report.* Presented to the California Department of Water Resources.

Parker, D., Zilberman, D. and Castillo, F. (1998) Offices of technology transfer and privatization of university innovations. *Choices*, first quarter, 19–25.

Pingali, P.L. (2001) *Milestones in Impact Assessment Research in the CGIAR, 1970–1999. With an Annotated Bibliography of Impact Assessment studies conducted in the CGIAR, 1970–1999*, prepared by M.P. Feldmann, Mexico, DF. Standing panel on Impact Assessment, Technical Advisory Committee of the Consultative Group on International Agricultural Research.

Rola, A. and Pingali, P. (1993) *Pesticides, Rice Productivity and Farmers' Health.* International Rice Research Institute, Malina and World Resource Institute, Washington, DC.

Rossi, P.H. and Freeman, H.E. (1993) *Evaluation: a Systematic Approach*, 5th edn. Sage Publications, Beverley Hills, California.

Ryan, J.G. (2001) *Some lessons from studies of the impact of policy research by IFPRI.* Paper presented at the DGIs/IFPRI workshop on Assessing the Impacts of Policy-Oriented Social Research held in The Hague, Netherlands, 12–13 November, 2001.

Smith, V.H. and Parday, P.G. (1997) Sizing up social science research. *American Journal of Agricultural Economics* 79(5), 1530–1533.

Smutylo, T. (2005) *Outcome mapping: a method for tracking behavioral changes in development programs.* ILCA Brief 7. International Livestock Research Institute, Addis Ababa.

Steiner, R.A., McLaughlin, L., Faeth, P. and Janke, R.R. (1995) Incorporating externality costs into productivity measures: A case study using US agriculture. In: Barnett, V., Payne, R. and Steiner, R. (eds) *Agricultural Sustainability: Economic, Environmental, and Statistical Considerations.* Wiley, Chichester, UK, pp. 209–230.

Technical Advisory Committee (TAC) Secretariat (2001) Summary of the sessions. In: *The Future of Impact Assessment in the CGIAR: Needs, Constraints and Options.* Proceedings of a Workshop Organized by the Standing Panel on Impact Assessment (SPIA) of the Technical Advisory Committee (TAC), 3–5 May 2000 FAO, Rome, pp. 7–27.

Weiss, C.H. and Bucavalas, M.J. (1980) *Social Science Research and Decision Making.* Colombia University Press, New York.

Williams, D. (1993) *Methods for assessing the socio-economic impacts of Government S&T.* Report prepared under contract. ARA Consulting Group, Inc. Canada. Ottawa, Canada.

Winpenny, J. (1991) *Values for the Environment: a Guide to Economic Appraisal.* HMSO/ODL, London.

Index